Universitext

Universitext is a series of textbooks that presents material from a wide variety of mathematical disciplines at master's level and beyond. The books, often well class-tested by their author, may have an informal, personal, or even experimental approach to their subject matter. Some of the most successful and established books in the series have evolved through several editions, always following the evolution of teaching curricula, into very polished texts.

Thus as research topics trickle down into graduate-level teaching, first textbooks written for new, cutting-edge courses may find their way into *Universitext*.

Masanori Morishita

Knots and Primes

An Introduction to Arithmetic Topology

Second Edition

 Springer

Masanori Morishita
Graduate School of Mathematics
Kyushu University
Fukuoka, Japan

ISSN 0172-5939   ISSN 2191-6675   (electronic)
Universitext
ISBN 978-981-99-9254-6   ISBN 978-981-99-9255-3   (eBook)
https://doi.org/10.1007/978-981-99-9255-3

English Language edition of 'Musubime to Sosu' Copyright © Springer Japan 2009

This Springer imprint is published by the registered company Springer Nature Singapore Pte Ltd.
The registered company address is: 152 Beach Road, #21-01/04 Gateway East, Singapore 189721, Singapore

Paper in this product is recyclable.

To the memory of my mother,
Chieko Morishita

Preface

The theme of the present book is the analogy between knot theory and number theory, based on the homotopical analogies between knots and primes, 3-manifolds and number rings. Thus the purpose of this book is to discuss and present, in a parallel and systematic manner, the analogies between the fundamental notions and theories of knot theory and number theory. To aid the reader, basic materials from each field are recollected in Chap. 2.

If we look back over the history of knot theory and number theory, an origin of the modern development of both fields may be found in the work of C.F. Gauss (1777–1855). The aim of this book may be rephrased as bridging the two subjects that branched out after Gauss and providing a foundation of *arithmetic topology*.

This volume is an English translation of my Japanese book [164] with some things added. I thank Professor Y. Matsumoto for recommending that I should write a book on this subject. The contents of this book grew out of my intensive lectures at some universities in Japan (Kyushu, Kyoto, Tohoku and Tokyo) on various occasions during 2002–2007 and at the University of Heidelberg in the fall of 2008. I take this opportunity to acknowledge my gratitude to Y. Taguchi, K. Kato, A. Yukie, T. Yamazaki, T. Oda and D. Vogel for inviting me to give lectures on arithmetic topology, and I thank Y. Terashima for useful communication and joint work on Chap. 12. I am thankful to H. Hida, M. Kaneko, M. Kato, M. Kurihara, Y. Mizusawa and S. Ohtani for useful communication in the course of writing this book, and to F. Amano, H. Niibo and Y. Takakura for pointing out some misprints in the Japanese version. I also thank the referees for their useful comments and L. Stoney, D. Akmanavičius and M. Nakamura at Springer for their help to publish this text. I would like to thank C. Deninger, M. Kapranov, T. Kohno, B. Mazur, J. Morava and T. Ono for their encouragement and interest in my work. I express my hearty thanks to J. Hillman and K. Murasugi for answering patiently my questions on knot/link theory over the years, especially to J. Hillman for his useful (both linguistic and mathematical) comments on the manuscript of this English version. Finally, I thank the Japan Society for the Promotion of Science for the support during the writing of this book.

Fukuoka, Japan   Masanori Morishita
July 2009

Preface to the Second Edition

The present second edition is a corrected and enlarged version of the first edition.

Since the publication of the first edition, some remarkable developments have taken place in arithmetic topology. The present edition includes two new chapters on them. One is Chap. 7 which deals with Idelic Class Field Theory for 3-Manifolds and Number Fields. It was an important problem in the foundation of arithmetic topology to construct a topological analogue for 3-manifolds of idelic class field theory. The works of Niibo–Ueki and Mihara [145], [179], [181] gave some answers to this problem. Chapter 7 treats their works as well as the (classical) arithmetic counterpart for number fields. The other is Chap. 16 which deals with Dijkgraaf–Witten Theory for 3-Manifolds and Number Rings. Since arithmetic topology started to be investigated, the connection with mathematical physics has been an intriguing theme to be explored. (See Introduction to the first edition.) The work of Minhyong Kim [103] on arithmetic Chern–Simons theory was the first concrete attempt to bridge arithmetic and quantum field theory, based on arithmetic topology. Chapter 16 treats his work and some subsequent works as well as the (original) topological counterpart, namely, Dijkgraaf–Witten theory for 3-manifolds. In accordance with these new added materials, I changed some expositions in the text of the first edition.

Other changes from the first edition include correcting misprints and mistakes as well as updating the references. I also changed the expositions at some points and added remarks (for example, multiple power residue symbols in Chap. 9), according to the results/references which appeared after the first edition. I thank Jonas Stelzig for pointing out mistakes I made in the first edition. I am thankful to Yuqi Deng for drawing the figures in the text and to Ibuki Hanada for pointing out misprints in Chap. 7. Some parts of the second edition were written during my stay as a Research Fellow at Münster University. I thank Christopher Deninger and the

Cluster of Excellence "Mathematics Münster" for their hospitality and support. I also acknowledge the support of the Japan Society for the Promotion of Science. Finally, I would like to thank Mr. Masayuki Nakamura at Springer for his help to publish the second edition.

Itoshima, Japan   Masanori Morishita
July 2023

Contents

Notation and Conventions

$\mathbb{N}$: the set of natural numbers (≥ 1).

$\mathbb{Z}$: the set of integers.

$\mathbb{Q}$: the set of rational numbers.

$\mathbb{R}$: the set of real numbers.

$\mathbb{C}$: the set of complex numbers.

$\#A$: the cardinality of a set A.

$R^{\times}$: the group of invertible elements in a ring R, called the unit group of R.

For $x = (x_1, \ldots, x_n) \in \mathbb{R}^n$, $||x|| := \sqrt{x_1^2 + \cdots + x_n^2}$.

For an integer $n \geq 0$, an n-manifold means an n-dimensional manifold.

For topological spaces X and Y, $X \approx Y$ means that X and Y are homeomorphic, and $X \simeq Y$ means that X and Y are homotopically equivalent.

For a topological space X in an ambient space Y, $\text{int}(X)$, ∂X and $\overline{X}$ denote the interior, the boundary and the closure of X respectively.

For objects A and B with an algebraic structure, $A \simeq B$ means that A and B are isomorphic.

For a topological space X, $H_*(X)$ stands for the homology group of X with coefficients in $\mathbb{Z}$.

For closed subgroups A, B of a topological group G, $[A, B]$ denotes the closed subgroup generated by commutators $[a, b] := aba^{-1}b^{-1}$ for $a \in A, b \in B$.

For a topological group G and $d \in \mathbb{N}$, $G^{(d)}$ denotes the d-th term of the lower central series of G defined by $G^{(1)} := G$, $G^{(d+1)} := [G^{(d)}, G]$.

Chapter 1
Introduction

1.1 Two Subjects that Branched Out from C.F. Gauss—Quadratic Residues and Linking Numbers

In his youth, C.F. Gauss proved the law of quadratic reciprocity and further created the theory of genera for binary quadratic forms ([64], 1801).

For an odd prime number p and an integer a prime to p, consider the quadratic equation modulo p:

$$x^2 \equiv a \bmod p.$$

According to whether this equation has an integral solution or not, the integer a is called a quadratic residue or quadratic non-residue mod p, and the Legendre symbol is defined by

$$\left(\frac{a}{p}\right) := \begin{cases} +1, & a \text{ is a quadratic residue mod } p \\ -1, & a \text{ is a quadratic non-residue mod } p. \end{cases}$$

For odd prime numbers p and q, Gauss proved the following relation between p being a quadratic residue mod q and q being a quadratic residue mod p:

$$\left(\frac{q}{p}\right) = \left(\frac{p}{q}\right)(-1)^{\frac{p-1}{2}\frac{q-1}{2}}.$$

In particular, the symmetric relation holds if if p or $q \equiv 1 \bmod 4$:

$$\left(\frac{q}{p}\right) = \left(\frac{p}{q}\right).$$

M. Morishita, *Knots and Primes*, Universitext,
https://doi.org/10.1007/978-981-99-9255-3_1

In terms of algebra today, Gauss' genus theory may be viewed as a classification theory of ideals of a quadratic field $k = \mathbb{Q}(\sqrt{m})$. Let $\mathcal{O}_k$ be the ring of integers of k. Non-zero fractional ideals $\mathfrak{a}$ and $\mathfrak{b}$ of $\mathcal{O}_k$ (i.e., finitely generated $\mathcal{O}_k$-submodules of k) are said to be in the same class in the narrow sense if there is $\alpha \in k$ such that

$$\mathfrak{b} = \mathfrak{a}(\alpha), \quad \alpha, \bar{\alpha} > 0,$$

where $\bar{\alpha}$ denotes the conjugate of α. Let $H^+(k)$ denote the set of these classes. For the sake of simplicity, we assume for the moment that $m = p_1 \cdots p_r$ ($p_1, \cdots, p_r$ being different prime factors) and $p_i \equiv 1 \bmod 4$ ($1 \le i \le r$). Note that in each class we may choose an ideal in $\mathcal{O}_k$ prime to m. Such ideals $\mathfrak{a}$ and $\mathfrak{b}$ are said to be in the same genus, written as $\mathfrak{a} \approx \mathfrak{b}$, if one has

$$\left(\frac{N\mathfrak{a}}{p_i}\right) = \left(\frac{N\mathfrak{b}}{p_i}\right) \quad (1 \le i \le r),$$

where $N\mathfrak{a} := \#(\mathcal{O}_k/\mathfrak{a})$. This gives a well-defined equivalence relation on $H^+(k)$ and we can classify $H^+(k)$ by the relation $\approx$. Gauss proved that $H^+(k)$ forms a finite Abelian group by the multiplication of fractional ideals, which is called the ideal class group in the narrow sense, that the correspondence $[\mathfrak{a}] \mapsto ((N\mathfrak{a}/p_1), \cdots, (N\mathfrak{a}/p_r))$ gives rise to the following isomorphism

$$H^+(k)/\approx \; \simeq \left\{ (\xi_i) \in \{\pm 1\}^r \;\middle|\; \prod_{i=1}^r \xi_i = 1 \right\} \simeq (\mathbb{Z}/2\mathbb{Z})^{r-1}$$

and hence that the number of genera is 2^{r-1}. Gauss' investigation on quadratic residues may be seen as an origin of the modern development of algebraic number theory.

On the other hand, in [63], 1833, Gauss discovered the notion of the linking number, together with its integral expression, in the course of his investigations of electrodynamics. Let K and L be disjoint, oriented simple closed curves in $\mathbb{R}^3$ (i.e., a 2-component link) with parametrizations given by smooth functions $a : [0, 1] \longrightarrow \mathbb{R}^3$ and $b : [0, 1] \longrightarrow \mathbb{R}^3$ respectively. Let us turn on an electric current with strength I in L so that the magnetic field $\mathbf{B}(x)$ ($x \in \mathbb{R}^3$) is generated. By the law of Biot–Savart, $\mathbf{B}(x)$ is given by

$$\mathbf{B}(x) = \frac{I\mu_0}{4\pi} \int_0^1 \frac{b'(t) \times (x - b(t))}{\|x - b(t)\|^3} \, dt,$$

where μ_0 stands for the magnetic permeability of a vacuum. Then Gauss showed the following integral formula:

$$\frac{1}{I\mu_0} \int_0^1 \mathbf{B}(a(s)) \cdot a'(s) \, ds = \mathrm{lk}(L, K),$$

namely,

$$\frac{1}{4\pi} \int_0^1 \int_0^1 \frac{(b'(t) \times (a(s) - b(t))) \cdot a'(s)}{||a(s) - b(t)||^3} \, ds \, dt = \mathrm{lk}(L, K).$$

Here $\mathrm{lk}(L, K)$ is an integer, called the linking number of K and L, which is defined as follows. Let Σ_L be an oriented surface with $\partial \Sigma_L = L$. We may assume that K crosses Σ_L at right angles. Let P be an intersection point of K and Σ_L. According to whether a tangent vector of K at P has the same or opposite direction to a normal vector of Σ_L at P, we assign a number $\varepsilon(P) := 1$ or -1 to each P:

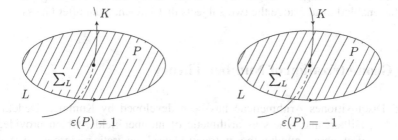

$$\varepsilon(P) = 1 \qquad\qquad \varepsilon(P) = -1$$

Let $P_1, \ldots, P_m$ be the set of intersection points of K and Σ_L. Then the linking number $\mathrm{lk}(L, K)$ is defined by

$$\mathrm{lk}(L, K) := \sum_{i=1}^m \varepsilon(P_i).$$

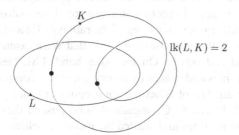

$$\mathrm{lk}(L, K) = 2$$

By this definition or by Gauss' integral formula, we easily see that the symmetric relation holds:

$$\mathrm{lk}(L, K) = \mathrm{lk}(K, L).$$

Gauss already recognized that the linking number is a topological invariant, a quantity which is invariant under continuous moves of K and L. Furthermore, it is remarkable that Gauss' integral formula has been overlooked; its first generalization was not studied until about 150 years later by E. Witten and M. Kontsevich in Chern–Simons gauge theory, again in connection with physics[1] [116], [244].

Although there seems no connection between the Legendre symbol and the linking number at first glance, as we shall show in Chap. 4, there is indeed a close analogy between both notions and in fact they are defined in an exactly analogous manner. Since Gauss took an interest in knots in his youth [50, XVII, p. 222], we may imagine that he already had a sense of the analogy between the Legendre symbol and the linking number. However, there was no mathematical language at his time to describe this analogy, and knot theory and number theory have grown up in separate ways for a century and a half after Gauss. It was the "geometric viewpoint and language" brought into number theory during this long period of time that enabled us to bridge the two subjects that branched out after Gauss.

1.2   Geometrization of Number Theory

Gauss' Disquisitiones Arithmeticae has been developed by Kummer, Dedekind, Kronecker, Hilbert and others as arithmetic of number fields, which provided a foundation of algebraic number theory today. Gauss' quadratic reciprocity was also generalized along this line of development and culminated in class field theory by T. Takagi and E. Artin (1927). A guiding principle leading to this development was the viewpoint of the analogy between number fields and function fields of one variable. Behind this thought there might be an influence from the theory of complex functions (Riemann surfaces) which was a major field in the 19th century mathematics. The basic idea comes from the well known analogy between integers and polynomials, and a maximal ideal of a number ring is regarded as an analogue of a point of an algebraic curve. In particular, there are close analogies between finite algebraic number fields (finite extensions of the rationals $\mathbb{Q}$) and algebraic function fields of one variable with finite constant fields that were extensively investigated by E. Artin, A. Weil and others. On the other hand, I.M. Gelfand clarified the equivalence between rings and spaces by showing that any commutative C^*-algebra R is obtained as the algebra of functions on a compact space given by the set of maximal ideals of R. It was A. Grothendieck who pushed these thoughts further for arbitrary commutative rings and created the theory of schemes. For instance, the prime spectrum $\mathrm{Spec}(\mathcal{O}_k)$, namely, the set of prime ideals of the ring $\mathcal{O}_k$ of integers of a finite algebraic number field k is a 1-dimensional scheme, called an "arithmetic curve". Grothendieck's thought unified number theory and algebraic geometry and led to arithmetical algebraic geometry today.

---

[1] Another origin of knot theory goes back to the work of William Thomson (Lord Kelvin; 1824–1907) on atomic theory. He considered atoms as knotted vortex tubes of ether and tried to classify atoms in the correspondence with knots. Although his theory on atoms was discarded in physics, the subsequent work by his collaborator, P. Tait, on the classification of knots led their theory (unexpectedly) to the development of knot theory in mathematics.

On the other hand, although there was work by J.B. Listing (a student of Gauss) and the physicist P.G. Tait, among others, there had not been remarkable progress in knot theory after Gauss, until the creation of topological notions such as fundamental groups and homology groups by H. Poincaré in the late 19th century. However, after Poincaré, the homology theory was rapidly developed by J.W. Alexander, S. Lefschetz and others, and subsequently knot theory was investigated by the homological and combinatorial group-theoretic methods (Alexander, M. Dehn, H. Seifert, K. Reidemeister et al.). Most notably, Alexander clarified the importance of knot theory in 3-dimensional topology by showing that any oriented connected closed 3-manifold is a finite covering of the 3-sphere ramified over a link, and also introduced the first polynomial invariant of a knot, called the Alexander polynomial (1928). It should also be noted that Reidemeister introduced the torsion of a CW complex and gave a homeomorphism classification of 3-dimensional lens spaces (1935).

The development of homology theory in topology had a favorable influence on algebraic number theory. Namely, T. Nakayama and J. Tate elaborated the theory of Galois cohomology and applied it to give a new proof of class field theory (1950s). On the other hand, motivated by Weil's conjecture asserting that there is a deep connection between arithmetic properties of an algebraic variety over a finite field and topological properties of the associated complex manifold, Grothendieck initiated the theory of étale topology and introduced the étale fundamental group and étale (co)homology group for schemes which enjoy properties similar to those of the topological fundamental group and singular (co)homology group. For example, class field theory for a finite algebraic number field k is stated as a sort of 3-dimensional Poincaré duality in the étale cohomology of $\mathrm{Spec}(\mathcal{O}_k)$ (M. Artin, J.-L. Verdier [136]). Furthermore, M. Artin and B. Mazur developed higher homotopy theory for schemes [9]. Along this line of thought, B. Mazur pointed out the analogy between a knot and prime as follows. We first note that the prime spectrum of the prime field $\mathbb{F}_p = \mathbb{Z}/p\mathbb{Z}$ for a prime number p has the following étale homotopy groups:

$$\pi_1^{\mathrm{\acute{e}t}}(\mathrm{Spec}(\mathbb{F}_p)) = \hat{\mathbb{Z}}, \quad \pi_i^{\mathrm{\acute{e}t}}(\mathrm{Spec}(\mathbb{F}_p)) = 0 \ (i \geq 2)$$

($\hat{\mathbb{Z}}$ being the pro-finite completion of $\mathbb{Z}$) and hence $\mathrm{Spec}(\mathbb{F}_p)$ is regarded as an arithmetic analogue of a circle S^1. Moreover, since $\mathrm{Spec}(\mathbb{Z})$ has the étale cohomological dimension 3 (up to 2-torsion) and $\pi_1^{\mathrm{\acute{e}t}}(\mathrm{Spec}(\mathbb{Z})) = 1$, the embedding

$$\mathrm{Spec}(\mathbb{F}_p) \hookrightarrow \mathrm{Spec}(\mathbb{Z})$$

is viewed as an analogue of an embedding, namely, a knot

$$S^1 \hookrightarrow \mathbb{R}^3.$$

The analogies between knots and primes and 3-manifolds and number rings were taken up later by M. Kapranov and A. Reznikov and the study of those analogies was christened *arithmetic topology* [98], [193], [194].

We note that in view of the analogy above, a knot group $G_K = \pi_1(\mathbb{R}^3 \setminus K)$ corresponds to a "prime group" $G_{\{(p)\}} = \pi_1^{\text{ét}}(\text{Spec}(\mathbb{Z}) \setminus \{(p)\})$. More generally, a link L corresponds to a finite set S of primes, and the link group $G_L = \pi_1(\mathbb{R}^3 \setminus L)$ corresponds to $G_S := \pi_1^{\text{ét}}(\text{Spec}(\mathbb{Z}) \setminus S)$, the Galois group of the maximal Galois extension of $\mathbb{Q}$ unramified outside S and the infinite prime.

1.3   The Outline of This Book

I started my own study of the analogy between knot theory and number theory, based on the analogy I noticed between the structures of a link group and a Galois group with restricted ramification (cf. Chap. 8). The purpose of this volume is to try to bridge two subjects that branched out after Gauss from this viewpoint by discussing in a parallel manner the analogies between the fundamental notions and theories of knot theory and number theory. The outline of this book is as follows. In Chap. 2, we recollect the basic materials from topology and number theory, which will be used in the subsequent chapters. In particular, we present a summary of fundamental groups, Galois theory for topological spaces and arithmetic rings, and arithmetic duality theorems in Galois and étale cohomology groups. In Chap. 3, we present basic analogies between 3-manifolds and number rings, knots and primes, which will be fundamental in later chapters. In Chaps. 4–6, from the viewpoint of the analogies between knots and primes, 3-manifolds and number rings in Chap. 3, we shall re-examine and unify Gauss' works on quadratic residues, genus theory and linking numbers in the context of Hilbert theory and genus theory for both 3-manifolds and number rings. In Chap. 7, we deal with idelic class field theory for 3-manifolds and finite algebraic number fields. For idelic theory for 3-manifolds, we introduce the notion of a stably generic link, the Chebotarev link, which consists of countably many knots. In Chap. 8, we present an analogy between Milnor's theorem on a link group and Koch's theorem on a pro-l Galois group with restricted ramification (l being a prime number). The analogy between linking numbers and power residue symbols is clearly explained by this group-theoretic point of view. Furthermore, in Chap. 9, we shall introduce arithmetic analogues of Milnor's higher linking numbers ($\overline{\mu}$-invariants). In particular, a triple symbol introduced by L. Rédei (1939) is interpreted as an arithmetic triple linking number. In Chap. 11, we shall describe, by using higher linking numbers, the Galois module structure of the l-part of the 1st homology of an l-fold cyclic ramified covering of S^3, and then we shall show, by using arithmetic higher linking numbers, an arithmetic analogue for the l-part of the ideal class group of a cyclic extension of $\mathbb{Q}$ of degree l, which may be regarded as a natural generalization of Gauss' genus theory. In Chaps. 10, 12 and 13, we discuss some analogies between Alexander–Fox theory and Iwasawa theory in a parallel manner, regarding the cyclotomic $\mathbb{Z}_p$-extension of a finite algebraic number field as an analogue of the infinite cyclic covering of a knot complement. Further, in Chap. 14, we present analogies between Alexander–Fox theory and Iwasawa theory and their non-Abelian generalization from the viewpoint of moduli and deformation of representations of knot and prime groups. For the case of 2-dimensional representations, in Chap. 15, we shall show some intriguing analogies between deformations of hyperbolic structures and of p-adic ordinary modular forms. Finally, in Chap. 16, we take up Dijkgraaf–Witten theory for 3-manifolds, which is a (2+1)-dimensional Chern–Simons topological quantum field theory with finite gauge group. We present its arithmetic analogue for number rings, called arithmetic Dijkgraaf–Witten theory.

As the history of mathematics tells us, pursuing analogies between different fields often raises new interesting problems and leads to development of both fields, and even opens a new field of study. As we explained above, the geometrization of number theory enabled us to pursue the analogy between knot theory and number theory, and it is the theme of this book. In the last few decades after the discovery of the Jones polynomial (1984), lots of knot invariants, called quantum invariants, have been constructed systematically in connection with mathematical physics. If one regards the classical electro-magnetic theory, from which Gauss' linking number originated, as Abelian gauge theory, this stream may be viewed as a development in the direction of non-Abelian gauge theory. On the other hand, Gauss' quadratic reciprocity is an origin of Abelian class field theory, and so non-Abelian class field theory may be seen as an arithmetic counterpart of non-Abelian gauge theory. Here the relation (Langlands conjecture) between the motivic L-functions associated to representations of Galois groups of number fields and the automorphic L-functions (zeta integrals over adèle groups) may correspond to the relation between the geometric invariants associated to representations of knot groups and 3-manifold groups and the path integral invariants (partition/correlation functions):

Abelian gauge theory $\longrightarrow$	Non-Abelian gauge theory
linking number	topological invariant
= Gaussian integral	= path integral invariant
Abelian class field theory $\longrightarrow$	Non-Abelian class field theory
Legendre symbol	motivic L-function
= Gaussian sum	= automorphic L-function

The aspect related to mathematical physics is an area of future investigation for arithmetic topology (cf. [48], [97], [131], [132]). The recent work initiated by M. Kim and his collaborators and successive works are the first concrete attempts to pursue the arithmetic analogues for number rings of Chern–Simons topological quantum field theory, based on arithmetic topology ([30], [31], [88], [102], [103]; see Chap. 16). I hope that pursuing further analogies between knot theory and number theory, in connection with mathematical physics, will raise new points of view and interesting problems, and lead to deeper understanding and progress of these fields.

Chapter 2
Preliminaries: Fundamental Groups and Galois Groups

The purpose of this chapter is to recollect the preliminary materials from topology and number theory, to aid the reader. In particular, we present a summary of fundamental groups and Galois theory for topological spaces and arithmetic rings in Sects. 2.1 and 2.2, since the analogies between topological and arithmetic fundamental/Galois groups are fundamental in this book. Sections 2.1 and 2.2 also contain basic concepts and examples in 3-dimensional topology and number fields which will be used in the subsequent chapters. In Sect. 2.3, we review arithmetic duality theorems in Galois and étale cohomology groups, from which we derive the main results and ingredients of local and global class field theory.

The reader who wants to know more or see precise proofs may consult [72], [133], [173] for fundamental groups and Galois theory, [10], [28], [73], [74], [146], [174], [176], [205], [206], [218] for Galois, étale cohomology and class field theory, and [24], [28], [83], [101], [127], [175], [186], [196] for the basic materials in knot theory and algebraic number theory. The author hopes that the forthcoming book [166] may not only cover this preliminary chapter together with proofs but also provide a sufficient foundation for the introduction to arithmetic topology.

2.1 The Case of Topological Spaces

Throughout this book, any topological space is assumed to be a PL-manifold and any map between topological spaces is assumed to be a PL-map (with obvious exceptions). For $d \in \mathbb{N}$, a d-manifold means a d-dimensional PL-manifold. Note that a manifold is arcwise-connected if and only if it is connected.

M. Morishita, *Knots and Primes*, Universitext,
https://doi.org/10.1007/978-981-99-9255-3_2

Let X be a connected topological space and fix a base point $x \in X$. For paths $\gamma, \gamma' : [0, 1] \to X$ with $\gamma(1) = \gamma'(0)$, we define a path $\gamma \vee \gamma' : [0, 1] \to X$ by $(\gamma \vee \gamma')(t) := \gamma(2t)$ if $0 \le t \le 1/2$ and $(\gamma \vee \gamma')(t) := \gamma'(2t - 1)$ if $1/2 \le t \le 1$. Let $\Omega(X, x)$ be the set of loops in X based at x. For $l, l' \in \Omega(X, x)$, we say that l and l' are homotopic, fixing the base point x, denoted by $l \simeq_x l'$, if there is a homotopy l_t connecting l and l' so that $l_t \in \Omega(X, x)$ for any $t \in I$. Let $\pi_1(X, x)$ be the set of equivalence classes, $\Omega(X, x)/\simeq_x$. We denote by $[l]$ the equivalence class of $l \in \Omega(X, x)$. Then $\pi_1(X, x)$ forms a group by the well-defined multiplication $[l] \cdot [l'] = [l \vee l']$. This is called the *fundamental group* of X with base point x. For another base point x', the correspondence $[l] \mapsto [\gamma^{-1} \vee l \vee \gamma]$ gives an isomorphism $\pi_1(X, x) \simeq \pi_1(X, x')$ where γ is a path from x to x'. Hence we sometimes omit the base point and write simply $\pi_1(X)$. A continuous map $f : X \to Y$ induces a homomorphism $f_* : \pi_1(X, x) \to \pi_1(Y, f(x))$ by $f_*([l]) := [f \circ l]$, and we have $f_* = g_*$ if $f, g : X \to Y$ are homotopic and $f(x) = g(x)$. Thus π_1 is a covariant functor from the homotopy category of based arcwise-connected topological spaces to the category of groups. We note that the Abelianization $\pi_1(X)/[\pi_1(X), \pi_1(X)]$ of $\pi_1(X)$ is isomorphic to the homology group $H_1(X)$ by sending $[l]$ to the homology class of l (Hurewicz theorem).

Example 2.1.1 (Circle) $S^1 := \{x \in \mathbb{R}^2 \mid ||x|| = 1\}$.

Let l be the loop based at $x \in S^1$ which goes once around the circle counterclockwise. Then $\pi_1(S^1, x)$ is an infinite cyclic group generated by $[l]$ (Fig. 2.1).

Example 2.1.2 (Solid Torus and the Boundary Torus) $V := D^2 \times S^1$, where $D^2 := \{x \in \mathbb{R}^2 \mid ||x|| \le 1\}$ is the unit 2-disk.

Since V is homotopically equivalent to S^1, one has $\pi_1(V) = \pi_1(S^1) = \langle[\beta]\rangle$, where $\beta = \{b\} \times S^1, b \in \partial D^2$.

The boundary ∂V of V is a 2-dimensional torus $T^2 := S^1 \times S^1 = \partial V$.

Define the projection $p_i : T^2 \to S^1$ for $i = 1, 2$ by $p_1(x, y) := x$, $p_2(x, y) := y$. Then $p_{1*} \times p_{2*}$ induces an isomorphism $\pi_1(T^2) \simeq \pi_1(S^1) \times \pi_1(S^1) = \langle[\alpha]\rangle \times \langle[\beta]\rangle$, where $\alpha = \partial D^2 \times \{a\}, a \in S^1$.

Two loops α and β on T^2 are called a *meridian* and a *longitude* respectively (Fig. 2.2).

Example 2.1.3 (n-Sphere) $S^n := \{x \in \mathbb{R}^{n+1} \mid ||x|| = 1\}$ $(n \ge 2)$.

Let $[l] \in \pi_1(S^n, x)$. We may take a point $y \in S^3 \setminus l$ so that $l \subset S^n \setminus \{y\}$. Since $S^n \setminus \{y\}$ is homotopically equivalent to $\mathbb{R}^{n-1}$ and so contractible, $[l] = 1$. Hence one has $\pi_1(S^n) = \{1\}$. A connected space X is called *simply-connected* if $\pi_1(X) = \{1\}$.

Fig. 2.1 A circle and the loop which goes once around it counterclockwise

Fig. 2.2 A solid torus, the
boundary torus, a meridian
and a longitude

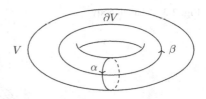

The Poincaré conjecture, which was proved by G. Perelman (2003), asserts that a simply-connected closed 3-manifold is homeomorphic to S^3.

The *van Kampen theorem* provides a useful method to present a fundamental group in terms of generators and relations. Let $F(x_1, \ldots, x_r)$ denote the free group on letters (or words) $x_1, \ldots, x_r$. For $R_1, \ldots, R_s \in F(x_1, \ldots, x_r)$, let $\langle\langle R_1, \ldots, R_s \rangle\rangle$ denote the smallest normal subgroup of $F(x_1, \ldots, x_r)$ containing $R_1, \ldots, R_s$. When a group G is isomorphic to the quotient group $F(x_1, \cdots, x_r)/\langle\langle R_1, \ldots, R_s \rangle\rangle$, we write G in the following form:

$$G = \langle x_1, \ldots, x_r \mid R_1 = \cdots = R_s = 1 \rangle$$

and call it a *presentation* of G in terms of generators and relations. Note that the choices of generators $x_1, \ldots, x_r$ and relators $R_1 = \cdots = R_s$ are not unique. If $r - s = k$, we say that G has a presentation of *deficiency* k. Now, let X be a topological space and suppose that there are two open subsets X_1 and X_2 of X such that $X = X_1 \cup X_2$ and $X_1 \cap X_2$ is non-empty. We assume that X, X_1, X_2 and $X_1 \cap X_2$ are arcwise-connected. Take a base point $x \in X_1 \cap X_2$ and suppose that we are given the following presentations:

$$\pi_1(X_1, x) = \langle x_1, \ldots, x_r \mid R_1 = \cdots = R_s = 1 \rangle,$$
$$\pi_1(X_2, x) = \langle y_1, \ldots, y_t \mid Q_1 = \cdots = Q_u = 1 \rangle,$$
$$\pi_1(X_1 \cap X_2, x) = \langle z_1, \ldots, z_v \mid P_1 = \cdots = P_w = 1 \rangle.$$

The inclusion maps $i_1 : X_1 \cap X_2 \hookrightarrow X_1$, $i_2 : X_1 \cap X_2 \hookrightarrow X_2$ induce the homomorphisms $i_{1*} : \pi_1(X_1 \cap X_2, x) \to \pi_1(X_1, x)$, $i_{2*} : \pi_1(X_1 \cap X_2, x) \to \pi_1(X_2, x)$. Then the van Kampen theorem asserts that $\pi_1(X, x)$ is given by amalgamating $\pi_1(X_1 \cap X_2, x)$ in $\pi_1(X_1, x)$ and $\pi_1(X_2, x)$, namely,

$$\pi_1(X, x) = \left\langle \begin{matrix} x_1, \ldots, x_r \\ y_1, \ldots, y_t \end{matrix} \,\middle|\, \begin{matrix} R_1 = \cdots = R_s = Q_1 = \cdots = Q_u = 1 \\ i_{1*}(z_1)i_{2*}(z_1)^{-1} = \cdots = i_{1*}(z_v)i_{2*}(z_v)^{-1} = 1 \end{matrix} \right\rangle.$$

Example 2.1.4 (Handlebody) Let us prepare g copies of a handle $D^2 \times D^1 = D^2 \times [0, 1]$ and a 3-ball D^3. For each handle, we fix a homeomorphism $D^2 \times \partial D^1 \to \partial D^3 = S^2$ and attach g handles to D^3 by identifying $x \in D^2 \times \partial D^2$ with $f(x)$. The resulting 3-manifold is called a *handlebody* of genus g and is denoted by H_g (Fig. 2.3).

Fig. 2.3 A handlebody

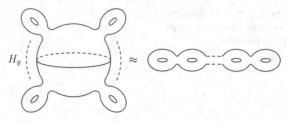

Fig. 2.4 A bouquet

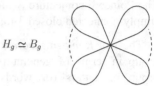

H_g is homotopically equivalent to a bouquet B_g obtained by attaching g copies of S^1 at one point b (Fig. 2.4).

Letting x_i be the loop starting from b and going once around the i-th S^1, the van Kampen theorem yields $\pi_1(H_g) = \pi_1(B_g) = F(x_1, \ldots, x_g)$.

Example 2.1.5 (Lens Space) Let V_1, V_2 be oriented solid tori and let $f : \partial V_2 \xrightarrow{\approx} \partial V_1$ be a given orientation-reversing homeomorphism. We then make an oriented connected closed 3-manifold $M = V_1 \cup_f V_2$ by identifying $x \in \partial V_2$ with $f(x) \in \partial V_1$ in the disjoint union of V_1 and V_2. Let α_i and β_i denote a meridian and a longitude on V_i respectively for each $i = 1, 2$. By Example 2.1.2, we may write

$$f_*([\alpha_2]) = p[\beta_1] + q[\alpha_1], \ (p, q) = 1$$

in a unique way. The topological type of the space M is determined by the pair (p, q) of integers above and so M is called the *lens space* of type (p, q) and denoted by $L(p, q)$. Let us calculate the fundamental group of $L(p, q)$. Let $i_1 : \partial V_2 \to V_1$ be the composite of f with the inclusion map $\partial V_1 \hookrightarrow V_1$ and let $i_2 : \partial V_2 \hookrightarrow V_2$ be the inclusion map. Noting $\pi_1(V_i) = \langle \beta_i \rangle$ and $\pi_1(\partial V_2) = \langle \alpha_2 \rangle \times \langle \beta_2 \rangle$ and applying the van Kampen theorem, we have

$$\begin{aligned}
\pi_1(L(p, q)) &= \langle \beta_1, \beta_2 \,|\, i_{1*}(\alpha_2) = i_{2*}(\alpha_2), i_{1*}(\beta_2) = i_{2*}(\beta_2) \rangle \\
&= \langle \beta_1, \beta_2 \,|\, \beta_1^p \alpha_1^q = 1, i_{1*}(\beta_2) = \beta_2 \rangle \\
&= \langle \beta_1 \,|\, \beta_1^p = 1 \rangle \\
&\simeq \mathbb{Z}/p\mathbb{Z}.
\end{aligned}$$

So $\pi_1(L(p, q))$ is a finite cyclic group except the case $p = 0$ for which we have $L(0, \pm 1) \approx S^2 \times S^1$.

More generally, for oriented handlebodies V_1, V_2 of genus g and an orientation-reversing homeomorphism $f : \partial V_2 \xrightarrow{\approx} \partial V_1$, we can make an oriented connected closed 3-manifold $M := V_1 \cup_f V_2$ in a similar manner. One calls $M = V_1 \cup_f V_2$ a *Heegaard splitting* of M and g the genus of the splitting. Conversely, it is known that any orientable connected closed 3-manifold has such a Heegaard splitting. For a proof of this, we refer to [77, Ch.2]. The fundamental group of a 3-manifold with a Heegaard splitting is computed in a similar way to the case of a lens space.

Example 2.1.6 (Knot Group, Link Group) A *knot* is the image of an embedding of S^1 into S^3. So, by our assumption, a knot is always assumed to be a simple closed polygon in this book. We denote by V_K a tubular neighborhood of K. The complement $X_K := S^3 \setminus \mathrm{int}(V_K)$ of an open tubular neighborhood $\mathrm{int}(V_K)$ in S^3 is called the *knot exterior*. It is a compact 3-manifold with a boundary being a 2-dimensional torus. A *meridian* of K is a closed (oriented) curve which is the boundary of a disk D^2 in V_K. A *longitude* of K is a closed curve on ∂X_K which intersects with a meridian at one point and is null-homologous in X_K (Fig. 2.5).

The fundamental group $\pi_1(X_K) = \pi_1(S^3 \setminus K)$ is called the *knot group* of K and is denoted by G_K. Firstly, let us explain how we can obtain a presentation of G_K. We may assume $K \subset \mathbb{R}^3$. A projection of a knot K onto a plane in $\mathbb{R}^3$ is called *regular* if there are only finitely many multiple points which are all double points and no vertex of K is mapped onto a double point. There are sufficiently many regular projections of a knot. We can draw a picture of a regular projection of a knot in the way that at each double point the overcrossing line is marked. So a knot can be reconstructed from its regular projection. Now let us explain how we can get a presentation of G_K from a regular projection of K, by taking a trefoil for K as an illustration.

(0) First give a regular projection of a knot K (Fig. 2.6):

Fig. 2.5 A meridian and a longitude of a knot

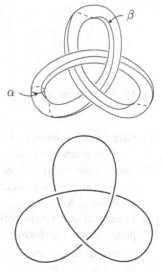

Fig. 2.6 A regular projection of a knot (trefoil)

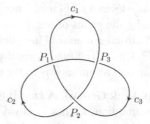

Fig. 2.7 Giving an orientation of the knot and dividing it into arcs $c_1, \ldots, c_n$

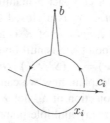

Fig. 2.8 Taking a base point b and a loop x_i coming down from b, going once around under each c_i from the right to the left, and returning to b

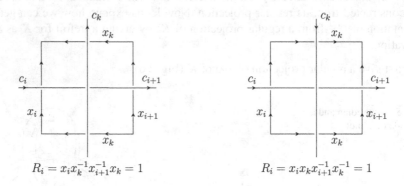

$$R_i = x_i x_k^{-1} x_{i+1}^{-1} x_k = 1 \qquad\qquad R_i = x_i x_k x_{i+1}^{-1} x_k^{-1} = 1$$

Fig. 2.9 The relations among x_i's at a double point on the regular projection

(1) Give an orientation to K and divide K into arcs $c_1, \cdots, c_n$ so that c_i ($1 \le i \le n-1$) is connected to c_{i+1} at a double point and c_n is connected to c_1 (Fig. 2.7):

(2) Take a base point b above K (for example $b = \infty$) and let x_i be a loop coming down from b, going once around under c_i from the right to the left, and returning to b (Fig. 2.8):

(3) In general one has the following two ways of crossing among c_i's at each double point. From the former case one derives the relation $R_i = x_i x_k^{-1} x_{i+1}^{-1} x_k = 1$, and from the latter case one derives the relation $R_i = x_i x_k x_{i+1}^{-1} x_k^{-1} = 1$ (Fig. 2.9):

Thus we have n relations $R_1 = \cdots = R_n = 1$ for n double points $P_1, \ldots, P_n$, which give a presentation of G_K, $G_K = \langle x_1, \ldots, x_n \mid R_1 = \cdots = R_n = 1 \rangle$ (Fig. 2.10).

Among these n relations we can derive any one from the other relations as follows. Let E be a plane, below K, on which we have a regular projection of K. Let C be an oriented circle such that a projection of K on E is lying inside C. Let γ be a path in X_K starting from the base point b to a fixed point Q on C and let $l := \gamma \vee C \vee \gamma^{-1}$. Note that $[l]$ is the identity in G_L. On the other hand, let l_i be a path in E starting from Q, going toward P_i and once around P_i with the same orientation as C, and returning Q. Then we see l is homotopic to $\prod\limits_{i=1}^{n} \gamma \vee l_i \vee \gamma^{-1}$ (Fig. 2.11):

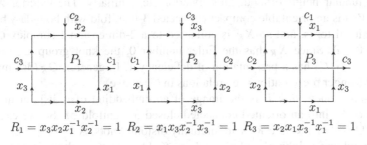

$$R_1 = x_3 x_2 x_1^{-1} x_2^{-1} = 1 \quad R_2 = x_1 x_3 x_2^{-1} x_3^{-1} = 1 \quad R_3 = x_2 x_1 x_3^{-1} x_1^{-1} = 1$$

Fig. 2.10 Giving the relations $R_i = 1$ ($i = 1, \ldots, n$) among x_j's at all double points P_i

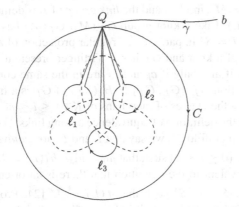

Fig. 2.11 Considering a circle C on the plane of the regular projection such that the regular projection of the knot is lying inside C. One then obtains one relation among R_i's and hence among x_j's

Since a small circle around P_i corresponds to R_i or R_i^{-1}, there are $z_i \in$ $F(x_1, \ldots, x_n)$ such that one has

$$\prod_{i=1}^{n} z_i R_i^{\pm 1} z_i^{-1} = 1. \tag{2.1}$$

Thus G_K has a presentation of deficiency 1.

A presentation of G_K obtained in the way described above is called a *Wirtinger presentation*. As we can see from the form of each relation in (3), $x_1, \ldots, x_n$ are conjugate to each other in G_K. Therefore the Abelianization $G_K/[G_K, G_K] \simeq H_1(X_K)$ of G_K is an infinite cyclic group generated by the class of a meridian of K.

We can, of course, consider a knot in any orientable connected 3-manifold and define a tubular neighborhood, knot exterior etc, similarly. The exterior $X_K = M \setminus \text{int}(V_K)$ is an orientable compact connected 3-manifold with boundary being a 2-dimensional torus, and so X_K is collapsed to a 2-dimensional complex C with a single 0-cell. Since X_K has the Euler number 0, the knot group $G_K(M) := \pi_1(X_K) = \pi_1(C)$ has a presentation of deficiency 1. In general, $G_K(M)$ may not have a Wirtinger presentation (i.e., relations in (3) above).

An r-component *link* L is the image of an embedding of a disjoint union of r copies of S^1 into an oriented connected closed 3-manifold M. So we can write $L = K_1 \cup \cdots \cup K_r$ where K_i's are mutually disjoint knots. A 1-component link is a knot. A tubular neighborhood V_L of $L = K_1 \cup \cdots \cup K_r$ is the union of tubular neighborhoods of K_i, $V_L = V_{K_1} \cup \cdots \cup V_{K_n}$ ($V_{K_i} \cap V_{K_j} = \emptyset$ for $i \neq j$). The *exterior* of L is $X_L := M \setminus \text{int}(V_L)$ and the *link group* of L is defined by $G_L(M) := \pi_1(X_L) = \pi_1(M \setminus L)$. Like a knot group $G_K(M)$, $G_L(M)$ has a presentation of deficiency 1. When $M = S^3$ in particular, a regular projection of a link L is defined similarly to the case of a knot and G_L has a Wirtinger presentation. Here loops x_i and x_j are conjugate if and only if c_i and c_j are in the same component of a link and so the Abelianization $G_L/[G_L, G_L] \simeq H_1(X_L)$ of G_L is a free Abelian group of rank r generated by the classes of meridians of K_i, $1 \leq i \leq r$.

Finally, let us give the definition of equivalence among links. For links L, L' in an oriented connected 3-manifold M, we say that L and L' are *equivalent* if there is an isotopy $h_t : M \xrightarrow{\approx} M$ ($0 \leq t \leq 1$) such that $h_0 = \text{id}_M, h_1(L) = L'$. For links in S^3, this condition is equivalent to the condition that there is an orientation-preserving homeomorphism $f : S^3 \xrightarrow{\approx} S^3$ such that $f(L) = L'$ [24, Proposition 1.10]. A quantity $\text{inv}(L)$ defined on the set of all links is called a *link invariant* if $\text{inv}(L) = \text{inv}(L')$ for any two equivalent links L and L'. Likewise a *knot invariant* is a quantity defined on the set of all knots, which takes the same for any two equivalent knots. For example, a knot group is a knot invariant and a link group is a link invariant.

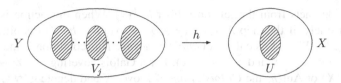

Fig. 2.12 A covering

Next, let us recall basic materials concerning covering spaces. Let X be a connected space. A continuous map $h : Y \to X$ is called an *(unramified) covering* if for any $x \in X$, there is an open neighborhood U of x such that

$$\begin{cases} (1) \ \ h^{-1}(U) = \bigsqcup_{j \in J} V_j, \ \ V_i \cap V_j = \emptyset \ (i \neq j), \\ (2) \ \ h|_{V_j} : V_j \xrightarrow{\approx} U \ \text{(homeomorphism)}, \end{cases}$$

where V_j is a connected component of $h^{-1}(U)$ and an open subset of Y (Fig. 2.12).

A covering $h' : Y' \to X$ is called a *subcovering* of $h : Y \to X$ if there is a continuous map $\varphi : Y \to Y'$ such that $h' \circ \varphi = h$. Then φ is also a covering, and we denote by $C_X(Y, Y')$ the set of all such φ. If there is a homeomorphic $\varphi \in C_X(Y, Y')$, Y and Y' are said to be isomorphic over X. The set of isomorphisms $\varphi \in C_X(Y, Y)$ forms a group, called the group of *covering transformations* of $h : Y \to X$, and is denoted by $\mathrm{Aut}(Y/X)$ or $\mathrm{Aut}(h)$.

The most basic fact in covering theory is the following lifting property of a path and its homotopy:

Proposition 2.1.7 *Let $h : Y \to X$ be a covering. For any path $\gamma : [0, 1] \to X$ and any $y \in h^{-1}(x)$ ($x = \gamma(0)$), there exists a unique lift $\hat{\gamma} : [0, 1] \to Y$ of γ (i.e, $h \circ \hat{\gamma} = \gamma$) with $\hat{\gamma}(0) = y$. Furthermore, for any homotopy γ_t ($t \in [0, 1]$) of γ with $\gamma_t(0) = \gamma(0)$ and $\gamma_t(1) = \gamma(1)$, there exists a unique lift of $\hat{\gamma}_t$ such that $\hat{\gamma}_t$ is the homotopy of $\hat{\gamma}$ with $\hat{\gamma}_t(0) = \hat{\gamma}(0)$ and $\hat{\gamma}_t(1) = \hat{\gamma}(1)$.*

In the following, we assume that any covering space is connected. By Proposition 2.1.7, the cardinality of the fiber $h^{-1}(x)$ is independent of $x \in X$. So we call $\#h^{-1}(x)$ the *degree* of $h : Y \to X$ which is denoted by $\deg(h)$ or $[Y : X]$. We define the right action of $\pi_1(X, x)$ on $h^{-1}(x)$ as follows. For $[l] \in \pi_1(X, x)$ and $y \in h^{-1}(x)$, we define $y.[l]$ to be the terminus $\hat{l}(1)$ where $\hat{l}$ is the lift of l with origin $\hat{l}(0) = y$. It is a transitive action such that the stabilizer of y is $h_*(\pi_1(Y, y))$ by Proposition 2.1.7 and hence one has a bijection $h^{-1}(x) \simeq h_*(\pi_1(Y, y)) \backslash \pi_1(X, x)$. The induced representation $\rho_x : \pi_1(X, x) \to \mathrm{Aut}(h^{-1}(x))$ is called the *monodromy permutation representation* of $\pi_1(X, x)$, where $\mathrm{Aut}(h^{-1}(x))$ denotes the group of permutations on $h^{-1}(x)$ so that the multiplication $\sigma_1 \cdot \sigma_2$ is defined by the composite of maps $\sigma_2 \circ \sigma_1$ for $\sigma_1, \sigma_2 \in \mathrm{Aut}(h^{-1}(x))$. The representation ρ_x induces an isomorphism $\mathrm{Im}(\rho_x) \simeq \pi_1(X, x) / \cap_{y \in h^{-1}(x)} h_*(\pi_1(Y, y))$. It can be shown that the isomorphism class of a covering is determined by the equivalence class of the monodromy representation. On the other hand, the group $\mathrm{Aut}(Y/X)$ of covering transformations acts from the left on a fiber $h^{-1}(x)$. When this action is simply-transitive, namely, if the map $\mathrm{Aut}(Y/X) \ni \sigma \mapsto \sigma(y) \in h^{-1}(x)$ is bijective for $y \in h^{-1}(x)$, $h : Y \to X$ is called a *Galois covering*. This condition is independent of the choice of $x \in X$ and $y \in h^{-1}(x)$. For a Galois covering $h : Y \to X$, we call $\mathrm{Aut}(Y/X)$ or $\mathrm{Aut}(h)$ the *Galois group* of Y over X and denote it by $\mathrm{Gal}(Y/X)$ or $\mathrm{Gal}(h)$, respectively. The following is the main theorem of the Galois theory for coverings.

Theorem 2.1.8 (Galois Correspondence) *The correspondence* $(h : Y \to X) \mapsto h_*(\pi_1(Y, y))$ $(y \in h^{-1}(x))$ *gives rise to the following bijection:*

$$\{connected\ covering\ h : Y \to X\}/isom.\ over\ X$$

$$\xrightarrow{\sim} \{subgroup\ of\ \pi_1(X, x)\}/conjugate.$$

Furthermore, this bijection satisfies the following properties:

- $h' : Y' \to X$ *is a subcovering of* $h : Y \to X$ $\Leftrightarrow$ $h'_*(\pi_1(Y', y'))$ $(y' \in h'^{-1}(x))$ *is a subgroup of* $h_*(\pi_1(Y, y))$ $(y \in h^{-1}(x))$ *up to conjugate.*
- $h : Y \to X$ *is a Galois covering* $\Leftrightarrow$ $h_*(\pi_1(Y, y))$ $(y \in h^{-1}(x))$ *is a normal subgroup of* $\pi_1(X, x)$. *Then one has* $\mathrm{Gal}(Y/X) \simeq \pi_1(X, x)/h_*(\pi_1(Y, y))$.

More generally, we can replace $\pi_1(X, x)$ *by* $\mathrm{Gal}(Z/X)$ *for a fixed Galois covering* $Z \to X$ *in the above bijection, and then we have a similar bijection:*

$$\{connected\ subcovering\ of\ Z \to X\}/isom.\ over\ X.$$

$$\xrightarrow{\sim} \{subgroup\ of\ \mathrm{Gal}(Z/X)\}/conjugate.$$

Thus the fundamental group of a space X may be viewed as a group which controls the symmetry of the set of coverings of X. In particular, the covering $\tilde{h} : \tilde{X} \to X$ (unique up to isom. over X) which corresponds to the identity group of $\pi_1(X, x)$ is called the *universal covering* of X. The universal covering has the following properties (U):

$$(\mathrm{U}) \begin{cases} \text{(i)} & \text{Fixing } \tilde{x} \in \tilde{X}, \text{ the map } C_X(\tilde{X}, Y) \ni \varphi \mapsto \varphi(\tilde{x}) \in h^{-1}(x) \\ & \text{is bijective for any covering } h : Y \longrightarrow X \ (x = \tilde{h}(\tilde{x})). \\ \text{(ii)} & \mathrm{Gal}(\tilde{X}/X) \simeq \pi_1(X, x) \ (x = \tilde{h}(\tilde{x})). \end{cases}$$

Example 2.1.9 The universal covering of S^1 is given by

$$\tilde{h} : \mathbb{R} \longrightarrow S^1; \quad \tilde{h}(\theta) := (\cos(2\pi\theta), \sin(2\pi\theta)).$$

Let l be a loop starting from a base point x and going once around S^1 counterclockwise. Define the covering transformation $\sigma \in \mathrm{Gal}(\mathbb{R}/S^1)$ by $\sigma(\theta) := \theta+1$. Then the correspondence $\sigma^n \mapsto [l^n]$ $(n \in \mathbb{Z})$ gives an isomorphism $\mathrm{Gal}(\mathbb{R}/S^1) \simeq \pi_1(S^1, x)$.

Any subgroup ($\neq \{1\}$) of $\pi_1(X, x) = \langle [l] \rangle$ is given by $\langle [l^n] \rangle$ for some $n \in \mathbb{N}$ and the corresponding covering is given by

$$h_n : \mathbb{R}/n\mathbb{Z} \longrightarrow S^1; \; h_n(\theta \bmod n\mathbb{Z}) := (\cos(2\pi\theta), \sin(2\pi\theta)).$$

Example 2.1.10 The universal covering of a 2-dimensional torus $T^2 = S^1 \times S^1$ is the product of two copies of the universal covering S^1, namely,

$$\tilde{h} : \mathbb{R}^2 \longrightarrow T^2; \; \tilde{h}(\theta_1, \theta_2) := ((\cos(2\pi\theta_1), \sin(2\pi\theta_1)), (\cos(2\pi\theta_2), \sin(2\pi\theta_2))).$$

Define the covering transformation $\sigma_1, \sigma_2 \in \mathrm{Gal}(\mathbb{R}^2/T^2)$ by $\sigma_1(\theta_1, \theta_2) := (\theta_1 + 1, \theta_2), \sigma_2(\theta_1, \theta_2) := (\theta_1, \theta_2 + 1)$. Then the correspondence $\sigma_1 \mapsto [\alpha]$ (meridian), $\sigma_2 \mapsto [\beta]$ (longitude) gives an isomorphism $\mathrm{Gal}(\mathbb{R}^2/T^2) \simeq \pi_1(T^2)$.

Example 2.1.11 Let $L(p, q)$ be a lens space of type (p, q) (Example 2.1.5), where p and q are coprime integers. When $p = 0$, $L(0, \pm 1) = S^2 \times S^1$ and so the universal covering is given by $S^2 \times \mathbb{R}$. Assume $p \neq 0$ and let us construct the universal covering $L(p, q)$.[1] We identify S^1 with $\mathbb{R}/\mathbb{Z}$, D^2 with $(\mathbb{R}/\mathbb{Z} \times (0, 1]) \cup \{(0, 0)\}$, and regard a solid torus V as $\mathbb{R}/\mathbb{Z} \times \mathbb{R}/\mathbb{Z} \times [0, 1]$, ∂V as $\mathbb{R}/\mathbb{Z} \times \mathbb{R}/\mathbb{Z}$. Let V_1, V_2, V_1', V_2' be copies of V. Let us consider the following map

$$f : \partial V_1 \longrightarrow \partial V_2; \; f(x, y) := \left(qx + \frac{y}{p}, px \right).$$

Since

$$\det \begin{pmatrix} q & p \\ \dfrac{1}{p} & 0 \end{pmatrix} = -1,$$

f is an orientation-reversing homeomorphism and $L(p, q)$ is obtained from the disjoint union of V_1 and V_2 by identifying ∂V_1 with ∂V_2 via f. Next, consider the following orientation-reversing homeomorphism:

$$g : \partial V_1' \longrightarrow \partial V_2'; \; g(x, y) := (y, x).$$

The space obtained from the disjoint union of V_1' and V_2' by identifying $\partial V_1'$ with $\partial V_2'$ via g is S^3. Now define the map $h : S^3 = V_1' \cup_g V_2' \to L(p, q) = V_1 \cup_f V_2$ by

$$h|_{V_1'} : V_1' \longrightarrow V_1; \; h|_{V_1'}(x, y, z) := (x, p(y - qx), z),$$
$$h|_{V_2'} : V_2' \longrightarrow V_2; \; h|_{V_2'}(x, y, z) := (x, py, z).$$

---

[1] The following argument is due to S. Miyasaka, a graduate student at Kyoto University (2005).

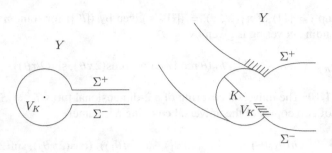

Fig. 2.13 Cutting the knot exterior along a Seifert surface

Then we see that h is well-defined and $h|_{V'_i}$ ($i = 1, 2$) are both p-fold cyclic coverings, and hence h is a p-fold cyclic covering. Since S^3 is simply connected, $h : S^3 \to L(p, q)$ defined as above is the universal covering.

Example 2.1.12 Let $K \subset S^3$ be a knot, V_K a tubular neighborhood, $X_K := S^3 \setminus \text{int}(V_K)$ the exterior of K, and $G_K := \pi_1(X_K)$ the knot group. Let α be a meridian of K. Since $G_K/[G_K, G_K]$ is the infinite cyclic group generated by the class of α, the map sending α to 1 defines a surjective homomorphism $\psi_\infty : G_K \to \mathbb{Z}$. Let $h_\infty : X_\infty \to X_K$ be the covering corresponding to $\text{Ker}(\psi_\infty)$ in Theorem 2.1.8. The covering space X_∞ is independent of the choice of α and called the *infinite cyclic covering* of X_K. Let τ be the generator of $\text{Gal}(X_\infty/X_K)$ corresponding to $1 \in \mathbb{Z}$. For each $n \in \mathbb{N}$, $\psi_n : G_K \to \mathbb{Z}/n\mathbb{Z}$ be the composite of ψ_∞ with the natural homomorphism $\mathbb{Z} \to \mathbb{Z}/n\mathbb{Z}$, and let $h_n : X_n \to X_K$ be the covering corresponding to $\text{Ker}(\psi_n)$. The space X_n is the unique subcovering of X_∞ such that $\text{Gal}(X_n/X_K) \simeq \mathbb{Z}/n\mathbb{Z}$. We denote by the same τ for the generator of $\text{Gal}(X_n/X_K)$ corresponding to 1 mod $n\mathbb{Z}$. The covering spaces X_n ($n \in \mathbb{N}$), X_∞ are constructed as follows. First, take a *Seifert surface* of K, an oriented connected surface Σ_K whose boundary is K. Let Y be the space obtained by cutting X_K along $X_K \cap \Sigma_K$. Let Σ^+, Σ^- be the surfaces, which are homeomorphic to $X_K \cap \Sigma_K$, as in the following picture (Fig. 2.13).

Let $Y_0, \ldots, Y_{n-1}$ be copies of Y and let X_n be the space obtained from the disjoint union of all Y_i's by identifying Σ_0^+ with Σ_1^-, $\ldots$, and Σ_{n-1}^+ with Σ_0^- (Fig. 2.14).

Define $h_n : X_n \to X_K$ as follows: If $y \in Y_i \setminus (\Sigma_i^+ \cup \Sigma_i^-)$, define $h_n(y)$ to be the corresponding point of Y via $Y_i = Y$. If $y \in \Sigma_i^+ \cup \Sigma_i^-$, define $h_n(y)$ to be the corresponding point of Σ_K via $\Sigma_i^+, \Sigma_i^- \subset \Sigma_K$. By the construction, $h_n : X_n \to X_K$ is an n-fold cyclic covering. The generating covering transformation $\tau \in \text{Gal}(X_n/X_K)$ is then given by the shift sending Y_i to Y_{i+1} ($i \in \mathbb{Z}/n\mathbb{Z}$). This construction is readily extended to the case $n = \infty$. Namely, taking copies Y_i ($i \in \mathbb{Z}$) of Y, let X_K^∞ be the space obtained from the disjoint union of all Y_i's by identifying Σ_i^+ with Σ_{i+1}^- ($i \in \mathbb{Z}$) (Fig. 2.15).

The generating covering transformation $\tau \in \text{Gal}(X_K^\infty/X_K)$ is given by the shift sending Y_i to Y_{i+1} ($i \in \mathbb{Z}$).

Fig. 2.14 The n-fold cyclic covering of a knot complement

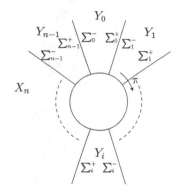

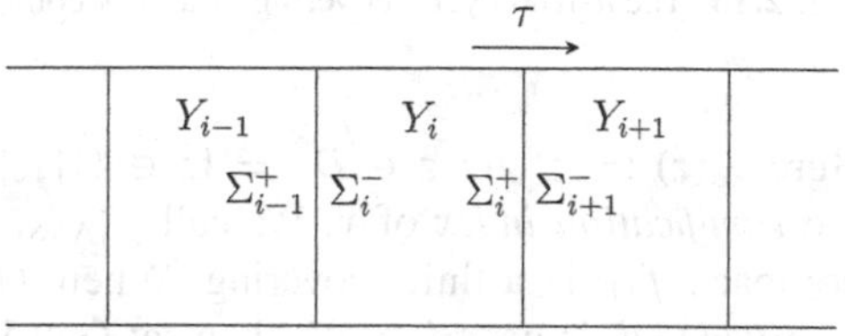

Fig. 2.15 The infinite cyclic covering of a knot complement

Y_{i-1}	Y_i	Y_{i+1}
Σ_{i-1}^+ Σ_i^-	Σ_i^+ Σ_{i+1}^-	

Example 2.1.13 The *Abelian fundamental group* of X is the Abelianization of $\pi_1(X)$, which we denote by $\pi_1^{ab}(X)$. By the Hurewicz theorem, $H_1(X) \simeq \pi_1^{ab}(X)$. The covering space corresponding to the commutator subgroup $[\pi_1(X), \pi_1(X)]$ in Theorem 2.1.8 is called the *maximal Abelian covering* of X, which we denote by X^{ab}. Since $\pi_1^{ab}(X) \simeq \mathrm{Gal}(X^{ab}/X)$, we have a canonical isomorphism

$$H_1(X) \simeq \mathrm{Gal}(X^{ab}/X),$$

which we call the *Hurewicz isomorphism* for the convenience. Therefore Abelian coverings of X are controlled by the homology group $H_1(X)$. This may be regarded as a topological analogue of unramified class field theory, which will be presented in Example 2.3.4.

Finally, we shall consider ramified coverings. Let M, N be n-manifolds ($n \geq 2$) and let $f : N \to M$ be a continuous map. Set $S_N := \{y \in N \mid f$ is not a homeomorphism in a neighborhood of $y\}$, $Y := N \setminus S_N$ and $S_M := f(S_N)$, $X := M \setminus S_M$. Let $D^k := \{x \in \mathbb{R}^k \mid \|x\| \leq 1\}$. Then $f : N \to M$ is called a *covering ramified over S_M* if the following conditions are satisfied:

$\begin{cases} (1) \ f|_Y : Y \setminus X \setminus S_M \text{ is a covering.} \\ (2) \ \text{For any } y \in S_N, \text{ there are a neighborhood } V \text{ of } y, \text{ a neighborhood } U \\ \quad \text{of } f(y), \text{ a homeomorphism } \varphi : V \xrightarrow{\approx} D^2 \times D^{n-2}, \ \psi : U \xrightarrow{\approx} D^2 \times D^{n-2} \\ \quad \text{and an integer } e = e(y)(> 1) \text{ such that } (f_e \times \mathrm{id}_{D^{n-2}}) \circ \varphi = \psi \circ f. \end{cases}$

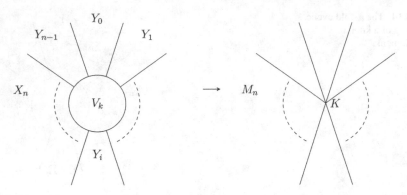

Fig. 2.16 The n-fold cyclic covering of a knot complement and its Fox completion

Here $g_e(z) := z^e$ for $z \in D^2 = \{z \in \mathbb{C} \,|\, |z| \leq 1\}$. The integer $e = e(y)$ is called the *ramification index* of y. We call $f|_{N \setminus S_N}$ the covering associated to f. If N is compact, $f|_Y$ is a finite covering. When $f|_Y$ is a Galois covering, f is called a *ramified Galois covering*. We then set $\mathrm{Gal}(N/M) := \mathrm{Gal}(Y/X)$, called the Galois group of $f : N \to M$.

Example 2.1.14 For a knot $K \subset S^3$, let V_K be a tubular neighborhood of K and $X_K = S^3 \setminus \mathrm{int}(V_K)$ the knot exterior. Let $h_n : X_n \to X_K$ be the n-fold cyclic covering defined in Example 2.1.12. Note that $h_n|_{\partial X_n} : \partial X_n \to \partial X_K$ is an n-fold cyclic covering of tori and a meridian of ∂X_n is given by $n\alpha$, where α is a meridian on ∂X_K. So we attach $V = D^2 \times S^1$ to X_n gluing ∂V with ∂X_n so that a meridian $\partial D^2 \times \{*\}$ coincides with $n\alpha$. Let M_n be the closed 3-manifold obtained in this way (Fig. 2.16).

Define $f_n : M_n \to S^3$ by $f_n|_{X_n} := h_n$ and $f_n|_V := f_n \times \mathrm{id}_{S^1}$. Then f_n is a covering ramified over K and the associated covering is h_n. $f_n : M_n \to S^3$ is called the completion of $h_n : X_n \to X_K$.

The completion given in Example 2.1.14 is called the *Fox completion* and such a completion can be constructed for any finite covering of a link exterior. In fact, the Fox completion can be defined for any covering (more generally, for a spread) of locally connected T_1-spaces [54]. Here let us explain an outline of the construction for a finite covering of a link exterior. Let M be an orientable connected closed 3-manifold and let L be a link in M. Let $X := M \setminus L$ and let $h : Y \to X$ be a given finite covering. Then there exists a unique covering $f : N \to M$ ramified over L such that the associated covering is $h : Y \to X$. Here the uniqueness means that if there are such coverings N, N', then there is a homeomorphism $N \xrightarrow{\approx} N'$ so that the restriction to Y is the identity map. The construction of $f : N \to M$ is given as follows. Let g be the composite of h with the inclusion $X \hookrightarrow M$: $g : Y \to M$. To each open neighborhood U of $x \in M$, we associate a connected component $y(U)$ of $g^{-1}(U)$ in a way that $y(U_1) \subset y(U_2)$ if $U_1 \subset U_2$. Let N_x be the set of all such correspondences y. Let $N := \cup_{x \in M} N_x$ and define $f : N \to M$ by $f(y) = x$ if $y \in N_x$, namely, $N_x = f^{-1}(x)$.

We give a topology on N so that the basis of open subsets of N are given by the subsets of the form $\{y \in N \mid y(U) = W\}$ where U ranges over all subsets of M and W ranges over all connected components of $f^{-1}(U)$. If $y \in Y$, we can associate to each open neighborhood U of $x = f(y)$ a unique connected component $y(U)$ of $g^{-1}(U)$ containing y and so we may regard $Y \subset N$. Intuitively, regarding $x \in L$ as the limit of its open neighborhood U as U smaller, $y \in N$ is defined as the limit of a connected component $y(U)$ of $g^{-1}(U)$. Let $V = D^2 \times D^1$ be a tubular neighborhood of L. Then it follows from the uniqueness of the Fox completion for the covering $h^{-1}(V \setminus L) \to V \setminus L$ that the condition (2) is satisfied in a neighborhood of $y \in f^{-1}(L)$.

Example 2.1.15 Let $L = K_1 \cup \cdots \cup K_r$ be a link in an orientable connected closed 3-manifold M, X_L the link exterior $G_L := \pi_1(X_L)$. Let α_i be a meridian of K_i ($1 \leq i \leq r$). The map sending all α_i to 1 defines a surjective homomorphism $\psi_\infty : G_L \to \mathbb{Z}$. The infinite covering of X_L corresponding to $\mathrm{Ker}(\psi_\infty)$ is called the *total linking number covering* of X_L. For each $n \in \mathbb{N}$, let ψ_n be the composite of ψ_∞ with the natural homomorphism $\mathbb{Z} \to \mathbb{Z}/m\mathbb{Z}$: $\psi_n : G_L \to \mathbb{Z}/n\mathbb{Z}$. For an n-fold cyclic covering of X_L corresponding to $\mathrm{Ker}(\psi_n)$, we have the Fox completion M_n, which is an n-fold cyclic covering of M ramified over L.

Example 2.1.16 Let L be a 2-*bridge link* $B(a, b)$ ($0 < b < a, (a, b) = 1$) presented by Schubert's normal form. If a is odd, L is a knot, and if a is even, L is a 2-component link (Fig. 2.17).

The double covering of S^3 ramified over L is given by the lens space $L(a, b)$ (Example 2.1.5). To see this, divide $B(a, b)$ into two parts, say B_1 and B_2, where B_1 consists of 2 bridges (line segment $PP' \cup$ line segment QQ') and B_2 consists of 2 arcs passing under B_1 (arc $PQ' \cup$ arc $P'Q$ if a is odd and b is odd, arc $PQ \cup$ arc $P'Q'$ if a is odd and b is even, arc $PP' \cup$ arc QQ' if a is even). We see B_1 and B_2 as arcs inside 3-balls D_1^3 and D_2^3 respectively (Fig. 2.18).

According to the Heegaard decomposition $S^3 = D_1^3 \cup D_2^3$, L is decomposed as $L = B_1 \cup B_2$. Since the double covering of each D_i^3 ramified over B_i is a solid torus V_i, the double covering M of S^3 ramified over L is a lens space. Further, we see that the image of a meridian α_1 on ∂V_1 (a lift of the bridge PP' to V_1) in ∂V_2 is given by $a[\beta_2] + b[\alpha_2]$ as a homology class and hence $M = L(a, b)$ (Fig. 2.19).

Fig. 2.17 The 2-bridge knot $B(3, 1)$

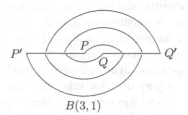

$$B(3, 1)$$

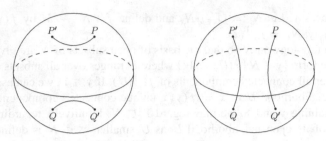

Fig. 2.18 Two 3-balls containing two parts of $B(3, 1)$

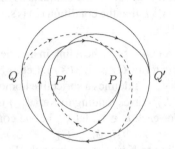

Fig. 2.19 The lens space $L(3, 1)$ obtained by gluing the two 3-balls

2.2 The Case of Arithmetic Rings

Throughout this section, any ring is assumed to be a commutative ring with identity element and any homomorphism between rings is assumed to send the identity element to the identity element.

For a commutative ring R, let $\operatorname{Spec}(R)$ the set of prime ideals of R, called the *prime spectrum* of R. For $a \in R$, let $U_a := \{\mathfrak{p} \in \operatorname{Spec}(R) \mid a \notin \mathfrak{p}\}$. The set $\operatorname{Spec}(R)$ is equipped with a topology, called the *Zariski topology*, whose open basis is given by $\mathcal{U} := \{U_a \mid a \in R\}$. On the topological space $\operatorname{Spec}(R)$, one has a sheaf of commutative rings $\mathcal{O}_{\operatorname{Spec}(R)}$ so that $\mathcal{O}_{\operatorname{Spec}(R)}(U_a) = R_a := \{\frac{r}{a^n} \mid r \in R, n \in \mathbb{Z} \geq 0\}$ $(a \neq 0)$. The pair $(\operatorname{Spec}(R), \mathcal{O}_{\operatorname{Spec}(R)})$ is called an *affine scheme*. A *scheme* is defined to be a topological space X equipped with a sheaf $\mathcal{O}_X$ of commutative rings such that locally $(U, \mathcal{O}_X|_U)$, U being an open subset of X, is given as an affine scheme. Hereafter, we simply call $\operatorname{Spec}(R)$ an affine scheme, omitting the sheaf $\mathcal{O}_{\operatorname{Spec}(R)}$. A homomorphism $\psi : A \to B$ of commutative rings gives a continuous map $\varphi : \operatorname{Spec}(B) \to \operatorname{Spec}(A)$ defined by $\varphi(\mathfrak{p}) := \psi^{-1}(\mathfrak{p})$ and a morphism $\psi^\# : \mathcal{O}_{\operatorname{Spec}(A)} \to \varphi_* \mathcal{O}_{\operatorname{Spec}(B)}$ of sheaves on $\operatorname{Spec}(A)$ defined by the natural homomorphism $A_a \to B_{\psi(a)}$ $(a \in A)$ induced by ψ. This correspondence gives rise to an anti-equivalence between the category of commutative rings and the category of affine schemes. Thus algebraic properties concerning a ring R can be expressed in terms of geometric properties concerning an affine $\operatorname{Spec}(R)$. However, as is easily seen, the Zariski topology is too coarse to define topological notions such as loops on $\mathrm{Spec}(R)$ etc. As explained in the previous section, the fundamental group of X controls the symmetry of the set of all coverings of X. So considering the fundamental group which describes the homotopy type of a space is equivalent to considering all coverings of the space. Similarly, we shall introduce the notion of an étale covering of $\mathrm{Spec}(R)$ which corresponds to a covering of a topological space and then define the étale fundamental group of $\mathrm{Spec}(R)$ following after the properties (U) of the pointed universal covering in the previous section.

For a commutative ring R and $\mathfrak{p} \in \mathrm{Spec}(R)$, let $R_{\mathfrak{p}}$ denote the localization of R at $\mathfrak{p}$: $R_{\mathfrak{p}} := \{ r/s \,|\, r \in R, s \in R \setminus \mathfrak{p} \}$. Let $\kappa(\mathfrak{p})$ denote the residue field of $R_{\mathfrak{p}}$: $\kappa(\mathfrak{p}) := R_{\mathfrak{p}}/\mathfrak{p}R_{\mathfrak{p}}$. A ring homomorphism $A \to B$ is said to be *finite étale* if

$$
\begin{cases}
(1) \ B \text{ is a finitely generated, flat } A\text{-module,} \\
(2) \ \text{for any } \mathfrak{p} \in \mathrm{Spec}(A), \\
\quad B \otimes_A \kappa(\mathfrak{p}) \simeq K_1 \times \cdots K_r \ (\kappa(\mathfrak{p})\text{-algebra isomorphism}),
\end{cases}
$$

where K_i is a finite separable extension of $\kappa(\mathfrak{p})$ $(1 \leq i \leq r)$.

In the rest of this section, a ring A shall denote an integrally closed domain and we let F be the quotient field of A. An A algebra B is called a *connected finite étale algebra* over A, if there is a finite separable extension K of F such that

$$
\begin{cases}
(1) \quad B \text{ is the integral closure of } A \text{ in } K, \\
(2) \quad \text{the inclusion map } A \hookrightarrow B \text{ is finite étale.}
\end{cases}
$$

An A-algebra B is called a *finite étale algebra* over A if B is isomorphic to the direct product $B_1 \times \cdots \times B_r$ of finite number of connected finite étale algebras $B_1, \ldots, B_r$ over A. An A-algebra B is called a *finite Galois algebra* over A if B is a connected finite étale algebra and if for any $\mathfrak{p} \in \mathrm{Spec}(A)$ and any algebraic closure Ω containing $\kappa(\mathfrak{p})$, the action of $\mathrm{Aut}(B/A) := \{ \sigma \,|\, A\text{-algebra automorphism of } B \}$ on $\mathrm{Hom}_{A\text{-alg}}(B, \Omega) := \{ \iota \,|\, A\text{-algebra homomorphism from } B \text{ to } \Omega \}$ defined by

$$
\mathrm{Aut}(B/A) \times \mathrm{Hom}_{A\text{-alg}}(B, \Omega) \longrightarrow \mathrm{Hom}_{A\text{-alg}}(B, \Omega); \ (\sigma, \iota) \longmapsto \iota \circ \sigma
$$

is simply transitive. This condition is independent of the choice of $\mathfrak{p}$ and Ω. If B is a finite Galois algebra over A, we write $\mathrm{Gal}(B/A)$ for $\mathrm{Aut}(B/A)$ and call it the *Galois group* of B over A. If K denotes the quotient field of B, B is a finite Galois algebra over A if and only if K/F is a finite Galois extension (see Example 2.2.1 below), and then $\mathrm{Gal}(B/A) = \mathrm{Gal}(K/F)$.

Example 2.2.1 (Field) Let F be a field. One has $\mathrm{Spec}(F) = \{(0)\}$. By definition, a connected finite étale algebra over F is nothing but a finite separable extension of F, and a étale algebra over F is an F-algebra which is isomorphic to the direct product of finite number of finite separable extensions of F. A finite Galois algebra over F is nothing but a finite Galois extension of F.

The most basic field in number theory is the *prime field* $\mathbb{F}_p := \mathbb{Z}/p\mathbb{Z}$ for a prime number p. More generally a *finite field* $\mathbb{F}_q$ consisting of q elements has a unique extension of degree n in a fixed separable closure $\overline{\mathbb{F}}_q$ for each $n \in \mathbb{N}$. On the other hand, over the field $\mathbb{Q}$ of rational numbers, there are infinitely many (non-isomorphic) quadratic extensions. Hence $\mathbb{Q}$ is much more complicated than $\mathbb{F}_q$ from the viewpoint of field extensions.

Example 2.2.2 (Complete Discrete Valuation Ring) A ring R is called a *discrete valuation ring* if the following conditions are satisfied:

$$\begin{cases} (1) & R \text{ is a principal ideal domain,} \\ (2) & R \text{ is a local ring with the maximal ideal } \mathfrak{p} \neq (0). \end{cases}$$

So $\mathrm{Spec}(R) = \{(0), \mathfrak{p}\}$. For $i \leq j$, let $f_{ij} : R/\mathfrak{p}^j \to R/\mathfrak{p}^i$ be the natural ring homomorphism. Then $\{R/\mathfrak{p}^i, f_{ij}\}$ is a projective system and the projective limit

$$\hat{R} := \varprojlim_{i \in \mathbb{N}} R/\mathfrak{p}^i := \left\{ (a_i) \in \prod_{i \in \mathbb{N}} R/\mathfrak{p}^i \,\middle|\, f_{ij}(a_j) = a_i \ (i \leq j) \right\}$$

forms a subring of the direct product ring $\prod_i R/\mathfrak{p}^i$. Giving $R/\mathfrak{p}^i$ the discrete topology, we endow $\hat{R}$ with the induced topology of the direct product space $\prod_i R/\mathfrak{p}^i$. Then $\hat{R}$ becomes a topological ring and is called the $\mathfrak{p}$-*adic completion* of R. By the injective map $x \mapsto (x \bmod \mathfrak{p}^i)$, R is regarded as a subring of $\hat{R}$. If $R = \hat{R}$, we call R a *complete discrete valuation ring*. Let K be the quotient field of R. Let us fix a prime element π so that $\mathfrak{p} = (\pi)$. Any element $x \in K^\times$ is then written as $x = u\pi^n$ ($u \in R^\times, n \in \mathbb{Z}$) uniquely and so we set $v(x) := n$. Then v is a discrete valuation on K (v is independent of the choice of π). Namely, $v : K^\times \to \mathbb{Z}$ is a surjective homomorphism such that $v(x + y) \geq \min(v(x), v(y))$ ($\forall x, y \in K$; $v(0) := \infty$). We call v the $\mathfrak{p}$-*adic (additive) valuation*. Take $c > 1$, define the $\mathfrak{p}$-adic multiplicative valuation by $|x| := c^{-v(x)}$. Then we have a metric d on K defined by $d(x, y) := |x - y|$. The topology on K defined in this way is independent of the choice of c. The completion $\hat{K}$ of the metric space (K, d) is called the $\mathfrak{p}$-*adic completion* of K. The metric d and the discrete valuation v are extended to those on $\hat{K}$ (denoted by the same d and v) so that $\hat{K}$ is a topological field. Now choose a system $S(\subset R)$ of complete representatives of $R/\mathfrak{p}$, where we choose 0 as a representative of the class 0 mod $\mathfrak{p}$). Then an element $x \in \hat{K}$ with $v(x) = n \in \mathbb{Z}$ is expanded uniquely as $x = a_n\pi^n + a_{n+1}\pi^{n+1} + \cdots$ ($a_i \in S$), called the $\mathfrak{p}$-*adic expansion* of x. By the correspondence $x \mapsto (x \bmod \mathfrak{p}^i)$, the valuation ring $\{x \in \hat{K} \mid v(x) \geq 0\}$ of $\hat{K}$ is identified with $\hat{R}$. The quotient field $\hat{K}$ of $\hat{R}$ is called a *complete discrete valuation field*. The maximal ideal of $\hat{R}$ is the valuation ideal $\hat{\mathfrak{p}} := \{x \in \hat{K} \mid v(x) > 0\}$ and the residue field $\hat{R}/\hat{\mathfrak{p}}$ is identified with $R/\mathfrak{p}$. For example, $\mathbb{Z}_{(p)}$ for a prime number p is a discrete valuation ring. The completions of $\mathbb{Z}_{(p)}$ and $\mathbb{Q}$ with respect to the associated p-adic valuation are called the ring of p-*adic integers* and the p-*adic field* respectively which are denoted by $\mathbb{Z}_p$ and $\mathbb{Q}_p$ respectively.

Let A be a complete discrete valuation ring and let F be the quotient field of A. Let K be a separable extension of F of degree n and let B be the integral closure of A in K. Then B is also a discrete valuation ring with the quotient field K. Furthermore B is a free A-module of rank n. Let $\mathfrak{p}$ and $\mathfrak{P}$ be the maximal ideals of A and B respectively. Then we can write $\mathfrak{p}B = \mathfrak{P}^e$ ($e \in \mathbb{N}$) uniquely. If $e = 1$, K/F is called an *unramified extension*, and if $e > 1$, K/F is called a *ramified extension*. The integer e is called the *ramification index* of K/F. If $e = n$, K/F is called a *totally ramified extension*. Since $B \otimes_A \kappa(\mathfrak{p}) \simeq B/\mathfrak{P}^e$, one has

B is a connected étale algebra over A

$\Longleftrightarrow K/F$ is an unramified extension

$\Longleftrightarrow \kappa(\mathfrak{P})/\kappa(\mathfrak{p})$ is a separable extension of degree n.

Thus the correspondences $K/F \mapsto B/A \mapsto \kappa(\mathfrak{P})/\kappa(\mathfrak{p})$ gives rise to the following bijections:

$$\{\text{finite unramified extension of } F \}/F\text{-isom.}$$

$$\xrightarrow{\sim} \{\text{connected finite étale algebra over } A \}/A\text{-isom.}$$

$$\xrightarrow{\sim} \{\text{finite separable extension of } \kappa(\mathfrak{p}) \}/\kappa(\mathfrak{p})\text{-isom.}$$

For the case that $A = \mathbb{Z}_p$ and $F = \mathbb{Q}_p$, B is called a *ring of $\mathfrak{p}$-adic integers* and K is called a *$\mathfrak{p}$-adic field* where $\mathfrak{p}$ stands for the maximal ideal of B.

Example 2.2.3 (Dedekind Domain) A ring R is called a *Dedekind domain* if the following conditions are satisfied

$$\begin{cases} (1) & R \text{ is a Noetherian integral domain (not a field),} \\ (2) & R \text{ is integrally closed,} \\ (3) & \text{any non-zero prime ideal of } R \text{ is a maximal ideal.} \end{cases}$$

For example, a principal ideal domain is a Dedekind domain. In the rest of this book, we denote by $\mathrm{Max}(R)$ the set of maximal ideals of R, called the *maximal spectrum* of R. The condition (3) is equivalent to the condition that $\mathrm{Spec}(R) = \mathrm{Max}(R) \cup \{(0)\}$. In terms of ideal theory, a Dedekind domain R is characterized as follows: "Any non-zero ideal $\mathfrak{a}$ of R is expressed uniquely (up to order) as $\mathfrak{a} = \mathfrak{p}_1^{e_1} \cdots \mathfrak{p}_r^{e_r}$ where $\mathfrak{p}_i$'s are distinct prime ideals of R and $e_i \in \mathbb{N}$. Let K be the quotient field of a Dedekind domain R. A finitely generated R-submodule($\neq (0)$) of K is called a *fractional ideal* of R. For a fractional ideal $\mathfrak{a}$, we let $\mathfrak{a}^{-1} := \{x \in K \mid x\mathfrak{a} \subset R\}$. Then $\mathfrak{a}^{-1}$ is a fractional ideal of R and one has $\mathfrak{a}\mathfrak{a}^{-1} = R$. So any non-zero ideal $\mathfrak{a}$ of R is expressed uniquely (up to order) as $\mathfrak{a} = \mathfrak{p}_1^{e_1} \cdots \mathfrak{p}_r^{e_r}$ where $\mathfrak{p}_i$'s are distinct prime ideals of R and $e_i \in \mathbb{Z}$. Hence the set of all fractional ideals forms a group by multiplication, called the *fractional ideal group* of R, which is the free Abelian group generated by $\mathrm{Max}(R)$. The quotient group of the fractional ideal group by the subgroup consisting of principal ideals $(a) = aR$ ($a \in K^\times$) is called the *ideal class group* of R.

Let R be a Dedekind domain. Since the localization $R_{\mathfrak{p}}$ of R at $\mathfrak{p} \in \mathrm{Max}(R)$ is a discrete valuation ring ([205, Ch.I, §3]), one has its completion $\hat{R}_{\mathfrak{p}}$ as in Example 2.2.2. The completed ring $\hat{R}_{\mathfrak{p}}$ is called the $\mathfrak{p}$-*adic completion* of R. The completion $K_{\mathfrak{p}}$ of the quotient field K of $R_{\mathfrak{p}}$ is defined similarly and is called the $\mathfrak{p}$-*adic completion* of K. We note that the localization $S^{-1}R$ of a Dedekind domain R with respect to any multiplicatively closed set $S (\neq R \setminus \{0\})$ is also a Dedekind domain.

Let A be a Dedekind domain and let F be the quotient field of A. Let K be a separable extension of F of degree n and let B be the integral closure of A in K. Then B is also a Dedekind domain with the quotient field K ([ibid, Ch.I, §4]). Since $B \otimes_A A_{\mathfrak{p}}$ is a finitely generated flat $A_{\mathfrak{p}}$-module for any $\mathfrak{p} \in \mathrm{Spec}(A)$, B is a finitely generated flat A-module. For $\mathfrak{p} \in \mathrm{Max}(A)$, we can write in a unique manner $\mathfrak{p}B = \mathfrak{P}_1^{e_1} \cdots \mathfrak{P}_r^{e_r}$ where $\mathfrak{P}_i$'s are distinct prime ideals of B and $e_i \in \mathbb{N}$. We then say that $\mathfrak{P}_i$ *lies over* $\mathfrak{p}$. We say that $\mathfrak{P}_i$ is *unramified* in K/F if $e_i = 1$, and we say that $\mathfrak{P}_i$ is *ramified* in K/F if $e_i > 1$. The integer e_i is called the *ramification index* of $\mathfrak{P}_i$ in K/F. We say that $\mathfrak{p}$ is *unramified* in K/F if $e_1 = \cdots = e_r = 1$, and we say that $\mathfrak{p}$ is *ramified* in K/F if $e_i > 1$ for some i. We say that $\mathfrak{p}$ is *totally ramified* in K/F if $r = 1$, $e_1 = n$, and we say $\mathfrak{p}$ is *completely decomposed* in K/F if $r = n$, $e_1 = \cdots = e_r = 1$. We also say that $\mathfrak{p}$ is *inert* in K/F if $r = e_1 = \cdots = e_r = 1$. If any $\mathfrak{p} \in \mathrm{Max}(A)$ is unramified in K/F, K/F is called an *unramified extension*, and if there is a $\mathfrak{p} \in \mathrm{Max}(A)$ which is ramified K/F, K/F is called a *ramified extension*. Since $B \otimes_A \kappa(\mathfrak{p}) \simeq B/\mathfrak{P}_1^{e_1} \times \cdots \times B/\mathfrak{P}_r^{e_r}$, one has

B is a connected finite étale algebra over A

$$\Longleftrightarrow K/F \text{ is an unramified extension.}$$

Since the étale fundamental group of a scheme is defined as a pro-finite group, we recall here some basic materials about pro-finite groups which will be used later on. Let (G_i, ψ_{ij}) $(i \in I)$ be a projective system consisting of finite groups G_i and homomorphisms $\psi_{ij} : G_j \to G_i$ $(i \leq j)$. Giving G_i the discrete topology, we endow the projective limit $\varprojlim_{i \in I} G_i$ with the induced topology as a subspace of the direct product space $\prod_{i \in I} G_i$. Then $\varprojlim_{i \in I} G_i$ becomes a topological group, called a *pro-finite group*. A pro-finite group is characterized as a topological group G which satisfies one of the following two properties: (1) G is a compact and totally disconnected, or (2) G has a fundamental system of neighborhoods of the identity consisting of compact and open subgroups of G. If each G_i is an l-group for a prime number l, the profinite $\varprojlim_i G_i$ is called a *pro-l group*.

Example 2.2.4 Let G be a group. Consider the set $\{N_i \mid i \in I\}$ of all normal subgroups of G with finite index and define $i \leq j$ if $N_j \subset N_i$. Let $\psi_{ij} : G/N_j \to G/N_i$ be the natural homomorphism for $i \leq j$. Then $(G/N_i, \psi_{ij})$ forms a projective system. The projective limit

$$\hat{G} := \varprojlim_i G/N_i$$

is called the *pro-finite completion* of G. If we consider only normal subgroups N_i of G such that each G/N_i is an l-group for a prime number l, the projective limit

$$\hat{G}(l) := \varprojlim_{G/N_i = l\text{-group}} G/N_i$$

is called the *pro-l completion* of G. If F is a free group on words $x_1, \ldots, x_r$, the pro-finite completion and pro-l completion of F (l being a prime number) is called a *free pro-finite group* and a *free pro-l group* on $x_1, \ldots, x_r$ respectively. For example, the pro-l completion $\varprojlim_n \mathbb{Z}/l^n\mathbb{Z}$ of the additive group is nothing but the additive group of the ring of l-adic integers $\mathbb{Z}_l$. The pro-finite completion $\varprojlim_n \mathbb{Z}/n\mathbb{Z}$ of $\mathbb{Z}$ is the direct product $\prod_l \mathbb{Z}_l$ (l running over all prime numbers) which is denoted by $\hat{\mathbb{Z}}$.

Let $\hat{F}$ be a free pro-finite group on words $x_1, \ldots, x_r$. For $R_1, \ldots, R_s \in \hat{F}$, we denote by $\langle\langle R_1, \ldots, R_s \rangle\rangle$ the smallest normal closed subgroup of F containing $R_1, \ldots, R_s$. If a pro-finite group $\mathfrak{G}$ is isomorphic to the quotient $\hat{F}/\langle\langle R_1, \ldots, R_s \rangle\rangle$, we write $\mathfrak{G}$ in the following form:

$$\mathfrak{G} = \langle x_1, \ldots, x_r \mid R_1 = \cdots = R_s = 1 \rangle$$

and call it a *presentation* of $\mathfrak{G}$ in terms of generators and relations. We also define a presentation of a pro-l group similarly as a quotient of a free pro-l group $\hat{F}(l)$. For a pro-l group $\mathfrak{G}$, one has the following [176, Ch.III, §9]:

Proposition 2.2.5 *A subset $\mathfrak{S}$ of $\mathfrak{G}$ generates G topologically if and only if the set of residue classes $\overline{\mathfrak{S}}$ mod $\mathfrak{G}^l[\mathfrak{G}, \mathfrak{G}]$ generates $\mathfrak{G}/\mathfrak{G}^l[\mathfrak{G}, \mathfrak{G}]$ topologically. The cardinality of a minimal generator system of $\mathfrak{G}$ is given by the dimension of the 1st group cohomology group $H^1(\mathfrak{G}, \mathbb{F}_l)$ over $\mathbb{F}_l$. Further, the cardinality of minimal relations in a minimal generator system is given by the dimension of the 2nd group cohomology group $H^2(\mathfrak{G}, \mathbb{F}_l)$ over $\mathbb{F}_l$.*

Example 2.2.6 Let $\mathfrak{G}$ be a pro-finite group and let l be a prime number. By Zorn's lemma, one has a minimal element $\mathfrak{N}_l$ with respect to the inclusion relation among all normal subgroups $\mathfrak{N}$ of $\mathfrak{G}$ such that $\mathfrak{G}/\mathfrak{N}$ is a pro-l group. In fact, $\mathfrak{N}_l$ is characterized by the following two properties: (1) $\mathfrak{G}/\mathfrak{N}_l$ is a pro-l group; (2) if $\mathfrak{G}/\mathfrak{N}$ is a pro-l group, then $\mathfrak{N}_l \subset \mathfrak{N}$. We call $\mathfrak{G}/\mathfrak{N}_l$ the *maximal pro-l quotient* of $\mathfrak{G}$ and denote it by $\mathfrak{G}(l)$. The pro-l completion $\hat{G}(l)$ of a group G is the maximal pro-l quotient of the pro-finite completion $\hat{G}$ of G. For instance, $\mathbb{Z}_l$ is the maximal pro-l quotient of $\hat{\mathbb{Z}}$.

Now let A be an integrally closed domain again and let $X := \text{Spec}(A)$. In order to define the étale fundamental group of X as a covariant functor, we need to consider all finite étale coverings of X including non-connected ones. We call a morphism $h : Y \to X$ of schemes a *finite étale covering* if there is a finite étale algebra $B = B_1 \times \cdots \times B_r$ (B_i being connected) over A such that $Y = \text{Spec}(B) = \bigsqcup_{i=1}^r \text{Spec}(B_i)$

(disjoint union of schemes) and h is the morphism associated to the inclusion $A \hookrightarrow B$. A finite étale covering $h' : Y' \to X$ is called a *subcovering* of $h : Y \to X$ if there is a morphism $\varphi : Y \to Y'$ such that $h' \circ \varphi = h$. We denote by $C_X(Y, Y')$ the set of such morphisms φ. If there is an isomorphism $\varphi \in C_X(Y, Y')$, we say that Y and Y' are isomorphic over X. The set of isomorphisms $\varphi \in C_X(Y, Y)$ forms a group, called the *group of covering transformations* of $h : Y \to X$, denoted by $\mathrm{Aut}(Y/X)$ or $\mathrm{Aut}(h)$.

Let $\mathfrak{p} \in X$ and fix an algebraically closed field Ω containing $\kappa(\mathfrak{p})$. It defines a morphism $\overline{x} : \mathrm{Spec}(\Omega) \to X$, called a *geometric base point* or simply a *base point* of X. For a finite étale covering $h : Y \to X$, we define the fiber of $\overline{x}$ by

$$F_{\overline{x}}(Y) := \mathrm{Hom}_X(\mathrm{Spec}(\Omega), Y) := \{\overline{y} : \mathrm{Spec}(\Omega) \longrightarrow Y \mid h \circ \overline{y} = \overline{x}\}$$
$$\simeq \mathrm{Hom}_{A\text{-alg}}(B, \Omega),$$

and, for $\varphi \in C_X(Y, Y')$, we define $F_{\overline{x}}(\varphi) : F_{\overline{x}}(Y) \to F_{\overline{x}}(Y')$ by $F_{\overline{x}}(\overline{y}) := \varphi \circ \overline{y}$ ($F_{\overline{x}}$ is called the *fiber functor* from the category of finite étale coverings of X to the category of sets). If Y is a connected (i.e., $Y = \mathrm{Spec}(B)$ for a connected finite étale algebra B over A), $\#F_{\overline{x}}(Y)$ is independent of the choice of $\overline{x}$. So we call $\#F_{\overline{x}}(Y)$ the *degree* of $h : Y \to X$, which is denoted by $\deg(h)$ or $[Y : X]$. A morphism $h : Y \to X$ is called a *finite Galois covering* if $Y = \mathrm{Spec}(B)$ for a finite Galois algebra B over A. In other words, Y is connected and the action of $\mathrm{Aut}(Y/X)$ on $F_{\overline{x}}(Y)$ defined by $(\sigma, \overline{y}) \mapsto \sigma \circ \overline{y}$ is simply transitive. This condition is independent of the choice of $\overline{x}$ (i.e., the choice of $\mathfrak{p}$ and Ω). For a finite Galois covering $h : Y \to X$, we call $\mathrm{Aut}(Y/X)$ the *Galois group* of Y over X and denote it by $\mathrm{Gal}(Y/X)$ or $\mathrm{Gal}(h)$.

A pair of a finite étale covering $h : Y \to X$ and $\overline{y} \in F_{\overline{x}}(Y)$ is called a *pointed finite étale covering*. A morphism between pointed finite étale coverings (Y, y) and (Y', y') over X is given by a $\varphi \in C_X(Y, Y')$ satisfying $\varphi \circ \overline{y} = \overline{y}'$. Then we have the following theorem, which is regarded as an analogue of the property (U)-(i) of the universal covering in Sect. 2.1.

Theorem 2.2.7 *There is a projective system* $((Y_i \xrightarrow{h_i} X, \overline{y}_i), \varphi_{ij})$ *of pointed finite Galois coverings such that for any finite étale covering* $h : Y \to X$, *the correspondence* $C_X(Y_i, Y) \ni \varphi \mapsto \varphi \circ \overline{y}_i \in F_{\overline{x}}(Y)$ *gives the following bijection:*

$$\varinjlim_i C_X(Y_i, Y) \simeq F_{\overline{x}}(Y).$$

Let $\tilde{X} = \varprojlim_i Y_i$ and $\tilde{x} = (\overline{y}_i)$. The pair $(\tilde{X}, \tilde{x})$ plays a role similar to the pointed universal covering of a manifold. Thus, as an analogue of (U)-(ii) in Sect. 2.1, we define the *étale fundamental group*[2] of X with base point $\overline{x}$ by

$$\pi_1(X, \overline{x}) := \operatorname{Gal}(\tilde{X}/X) := \varprojlim_i \operatorname{Gal}(Y_i/X),$$

where the projective limit is taken with respect to the composite

$$\operatorname{Gal}(Y_j/X) \simeq F_{\overline{x}}(Y_j) \overset{F_{\overline{x}}(\varphi_{ij})}{\longrightarrow} F_{\overline{x}}(Y_i) \simeq \operatorname{Gal}(Y_i/X) \ (i \leq j).$$

The group structure of $\pi_1(X, \overline{x})$ is independent of the choice of $\overline{x}$ (non-canonically isomorphic). Thus we often write simply $\pi_1(X)$, omitting a base point, and call it the étale fundamental group of X. By Theorem 2.2.7, for any finite étale covering Y over X, $\pi_1(X, \tilde{x})$ acts on $F_{\overline{x}}(Y)$ continuously from the right. We write this action as $\overline{y}.\sigma$ $(\sigma \in \pi_1(X, \overline{x}), \overline{y} \in F_{\tilde{x}}(Y))$.

Let A' be an integrally closed domain and let $A \to A'$ be a ring homomorphism. Let $f : X' := \operatorname{Spec}(A') \to X$ be the associated morphism of affine schemes. We fix an algebraic closure Ω' of $\kappa(\mathfrak{p}')$ and let $\overline{x}' : \operatorname{Spec}(\Omega') \to X'$ be the corresponding base point of X'. The composite $\overline{x} := f \circ \tilde{x}' : \operatorname{Spec}(\Omega') \to X$ gives a base point of X. Then, for any finite étale covering $h : Y \to X$, one has the bijection

$$F_{\overline{x}'}(Y \times_X X') = \operatorname{Hom}_{X'}(\operatorname{Spec}(\Omega'), Y \times_X X') \simeq \operatorname{Hom}_X(\operatorname{Spec}(\Omega'), Y) = F_{\overline{x}}(Y).$$

Here we note that $Y \times_X X'$ may not be connected, even though Y is connected. In the above bijection, let us take Y to be Y_i in Theorem 2.2.7 and let $\overline{y}'_i$ be the point in $F_{\overline{x}'}(Y_i \times_X X')$ corresponding to $\overline{y}_i \in F_{\tilde{x}}(Y_i)$. Then for $\sigma' \in \pi_1(Y', \overline{y}')$, we have the unique $\sigma_i \in \operatorname{Gal}(Y_i/X)$ such that $\overline{y}'_i.\sigma' = \sigma_i \circ \overline{y}_i$. So letting $f_*(\sigma') := (\sigma_i)$, we obtain a continuous homomorphism $f_* : \pi_1(X', \overline{x}') \to \pi_1(X, \overline{x})$.

For a projective system $(Y_i \overset{h_i}{\to} X, \varphi_{ij})$ of (connected) finite étale coverings of X, the projective limit $Y = \varprojlim_i Y_i$ is called a (connected) *pro-finite étale covering*, and we let $F_{\overline{x}}(Y) := \{(\overline{y}_i) \mid \overline{y}_i \in F_{\overline{x}}(Y_i), \ \varphi_{ij} \circ \overline{y}_j = \overline{y}_i \ (i \leq j)\}$. For $\overline{y} \in F_{\overline{x}}(Y)$, we set $h_*(\pi_1(Y, \overline{y})) := \bigcap_i h_{i*}(\pi_1(Y_i, \overline{y}_i))$. If each Y_i is a Galois covering of X, we call Y a *pro-finite Galois covering* and define the *Galois group* of Y over X by $\operatorname{Gal}(Y/X) := \varprojlim_i \operatorname{Gal}(Y_i/X)$. The main theorem of the Galois theory (Galois correspondence) over X is stated as follows.

---

[2]Although the étale fundamental group is often denoted by $\pi_1^{\text{ét}}(X, \overline{x})$, we write it by $\pi_1(X, \overline{x})$ or $\pi_1(X)$ for simplicity.

Theorem 2.2.8 (Galois Correspondence) *The correspondence* $(h : Y \to X) \mapsto$ $h_*(\pi_1(Y, \overline{y}))$ $(\overline{y} \in F_{\overline{x}}(Y))$ *gives rise to the following bijection:*

{connected pro-finite étale covering $h : Y \to X$}/isom. over X
$$\xrightarrow{\sim} \text{\textit{{closed subgroup of} }} \pi_1(X, \overline{x})\text{\textit{}/conjugate.}$$
Furthermore, this bijection satisfies the following:

- $h : Y \to X$ *is a connected finite étale covering* $\Leftrightarrow$ $h_*(\pi_1(Y, \overline{y}))$ *is an open subgroup.*
- $h' : Y' \to X$ *is a subcovering of* $h : Y \to X$ $\Leftrightarrow$ $h_*(\pi_1(Y, \overline{y}))$ $(\overline{y} \in F_{\overline{x}}(Y))$ *is a subgroup of* $h'_*(\pi_1(Y', \overline{y}'))$ $(\overline{y}' \in F_{\overline{x}}(Y'))$ *up to conjugate.*
- $h : Y \to X$ *is a Galois covering* $\Leftrightarrow$ $h_*(\pi_1(Y, \overline{y}))$ $(\overline{y} \in F_{\overline{x}}(Y))$ *is a normal subgroup of* $\pi_1(X, \overline{x})$. *Then one has* $\mathrm{Gal}(Y/X) \simeq \pi_1(X, \overline{x})/h_*(\pi_1(Y, \overline{y}))$.

More generally, we can replace $\pi_1(X, \overline{x})$ *by* $\mathrm{Gal}(Z/X)$ *for a fixed pro-finite Galois covering* $Z \to X$ *in the above, and then we have a similar bijection:*

{connected subcovering of $Z \to X$}/isom. over X
$$\xrightarrow{\sim} \text{\textit{{closed subgroup of} }} \mathrm{Gal}(Z/X)\text{\textit{}/conjugate.}$$

Example 2.2.9 Let F be a field. Choose an algebraically closed field Ω containing F which defines a base point $\overline{x} : \mathrm{Spec}(\Omega) \to \mathrm{Spec}(F)$. Let $\overline{F}$ be the separable closure of F in Ω. The set of all finite Galois extensions of $K_i \subset \Omega$ of F is inductively ordered with respect to the inclusion relation and one has $\overline{F} = \varinjlim_i K_i$, the composite field of K_i's. Therefore we can take $\mathrm{Spec}(K_i)$ for Y_i in Theorem 2.2.7 and hence

$$\pi_1(\mathrm{Spec}(F), \overline{x}) = \varprojlim_i \mathrm{Gal}(K_i/F) = \mathrm{Gal}(\overline{F}/F).$$

Let F be a finite field $\mathbb{F}_q$. For each $n \in \mathbb{N}$, there is the unique subfield $\mathbb{F}_{q^n} \subset \overline{\mathbb{F}}_q$ of degree n over $\mathbb{F}_q$ and so $\overline{\mathbb{F}}_q = \varinjlim_n \mathbb{F}_{q^n}$. Define the *Frobenius automorphism* $\sigma \in \mathrm{Gal}((\overline{\mathbb{F}}_q/\mathbb{F}_q)$ by

$$\sigma(x) = x^q \quad (x \in \overline{\mathbb{F}}_q).$$

For each $n \in \mathbb{N}$, the correspondence $\sigma|_{\mathbb{F}_{q^n}} \mapsto 1 \mod n$ gives an isomorphism $\mathrm{Gal}(\mathbb{F}_{q^n}/\mathbb{F}_q) \simeq \mathbb{Z}/n\mathbb{Z}$. Hence

$$\pi_1(\mathrm{Spec}(\mathbb{F}_q)) = \varprojlim_n \mathrm{Gal}(\mathbb{F}_{q^n}/\mathbb{F}_q) \simeq \varprojlim_n \mathbb{Z}/n\mathbb{Z} = \hat{\mathbb{Z}}.$$

Here the Frobenius automorphism σ corresponds to $1 \in \hat{\mathbb{Z}}$.

Example 2.2.10 Let A be a complete discrete valuation ring with the quotient field F. Choose an algebraically closed field Ω containing F which defines a base point

$\overline{x}$: Spec(Ω) $\to$ Spec(A). Consider the set of all finite Galois algebras B_i over A in Ω, which is inductively ordered. Let K_i be the quotient field of B_i and let $\tilde{F} = \varinjlim_i K_i$, the composite field of K_i's. The field $\tilde{F}$ is called the *maximal unramified extension* of F in Ω. Then we can take Spec(B_i) for Y_i in Theorem 2.2.7 and hence

$$\pi_1(\mathrm{Spec}(A), \overline{x}) = \varprojlim_i \mathrm{Gal}(B_i/A) = \varprojlim_i \mathrm{Gal}(K_i/F) = \mathrm{Gal}(\tilde{F}/F).$$

Let $\mathfrak{p}$ be the maximal ideal of A. Let $f : \mathrm{Spec}(\kappa(\mathfrak{p})) \to \mathrm{Spec}(A)$ be the morphism associated to the natural homomorphism $A \to \kappa(\mathfrak{p})$. Choose an algebraically closed field Ω' containing $\kappa(\mathfrak{p})$. Let $\overline{x}' : \mathrm{Spec}(\Omega') \to \mathrm{Spec}(\kappa(\mathfrak{p}))$ be the associated base point and let $\overline{x} := f \circ \overline{x}'$. Since there is the bijection between the set of $\kappa(\mathfrak{p})$-isomorphism classes of finite separable extensions of $\kappa(\mathfrak{p})$ and the set of F-isomorphism classes of finite unramified extensions of F (Example 2.2.2), f induces the isomorphism $f_* : \pi_1(\mathrm{Spec}(\kappa(\mathfrak{p})), \overline{x}') \simeq \pi_1(\mathrm{Spec}(A), \overline{x})$.

Example 2.2.11 Let A a Dedekind domain with the quotient field F. Choose an algebraically closed field Ω containing F which defines a base point Spec(Ω) $\to$ Spec(A). Consider the set of all finite Galois algebras B_i over A in Ω, which is inductively ordered. Let K_i be the quotient field of B_i and let $\tilde{F} = \varinjlim_i K_i$, the composite of K_i's. The field $\tilde{F}$ is called the *maximal unramified extension* of F in Ω. Then we can take Spec(B_i) for Y_i in Theorem 2.2.7 and hence

$$\pi_1(\mathrm{Spec}(A), \overline{x}) = \varprojlim_i \mathrm{Gal}(B_i/A) = \varprojlim_i \mathrm{Gal}(K_i/F) = \mathrm{Gal}(\tilde{F}/F).$$

Example 2.2.12 Let A be an integrally closed domain and let $X = \mathrm{Spec}(A)$. Let Y_i's be finite Galois coverings of X in Theorem 2.2.7. Now let us consider only $Y_i \to X$ whose degree is a power of a fixed prime number l. We then define the *pro-l étale fundamental group* of X by

$$\pi_1(X, \overline{x})(l) := \varprojlim_{[Y:X] = \text{a power of } l} \mathrm{Gal}(Y_i/X).$$

In fact, $\pi_1(X, \overline{x})(l)$ is the maximal pro-l quotient of $\pi_1(X, \overline{x})$ (Example 0.2.6). Suppose A is a field F. Let $F(l)$ be the composite field of all finite l-extensions K_i of F (a finite l-extension means a finite Galois extension whose degree is a power of l) in Ω, called the *maximal l-extension* of F. Then one has $\pi_1(X, \overline{x})(l) = \mathrm{Gal}(F(l)/F)$. Suppose A is a Dedekind domain. Let $\tilde{F}(l)$ be the composite field of all finite unramified l-extensions of F in Ω, called the *maximal unramified l-extension* of F. Then one has $\pi_1(X, \overline{x})(l) = \mathrm{Gal}(\tilde{F}(l)/F)$.

A typical example of a Dedekind domain is the ring of integers of a number field and its localizations. Here we recall some basic materials concerning number fields which shall be used later. A *number field* is an algebraic extension of the field of rational numbers $\mathbb{Q}$. The *ring of integers* of a number field k is the integral closure of $\mathbb{Z}$ in k and is denoted by $\mathcal{O}_k$. We also call a number field of finite degree over $\mathbb{Q}$ a *finite algebraic number field*. In the following, we assume k is a number field of finite degree over $\mathbb{Q}$ and set $n := [k : \mathbb{Q}]$ is finite. Since $\mathbb{Z}$ is a principal ideal domain, $\mathcal{O}_k$ is a Dedekind domain ([205, Ch.I, §4]). Further, $\mathcal{O}_k$ is a free $\mathbb{Z}$-module of rank n. For $\mathfrak{p} \in \mathrm{Max}(\mathcal{O}_k)$, $\mathfrak{p} \cap \mathbb{Z}$ is an ideal of $\mathbb{Z}$ generated by a prime number p and the residue field $\kappa(\mathfrak{p}) = \mathcal{O}_k/\mathfrak{p}$ is a finite extension of $\mathbb{F}_p$. In this book, we shall often write $\mathbb{F}_\mathfrak{p}$ instead of $\kappa(\mathfrak{p})$ to indicate that it is a finite field. For an ideal $\mathfrak{a}(\neq (0))$, the quotient ring $\mathcal{O}_k/\mathfrak{a}$ is finite. The order $\#(\mathcal{O}_k/\mathfrak{a})$ is called the *norm* of $\mathfrak{a}$ and is denoted by $\mathrm{N}\mathfrak{a}$. For a fractional ideal $\mathfrak{a} = \prod_\mathfrak{p} \mathfrak{p}^{e_\mathfrak{p}}$ $(e_\mathfrak{p} \in \mathbb{Z})$, the norm $\mathrm{N}\mathfrak{a}$ is defined by $\prod_\mathfrak{p}(\mathrm{N}\mathfrak{p})^{e_\mathfrak{p}}$. For a principal ideal (α) $(\alpha \in k^\times)$, one has $\mathrm{N}(\alpha) = |\mathrm{N}_{k/\mathbb{Q}}(\alpha)|$, where $\mathrm{N}_{k/\mathbb{Q}}(\alpha) := \prod_{i=1}^n \alpha_i$ (α_i running over conjugates of α over $\mathbb{Q}$). The group of fractional ideals of $\mathcal{O}_k$ is called the *ideal group* of k, which we denote by $I(k)$. It is a free Abelian group generated by $\mathrm{Max}(\mathcal{O}_k)$. The subgroup $P(k)$ consisting of principal fractional ideals is called the *principal ideal group* of k. The quotient group $I(k)/P(k)$ is called the *ideal class group* of k, which we denote by $H(k)$.

For $\mathfrak{p} \in \mathrm{Max}(\mathcal{O}_k)$, we denote by $\mathcal{O}_\mathfrak{p}$ the $\mathfrak{p}$-adic completion of $\mathcal{O}_k$ and by $k_\mathfrak{p}$ the $\mathfrak{p}$-adic completion of k, which are a ring of $\mathfrak{p}$-adic integers and a $\mathfrak{p}$-adic field in the sense of Example 2.2.2 respectively. So $k_\mathfrak{p}$ is equipped with the topology defined by the $\mathfrak{p}$-adic valuation. Since the residue field $\mathbb{F}_\mathfrak{p}$ of $\mathcal{O}_\mathfrak{p}$ is finite, $\mathcal{O}_\mathfrak{p}$ is a compact topological ring and $k_\mathfrak{p}$ is a locally compact topological field. An embedding of k into a locally compact topological field is given as one of the embeddings $k \hookrightarrow k_\mathfrak{p}$ for some $\mathfrak{p} \in \mathrm{Max}(\mathcal{O}_k)$, $k \hookrightarrow \mathbb{R}$ or $k \hookrightarrow \mathbb{C}$. Among the conjugate fields of k (i.e., the images of embeddings $k \hookrightarrow \mathbb{C}$), let $\iota_i : k \simeq k^{(i)} \subset \mathbb{R}$ ($1 \leq i \leq r_1$) be the real embeddings, and let $\iota_{r_1+j} : k \simeq k^{(r_1+j)} \subset \mathbb{C}$, $\bar{\iota}_{r_1+j} : k \simeq \overline{k}^{(r_1+j)} \subset \mathbb{C}$ ($1 \leq j \leq r_2$) be the complex but not real embeddings, where $\bar{\iota}_{r_1+j}$ and $\overline{k}^{(r_1+j)}$ mean the complex conjugate of ι_{r_1+j} and $k^{(r_1+j)}$ respectively and so $r_1 + 2r_2 = n$. For $a \in k$, we set $|a|_\mathfrak{p} := \mathrm{N}\mathfrak{p}^{-v_\mathfrak{p}(a)}$ ($\mathfrak{p} \in \mathrm{Max}(\mathcal{O}_k)$, $v_\mathfrak{p}$ is a $\mathfrak{p}$-adic additive valuation), $|a|_{\infty_i} := |\iota_j(a)|$ ($1 \leq j \leq r_1$), $|a|_{\infty_{r_1+j}} := |\iota_{r_1+j}(a)|^2 = \iota_{r_1+j}(a)\bar{\iota}_{r_1+j}(a)$ ($1 \leq j \leq r_2$). These give all non-trivial multiplicative valuations on k up to equivalence. We identify an embedding ι_j with the valuation $|\cdot|_{\infty_j}$ and call it an *infinite prime* of k and denote it by ∞_j or v_{∞_j} simply. The infinite primes $v_{\infty_1}, \ldots, v_{\infty_{r_1}}$ are called *real primes*, and $v_{\infty_{r_1+1}}, \ldots, v_{\infty_{r_1+r_2}}$ are called *complex primes*. We denote the set of infinite primes by $S_k^\infty := \{v_{\infty_1}, \ldots v_{\infty_{r_1+r_2}}\}$ and we set $S_k := \mathrm{Max}(\mathcal{O}_k) \cup S_k^\infty$. Then for $a \in k^\times$, the following product formula holds:

$$\prod_{v \in S_k} |a|_v = 1 \quad (a \in k^\times).$$

Intuitively, a scheme $\mathrm{Spec}(\mathcal{O}_k)$ is "compactified" by adding S_k^∞. We thus write

$$\overline{\mathrm{Spec}(\mathcal{O}_k)} := \mathrm{Spec}(\mathcal{O}_k) \cup S_k^\infty.$$

An element $a \in k^\times$ is said to be *totally positive* if $\iota_j(a) > 0$ ($1 \leq j \leq r_1$). We denote by $P^+(k)$ the group of principal fractional ideals generated by totally positive elements in k. The quotient group $I(k)/P^+(k)$ is called the *ideal class group in the narrow sense* or simply the *narrow ideal class group* of k, which we denote by $H^+(k)$. For a $\mathbb{Z}$-basis $\omega_1, \ldots, \omega_n$ of $\mathcal{O}_k$, we define the *discriminant* of k by $d_k := \det(\iota_i((\omega_j)))^2$. It is independent of the choice of basis $\omega_1, \ldots, \omega_n$. It can be shown that a prime number p is ramified in $k/\mathbb{Q}$ if and only if p divides d_k (see (2.2) below).

Let K/k be a finite extension. For an infinite prime $v \in S_k^\infty$, we say that v is *ramified* in K/k if v is a real prime and is extended to a complex prime of K. Otherwise, namely, if v is a complex prime of k, or if v is a real prime and any extension of v to K is a real prime, then we say that v is *unramified* in K/k. According to the convention in algebraic number theory, we say that K/k is an *unramified extension*, if all $\mathfrak{p} \in \text{Max}(\mathcal{O}_k)$ and all $v \in S_k^\infty$ are unramified in K/k. When all $\mathfrak{p} \in \text{Max}(\mathcal{O}_k)$ are unramified and some infinite prime may be ramified in K/k, we say that K/k is an *unramified extension in the narrow sense* or simply a *narrow unramified extension*.

Since a finite algebraic number field k is embedded into $\mathbb{C}$ (or $\mathfrak{p}$-adic field) as we have seen above, the ring of integers $\mathcal{O}_k$ is not only a Dedekind domain but also enjoys some analytic properties. Here are the most notable properties of a finite algebraic number field k. Notations are as above:

Minkowski's Theorem 2.2.13 *If $k \neq \mathbb{Q}$, then $|d_k| > 1$.*

The Finiteness of Ideal Classes 2.2.14 *The (narrow) ideal class group $H(k)$ (or $H^+(k)$) is a finite Abelian group.*

Dirichlet's Unit Theorem 2.2.15 *The unit group $\mathcal{O}_k^\times$ is the direct product of the cyclic group of roots of unity in k and a free Abelian group of rank $r_1 + r_2 - 1$.*

Example 2.2.16 (Quadratic Number Field) Let m be a square-free integer ($\neq 1$) and let $k := \mathbb{Q}(\sqrt{m})$, a *quadratic number field*. Then one has

$$\mathcal{O}_k = \begin{cases} \mathbb{Z}[\dfrac{1+\sqrt{m}}{2}] & m \equiv 1 \bmod 4 \\ \mathbb{Z}[\sqrt{m}] & m \equiv 2, 3 \bmod 4, \end{cases} \qquad d_k = \begin{cases} m & m \equiv 1 \bmod 4 \\ 4m & m \equiv 2, 3 \bmod 4. \end{cases}$$

$$\mathcal{O}_k^\times \simeq \begin{cases} \{\pm 1\} \times \mathbb{Z} & m > 0 \\ \{\pm 1, \pm\sqrt{-1}\} & m = -1 \\ \{\pm 1, \pm\omega, \pm\omega^2\} \, (\omega := \dfrac{1+\sqrt{-3}}{2}) & m = -3 \\ \{\pm 1\} & m = -2, \ m < -3. \end{cases}$$

Example 2.2.17 (Cyclotomic Field) Let n be an integer ≥ 3 and let $\zeta_n := \exp(\frac{2\pi\sqrt{-1}}{n})$. Let $k := \mathbb{Q}(\zeta_n)$, a *cyclotomic field*. Then k is a finite Abelian extension of $\mathbb{Q}$ whose Galois group $\text{Gal}(k/\mathbb{Q})$ is isomorphic to $(\mathbb{Z}/n\mathbb{Z})^\times$. This isomorphism is given as follows: For $g \in \text{Gal}(k/\mathbb{Q})$, define $m(g)$ by $g(\zeta_n) = \zeta_n^{m(g)}$. Then the map $g \mapsto m(g)$ gives an isomorphism $\text{Gal}(k/\mathbb{Q}) \simeq (\mathbb{Z}/n\mathbb{Z})^\times$. Hence $[k : \mathbb{Q}] =$

$\phi(n)$ (Euler function). One has $\mathcal{O}_k = \mathbb{Z}[\zeta_n]$ and $\mathcal{O}_k^{\times} \simeq \langle \pm \zeta_n \rangle \times \mathbb{Z}^{\phi(n)/2-1}$. The discriminant of k is given as follows: If $n = p^e$ for a prime number p,

$$d_k = \begin{cases} -p^{p^{e-1}(pe-e-1)} & p \equiv 3 \bmod 4 \text{ or } p = e = 2 \\ p^{p^{e-1}(pe-e-1)} & \text{otherwise.} \end{cases}$$

In general, for $n = p_1^{e_1} \cdots p_r^{e_r}$ the decomposition of prime factors of n, we have $d_k = d_{k_1}^{\frac{\phi(n)}{\phi(p_1^{e_1})}} \cdots d_{k_r}^{\frac{\phi(n)}{\phi(p_r^{e_r})}}$ where $k_i = \mathbb{Q}(\zeta_{p_i^{e_i}})$.

Example 2.2.18 Let $\mathcal{O}_{\mathfrak{p}}$ be a ring of $\mathfrak{p}$-adic integers and $k_{\mathfrak{p}}$ be its quotient field. By Example 2.2.10, one has

$$\pi_1(\text{Spec}(\mathcal{O}_{\mathfrak{p}})) \simeq \pi_1(\text{Spec}(\mathbb{F}_{\mathfrak{p}})) \simeq \hat{\mathbb{Z}}.$$

Since a separable closure of $\mathbb{F}_{\mathfrak{p}}$ is obtained by adjoining n-th roots of unity to $\mathbb{F}_{\mathfrak{p}}$ for all natural numbers n prime to $q := N\mathfrak{p}$, the maximal unramified extension $k_{\mathfrak{p}}^{\text{ur}}$ of $k_{\mathfrak{p}}$ is given by

$$k_{\mathfrak{p}}^{\text{ur}} = k_{\mathfrak{p}}(\zeta_n \mid (n, q) = 1),$$

where ζ_n is a primitive n-th root of unity in $\overline{k}_{\mathfrak{p}}$. The element of $\pi_1(\text{Spec}(\mathcal{O}_{\mathfrak{p}})) = \text{Gal}(k_{\mathfrak{p}}^{\text{ur}}/k_{\mathfrak{p}})$ corresponding to the Frobenius automorphism $\sigma \in \pi_1(\text{Spec}(\mathbb{F}_{\mathfrak{p}})) = \text{Gal}(\overline{\mathbb{F}}_{\mathfrak{p}}/\mathbb{F}_{\mathfrak{p}})$ under the above isomorphism is also called the *Frobenius automorphism*, denoted by the same σ, which is given by $\sigma(\zeta_n) = \zeta_n^q$.

Example 2.2.19 By Minkowski's theorem 2.2.13, there is no non-trivial connected finite étale algebra over $\mathbb{Z}$. Hence we have

$$\pi_1(\text{Spec}(\mathbb{Z})) = \{1\}.$$

Example 2.2.20 Let k be a number field of finite degree over $\mathbb{Q}$ and let $\mathcal{O}_k$ be the ring of integers of k. Let S be a finite set of maximal ideals of $\mathcal{O}_k$: $S = \{\mathfrak{p}_1, \ldots, \mathfrak{p}_n\}$. By the finiteness of ideal classes (2.2.14), one finds $n_i \in \mathbb{N}$ for each i so that $\mathfrak{p}_i^{n_i} = (a_i)$, $a_i \in \mathcal{O}_k$. Set $A = \mathcal{O}_k[\frac{1}{a_1 \cdots a_n}]$. Since A is a localization of $\mathcal{O}_k$, A is a Dedekind domain and $\text{Spec}(A) = \text{Spec}(\mathcal{O}_k) \setminus S$. Choose an algebraically closed field Ω containing k which defines a base point $\overline{x} : \text{Spec}(\Omega) \to \text{Spec}(A)$. For a finite extension K/k in Ω, if any maximal ideal which is not contained in S is unramified in K/k, we say that K/k is *unramified outside* $\overline{S} := S \cup S_k^{\infty}$, where S_k^{∞} is the set of infinite primes of k. Let $k_{\overline{S}} = \varinjlim_i k_{i,\overline{S}}$ be the composite field of all finite Galois extensions $k_{i,\overline{S}}$ of k in Ω which are unramified outside $\overline{S}$. The field $k_{\overline{S}}$ is called the *maximal Galois extension of k unramified outside $\overline{S}$*. We can take $\text{Spec}(k_i)$ for Y_i in Theorem 2.2.7 and hence

$$\pi_1(\text{Spec}(\mathcal{O}_k) \setminus S, \overline{x}) = \text{Gal}(k_{\overline{S}}/k) = \varprojlim_i \text{Gal}(k_{i,\overline{S}}/k).$$

We denote this pro-finite group by $G_S(k)$:

$$G_S(k) := \pi_1(\mathrm{Spec}(\mathcal{O}_k) \setminus S, \overline{x}).$$

In the case $k = \mathbb{Q}$, we shall simply write G_S. For a prime number l, let $k_{\overline{S}}(l)$ be *the maximal l-extension of k unramified outside $\overline{S}$*. We then have $G_S(k)(l) = \mathrm{Gal}(k_{\overline{S}}(l)/k)$.

More generally, let T be a finite set of primes of k, which may contain infinite primes. So T is a finite subset of $\overline{\mathrm{Spec}(\mathcal{O}_k)}$. Let $k_T = \varinjlim_i k_{i,T}$ be the maximal Galois extension of k unramified outside T, where $k_{i,T}$'s are all finite Galois extensions of k in Ω which are unramified outside T. Following the above construction, we define the *modified étale fundamental group* of $\overline{\mathrm{Spec}(\mathcal{O}_k)} \setminus T$ with base point $\overline{x}$ by

$$\pi_1(\overline{\mathrm{Spec}(\mathcal{O}_k)} \setminus T, \overline{x}) = \mathrm{Gal}(k_T/k) = \varprojlim_i \mathrm{Gal}(k_{i,T}/k).$$

In particular, $\pi_1(\overline{\mathrm{Spec}(\mathcal{O}_k)})$ is the Galois group of the maximal extension of k such that all (finite and infinite) primes of k are unramified.

Finally, we shall review some materials about ramified extensions over a Dedekind domain. Let A be a Dedekind domain with the quotient field F. Let K be a finite separable extension of F and let B be the integral closure of A in K. The morphism $f : N := \mathrm{Spec}(B) \to M := \mathrm{Spec}(A)$ induced from the inclusion $A \hookrightarrow B$ is called a *ramified covering* if K/F a ramified extension. The information on which $\mathfrak{P} \in \mathrm{Max}(B)$ or $\mathfrak{p} \in \mathrm{Max}(A)$ is ramified is detected by the different or the relative discriminant for B/A. Let $\iota_j : K \to \overline{F}$ $(1 \leq j \leq n)$ be all embeddings of K into a separable closure $\overline{F}$ of F. The trace $\mathrm{Tr}_{K/F}$ and the norm $\mathrm{N}_{K/F}$ are defined by $\mathrm{Tr}_{K/F}(a) := \iota_1(a) + \cdots + \iota_n(a)$ and $\mathrm{N}_{K/F}(a) := \iota_1(a) \cdots \iota_n(a)$ respectively. Let $\mathfrak{b} := \{b \in K \mid \mathrm{Tr}_{K/F}(ab) \in A \; \forall a \in B\}$. We easily see $\mathfrak{b}$ is a fractional ideal containing B. We then define the *different* of B/A by $\mathfrak{d}_{B/A} := \mathfrak{b}^{-1}$ and the *relative discriminant* of B/A by $d_{B/A} := \mathrm{N}_{K/F}(\mathfrak{d}_{B/A})$. If K/F is a finite extension of number fields of finite degree over $\mathbb{Q}$, we denote simply by $d_{K/F}$ the relative discriminant $d_{\mathcal{O}_K/\mathcal{O}_F}$ and call it the *relative discriminant* of K/F. In particular, $d_{k/\mathbb{Q}}$ coincides with the ideal of $\mathbb{Z}$ generated by the discriminant d_k. Now, as for the ramification, we have the following:

$$\begin{aligned} \mathfrak{P} \text{ is ramified in } K/F &\iff \mathfrak{P} | \mathfrak{d}_{B/A}, \\ \mathfrak{p} \text{ is ramified in } K/F &\iff \mathfrak{p} | d_{B/A}. \end{aligned} \tag{2.2}$$

Therefore only finitely many $\mathfrak{p} \in \mathrm{Max}(A)$ are ramified in K/F. Let S_F be the set of $\mathfrak{p} \in \mathrm{Max}(A)$ ramified in K/F and let $S_K := f^{-1}(S_F)$. We call $f|_{N \setminus S_K} : N \setminus S_K \to M \setminus S_F$ the *associated finite étale covering*. If $f|_{N \setminus S_K}$ is a Galois covering, f is called a *ramified Galois covering*. This condition amounts to K/F being a Galois extension. Finally, K/F is called a *tamely ramified extension*, if for any $\mathfrak{P} \in \mathrm{Max}(B)$ ramified in K/F, the ramification index of $\mathfrak{P}$ is prime to the characteristic of the residue field $\kappa(\mathfrak{P})$. Here if the characteristic of $\kappa(\mathfrak{P})$ is zero, no condition is meant. Let Ω be an algebraically closed field containing F which defines a base point $\overline{x} : \mathrm{Spec}(\Omega) \to X := \mathrm{Spec}(F)$, and let F^t be the composite field of all finite tamely ramified extensions K_i of F in Ω. The field F^t is called the *maximal tamely ramified extension* of F. Then we define the *tame fundamental group* of X by

$$\pi_1^t(X, \overline{x}) = \varprojlim_{K_i} \mathrm{Gal}(K_i/F) = \mathrm{Gal}(F^t/F),$$

where the projective limit is taken over all finite tamely ramified extensions K_i/F in Ω.

Example 2.2.21 Let k be a number field of finite degree over $\mathbb{Q}$ and let d_k be the discriminant of k. By (2.2), $\mathrm{Spec}(\mathcal{O}_k) \to \mathrm{Spec}(\mathbb{Z})$ is a finite covering which is ramified over primes (p), $p | d_k$, and $\mathrm{Spec}(\mathcal{O}_k[1/d_k]) \to \mathrm{Spec}(\mathbb{Z}[1/d_k])$ is the associated étale covering.

Example 2.2.22 Let p be a fixed prime number. For $n \in \mathbb{N}$, let $\zeta_{p^n} := \exp(\frac{2\pi\sqrt{-1}}{p^n})$ and $k_n := \mathbb{Q}(\zeta_{p^n})$. Set $\mathcal{O}_n := \mathcal{O}_{k_n}$, $M_n := \mathrm{Spec}(\mathcal{O}_n)$ and $X_n := \mathrm{Spec}(\mathcal{O}_n[\frac{1}{p}])$ for simplicity. By Example 2.2.17 and (2.2), the natural map $M_n \to M_0 = \mathrm{Spec}(\mathbb{Z})$ is a Galois covering ramified over (p), and $X_n \to X_0 = \mathrm{Spec}(\mathbb{Z}[\frac{1}{p}])$ is the associated étale covering. The Galois group is given by

$$\mathrm{Gal}(M_n/M_0) = \mathrm{Gal}(X_n/X_0) = \mathrm{Gal}(k_n/k_0) \simeq (\mathbb{Z}/p^n\mathbb{Z})^\times.$$

By the natural maps $M_{n+1} \to M_n$ and $X_{n+1} \to X_n$, $M_\infty := \varprojlim_n M_n$ is a pro-finite ramified Galois covering over M_0 and $X_\infty := \varprojlim_n X_n$ is a pro-finite Galois covering over X_0. Let $k_\infty := \varinjlim_n k_n = \mathbb{Q}(\zeta_{p^n} \,|\, n \geq 1)$. Then the Galois group of M_∞ over M_0 is given by

$$\mathrm{Gal}(M_\infty/M_0) = \mathrm{Gal}(X_\infty/X_0) = \mathrm{Gal}(k_\infty/\mathbb{Q}) \simeq \varprojlim_n (\mathbb{Z}/p^n\mathbb{Z})^\times = \mathbb{Z}_p^\times.$$

The ramification of (p) is as follows: Since the minimal polynomial of ζ_{p^n} over $\mathbb{Q}$ is $f(X) = \frac{X^{p^n}-1}{X^{p^{n-1}}-1}$ and $f(1+X) \equiv X^{p^{n-1}(p-1)} \bmod p$, we have $p\mathcal{O}_n = \mathfrak{p}^{p^{n-1}(p-1)}$, where $\mathfrak{p} = (\zeta_{p^n} - 1)$. The ramification index $p^{n-1}(p-1)$ is same as the covering degree $[k_n : \mathbb{Q}] = \phi(p^n)$ and so (p) is totally ramified in $M_n \to M_0$.

Example 2.2.23 Let $k_\mathfrak{p}$ be a $\mathfrak{p}$-adic field with $q = N\mathfrak{p}$ and let $X = \mathrm{Spec}(k_\mathfrak{p})$. Choose an algebraically closed field Ω containing $k_\mathfrak{p}$ and let $\overline{k}_\mathfrak{p}$ be the algebraic closure of $k_\mathfrak{p}$ in Ω. By Example 2.2.18, the maximal unramified extension $k_\mathfrak{p}^{\mathrm{ur}}$ of $k_\mathfrak{p}$ is given by $k_\mathfrak{p}(\zeta_n \,|\, (n, q) = 1)$, where ζ_n is a primitive n-th root of unity in $\overline{k}_\mathfrak{p}$ so that $\zeta_n^m = \zeta_{n/m}$ for $m | n$. The kernel of the natural homomorphism $\pi_1(X) = $

$\mathrm{Gal}(\overline{k}_{\mathfrak{p}}/k_{\mathfrak{p}}) \rightarrow \pi_1(\mathrm{Spec}(\mathcal{O}_{\mathfrak{p}})) = \mathrm{Gal}(k_{\mathfrak{p}}^{\mathrm{ur}}/k_{\mathfrak{p}})$ induced by the inclusion $\mathcal{O}_{\mathfrak{p}} \hookrightarrow k_{\mathfrak{p}}$ is called the *inertia group* of $k_{\mathfrak{p}}$, which we denote by $I_{k_{\mathfrak{p}}}$. The tame fundamental group $\pi_1^{\mathrm{t}}(X)$ will be described as an extension of $\mathrm{Gal}(k_{\mathfrak{p}}^{\mathrm{ur}}/k_{\mathfrak{p}})$ by the maximal tame quotient $I_{k_{\mathfrak{p}}}^{\mathrm{t}}$ of $I_{k_{\mathfrak{p}}}$ as follows. Let ϖ be a prime element of $k_{\mathfrak{p}}$. Then the maximal tamely ramified extension $k_{\mathfrak{p}}^{\mathrm{t}}$ of $k_{\mathfrak{p}}$ is given by

$$k_{\mathfrak{p}}^{\mathrm{t}} = k_{\mathfrak{p}}^{\mathrm{ur}}(\sqrt[n]{\varpi} \mid (n, q) = 1).$$

We define the monodromy $\tau \in \mathrm{Gal}(k_{\mathfrak{p}}^{\mathrm{t}}/k_{\mathfrak{p}})$ by

$$\tau(\zeta_n) = \zeta_n, \quad \tau(\sqrt[n]{\varpi}) = \zeta_n \sqrt[n]{\varpi}.$$

Then τ is a topological generator of $I_{k_{\mathfrak{p}}}^{\mathrm{t}} := \mathrm{Gal}(k_{\mathfrak{p}}^{\mathrm{t}}/k_{\mathfrak{p}})$, the maximal tame quotient of $I_{k_{\mathfrak{p}}}$, and gives the following isomorphism:

$$\mathrm{Gal}(k_{\mathfrak{p}}^{\mathrm{t}}/k_{\mathfrak{p}}^{\mathrm{ur}}) \simeq \varprojlim_{(n,q)=1} \mathbb{Z}/n\mathbb{Z} =: \hat{\mathbb{Z}}^{(q')},$$

where τ corresponds to $1 \in \hat{\mathbb{Z}}^{(q')}$. Hence we have the following short exact sequence:

$$1 \longrightarrow \mathrm{Gal}(k_{\mathfrak{p}}^{\mathrm{t}}/k_{\mathfrak{p}}^{\mathrm{ur}}) \longrightarrow \mathrm{Gal}(k_{\mathfrak{p}}^{\mathrm{t}}/k_{\mathfrak{p}}) \longrightarrow \mathrm{Gal}(k_{\mathfrak{p}}^{\mathrm{ur}}/k_{\mathfrak{p}}) \longrightarrow 1.$$
$$\wr\vert \qquad\qquad\qquad\qquad\qquad\qquad\quad \wr\vert$$
$$\hat{\mathbb{Z}}^{(q')} \qquad\qquad\qquad\qquad\qquad\qquad\quad \hat{\mathbb{Z}}$$

We define an extension of the Frobenius automorphism $\sigma \in \mathrm{Gal}(k_{\mathfrak{p}}^{\mathrm{ur}}/k_{\mathfrak{p}})$ to $\mathrm{Gal}(k_{\mathfrak{p}}^{\mathrm{t}}/k_{\mathfrak{p}})$, denoted by the same σ, by

$$\sigma(\zeta_n) = \zeta_n^q, \quad \sigma(\sqrt[n]{\varpi}) = \sqrt[n]{\pi}.$$

Then τ and σ are subject to the relation

$$\sigma\tau = \tau^q\sigma.$$

Thus

$$\pi_1^{\mathrm{t}}(X) = \mathrm{Gal}(k_{\mathfrak{p}}^{\mathrm{t}}/k_{\mathfrak{p}}) = \langle \tau, \sigma \mid \tau^{q-1}[\tau, \sigma] = 1 \rangle.$$

We note that for a prime number l prime to q, the pro-l fundamental group $\pi_1(\mathrm{Spec}(k_{\mathfrak{p}}))(l)$ has a similar presentation.

Example 2.2.24 Let k be a number field of finite degree over $\mathbb{Q}$. Let S be a finite subset of $\mathrm{Max}(\mathcal{O}_k)$ and let $X_S := \mathrm{Spec}(\mathcal{O}_k) \setminus S$. Let $k_{\overline{S}}$ be the maximal Galois extension of k unramified outside $\overline{S} := S \cup S_k^\infty$ so that $G_S(k) = \pi_1(X_S) =$ $\mathrm{Gal}(k_{\overline{S}}/k)$ (Example 2.2.20). Take a $\mathfrak{p} \in \mathrm{Max}(\mathcal{O}_k)$ and let $k_\mathfrak{p}$ be the $\mathfrak{p}$-adic field. Choose an algebraic closure $\overline{k}_\mathfrak{p}$ of $k_\mathfrak{p}$ and hence a base point $\overline{x} : \mathrm{Spec}(\overline{k}_\mathfrak{p}) \to$ $\mathrm{Spec}(k_\mathfrak{p})$. Combining $\overline{x}$ with the natural morphism $\mathrm{Spec}(k_\mathfrak{p}) \to X_S$, we have a base point of X_S, $\overline{y} : \mathrm{Spec}(\overline{k}_\mathfrak{p}) \to X_S$. This defines an embedding $k_{\overline{S}} \hookrightarrow \overline{k}_\mathfrak{p}$ over k and induces the homomorphism

$$\varphi_\mathfrak{p} : \pi_1(\mathrm{Spec}(k_\mathfrak{p}), \overline{x}) = \mathrm{Gal}(\overline{k}_\mathfrak{p}/k_\mathfrak{p}) \longrightarrow \pi_1(X_S, \overline{y}) = G_S(k).$$

The embedding $k_{\overline{S}} \hookrightarrow \overline{k}_\mathfrak{p}$ over k defines a prime $\overline{\mathfrak{p}}$ in $k_{\overline{S}}$ over $\mathfrak{p}$. Then the image of $\varphi_\mathfrak{p}$ coincides with the decomposition group of $\overline{\mathfrak{p}}$

$$D_{\overline{\mathfrak{p}}} := \{g \in G_S(k) \mid g(\overline{\mathfrak{p}}) = \overline{\mathfrak{p}}\}.$$

Hereafter, we suppose an embedding $k_{\overline{S}} \hookrightarrow \overline{k}_\mathfrak{p}$ and hence $\overline{\mathfrak{p}}$ is fixed, and we call $D_{\overline{\mathfrak{p}}}$ the *decomposition group over* $\mathfrak{p}$ in $k_{\overline{S}}/k$ and denote it by $D_\mathfrak{p}$. Similarly, we call the image of the inertia group $I_{k_\mathfrak{p}}$ under $\varphi_\mathfrak{p}$ the *inertia group over* $\mathfrak{p}$ in $k_{\overline{S}}/k$ and denote it by $I_\mathfrak{p}$. If we replace an embedding $k_{\overline{S}} \hookrightarrow \overline{k}_\mathfrak{p}$ by another one, $D_\mathfrak{p}$ and $I_\mathfrak{p}$ are changed to some conjugate subgroups in $G_S(k)$.

Suppose $\mathfrak{p} \notin S$. Then $\mathfrak{p}$ is unramified in $k_{\overline{S}}/k$, namely, $I_\mathfrak{p} = 1$. So $\varphi_\mathfrak{p}$ factors through $\mathrm{Gal}(k_\mathfrak{p}^{\mathrm{ur}}/k_\mathfrak{p})$. We call the image $\sigma_\mathfrak{p} := \varphi_\mathfrak{p}(\sigma) \in G_S(k)$ of the Frobenius automorphism $\sigma \in \mathrm{Gal}(k_\mathfrak{p}^{\mathrm{ur}}/k_\mathfrak{p})$ the *Frobenius automorphism over* $\mathfrak{p}$. If we replace an embedding $k_{\overline{S}} \hookrightarrow \overline{k}_\mathfrak{p}$ by another one, $\sigma_\mathfrak{p}$ is changed to a conjugate in $G_S(k)$. Therefore in an Abelian quotient of $G_S(k)$ the image of $\sigma_\mathfrak{p}$ is uniquely determined.

Although we have dealt with étale fundamental groups in this section, one has also the theories of étale (co)homology and higher homotopy groups for schemes which are defined by a simplicial method, similar to the method in topology (cf. [9, 57, 73]). For example, $\mathrm{Spec}(\mathcal{O}_\mathfrak{p})$ and $\mathrm{Spec}(\mathbb{F}_\mathfrak{p})$ are étale homotopically equivalent. For recent investigations on the subject, we refer to [200] and references therein.

2.3 Arithmetic Duality Theorems

In this section, we review the arithmetic duality theorems for local fields and number rings. In what follows, we shall consider some étale cohomology groups of $X = \mathrm{Spec}(A)$ with coefficients in locally constant étale sheaves on X defined by Abelian groups on which $\pi_1(X, \overline{x})$ acts continuously. Here an étale sheaf M on X is called *locally constant* if there is a connected finite étale covering $Y \to X$ such that $M|_Y$ is a constant sheaf of an Abelian group on Y. A finite $\pi_1(X, \overline{x})$-module M gives rise to a locally constant étale sheaf on X which is defined by associating to a connected finite étale covering $Y \to X$ the $\pi_1(Y, \overline{y})$-invariant subgroup $M^{\pi_1(Y,\overline{y})}$ ($\overline{y} \in F_{\overline{x}}(X)$) of M. Conversely, a locally constant, finite étale sheaf M gives rise to a finite $\pi_1(X, \overline{x})$-module $M_{\overline{x}}$, the stalk of M at $\overline{x}$. Thus we identify a locally constant étale sheaf on X with the associated finite $\pi_1(X, \overline{x})$-module. For the case that A is a field F, an étale sheaf of finite Abelian group on $\mathrm{Spec}(F)$ is same as a finite Abelian group on which $\pi_1(\mathrm{Spec}(F)) = \mathrm{Gal}(\overline{F}/F)$ acts continuously. So the étale cohomology group $H^i(\mathrm{Spec}(F), M)$ is identified with the Galois cohomology group $H^i(\mathrm{Gal}(\overline{F}/F), M)$, which we denote by $H^i(F, M)$ for simplicity. The *étale cohomological dimension* of $X = \mathrm{Spec}(A)$ is defined by the smallest integer n (or ∞) such that $H^i(X, M) = 0$ for $i > n$ and any torsion étale sheaf M of Abelian groups on X.

As applications of the arithmetic duality theorems, we obtain the main results in local and global class field theory, which describe the Abelian fundamental groups for local fields and number rings. The *Abelian fundamental group* of $X = \mathrm{Spec}(A)$ is the Abelianization of the étale fundamental group $\pi_1(X)$ and is denoted by $\pi_1^{\mathrm{ab}}(X)$. The pro-finite covering of X corresponding to the closed commutator subgroup $[\pi_1(X), \pi_1(X)]$ is called the *maximal Abelian covering* of X, which we denote by X^{ab}. So $\pi_1^{\mathrm{ab}}(X) = \mathrm{Gal}(X^{\mathrm{ab}}/X)$. If A is a field F, one has $\pi_1^{\mathrm{ab}}(\mathrm{Spec}(F)) = \mathrm{Gal}(F^{\mathrm{ab}}/F)$, where F^{ab} is the *maximal Abelian extension* of F, the composite field of all finite Abelian extensions of F. For a Dedekind domain A, let F be the quotient field of A. Then one has $\pi_1^{\mathrm{ab}}(\mathrm{Spec}(A)) = \mathrm{Gal}(\tilde{F}^{\mathrm{ab}}/F)$, where $\tilde{F}^{\mathrm{ab}}$ is the *maximal unramified Abelian extension* of F, the composite field of all finite unramified Abelian extensions of F. *Class field theory* for a finite algebraic number field k describes the Abelian fundamental group $\pi_1^{\mathrm{ab}}(\mathrm{Spec}(k)) = \mathrm{Gal}(k^{\mathrm{ab}}/k)$ in terms of the base field k. Its local version for a $\mathfrak{p}$-adic field $k_{\mathfrak{p}}$, the theory describing $\pi_1^{\mathrm{ab}}(\mathrm{Spec}(k_{\mathfrak{p}})) = \mathrm{Gal}(k_{\mathfrak{p}}^{\mathrm{ab}}/k_{\mathfrak{p}})$ in terms of the base field $k_{\mathfrak{p}}$, is called *local class field theory*. Since for a finite algebraic number field k, $\pi_1^{\mathrm{ab}}(\mathrm{Spec}(k)) = \varprojlim_S \pi_1^{\mathrm{ab}}(\mathrm{Spec}(\mathcal{O}_k) \setminus S) = \varprojlim_S \mathrm{Gal}(k_S^{\mathrm{ab}}/k)$ (S running over finite subsets of $\mathrm{Max}(\mathcal{O}_k)$), class field theory amounts to describing $G_S(k)^{\mathrm{ab}} = \pi_1^{\mathrm{ab}}(\mathrm{Spec}(\mathcal{O}_k) \setminus S) = \mathrm{Gal}(k_S^{\mathrm{ab}}/k)$ in terms of k and S, where k_S^{ab} is the maximal Abelian extension of k unramified outside $\overline{S} = S \cup S_k^{\infty}$ (Example 2.2.20). These descriptions are obtained as duality theorems in the étale cohomology of $\mathrm{Spec}(k_{\mathfrak{p}})$ and $\mathrm{Spec}(\mathcal{O}_k) \setminus S$.

In the following, for a locally compact Abelian group G, we denote by G^* the Pontryagin dual of G, the locally compact Abelian group consisting of continuous homomorphisms $G \to \mathbb{R}/\mathbb{Z}$.

2.3.1 Finite Fields

Let F be a finite field $\mathbb{F}_q$. For a finite $\mathrm{Gal}(\overline{\mathbb{F}}_q/\mathbb{F}_q)$-module M, let $M^* = \mathrm{Hom}(M, \mathbb{Q}/\mathbb{Z})$. The action of $\mathrm{Gal}(\overline{\mathbb{F}}_q/\mathbb{F}_q)$ on M^* is defined by $(g\varphi)(x) = \varphi(g^{-1}x)$ ($g \in \mathrm{Gal}(\overline{\mathbb{F}}_q/\mathbb{F}_q), \varphi \in M^*, x \in M$). Then the cup product

$$H^i(\mathbb{F}_q, M^*) \times H^{1-i}(\mathbb{F}_q, M) \longrightarrow H^1(\mathbb{F}_q, \mathbb{Q}/\mathbb{Z}) \simeq \mathbb{Q}/\mathbb{Z} \ \ (i = 0, 1)$$

gives a non-degenerate pairing of finite Abelian groups, and $\mathrm{Spec}(\mathbb{F}_q)$ has the étale cohomological dimension 1. In particular, if $\mathrm{Gal}(\overline{\mathbb{F}}_q/\mathbb{F}_q)$ acts on M trivially, by using $\mathrm{Gal}(\overline{\mathbb{F}}_q/\mathbb{F}_q) = \hat{\mathbb{Z}}$, this pairing reduces to the duality $M \simeq M^{**}$.

2.3.2 $\mathfrak{p}$-*Adic Fields*

Let $k_{\mathfrak{p}}$ be a $\mathfrak{p}$-adic field. Let $\mathcal{O}_{\mathfrak{p}}$ be the ring of $\mathfrak{p}$-adic integers, ϖ a prime element of $\mathcal{O}_{\mathfrak{p}}$ and $v_{\mathfrak{p}}$ the $\mathfrak{p}$-adic additive valuation with $v_{\mathfrak{p}}(\varpi) = 1$. For a finite $\mathrm{Gal}(\overline{k}_{\mathfrak{p}}/k_{\mathfrak{p}})$-module M, let $M' := \mathrm{Hom}(M, \overline{k}_{\mathfrak{p}}^{\times})$. The action of $\mathrm{Gal}(\overline{k}_{\mathfrak{p}}/k_{\mathfrak{p}})$ on M' is defined by $(g\varphi)(x) = g\varphi(g^{-1}x)$ $(g \in \mathrm{Gal}(\overline{k}_{\mathfrak{p}}/k_{\mathfrak{p}}), \varphi \in M', x \in M)$.

Tate Local Duality 2.3.1 *There is a canonical isomorphism* $\mathrm{inv}_{\mathfrak{p}} : H^2(k_{\mathfrak{p}}, \overline{k}_{\mathfrak{p}}^{\times}) \xrightarrow{\sim} \mathbb{Q}/\mathbb{Z}$, *called the invariant isomorphism for* $k_{\mathfrak{p}}$, *and the cup product*

$$H^i(k_{\mathfrak{p}}, M') \times H^{2-i}(k_{\mathfrak{p}}, M) \longrightarrow H^2(k_{\mathfrak{p}}, \overline{k}_{\mathfrak{p}}^{\times}) \simeq \mathbb{Q}/\mathbb{Z} \ \ (0 \leq i \leq 2)$$

gives a non-degenerate pairing of finite Abelian groups. The étale cohomological dimension of $\mathrm{Spec}(k_{\mathfrak{p}})$ *is* 2.

Now consider the case $i = 1$ and $M = \mathbb{Z}/n\mathbb{Z}$. Then one has $M' = \mu_n$, the group of n-th roots of unity, $H^1(k_{\mathfrak{p}}, \mathbb{Z}/n\mathbb{Z}) = \mathrm{Hom}(\mathrm{Gal}(k_{\mathfrak{p}}^{\mathrm{ab}}/k_{\mathfrak{p}}), \mathbb{Z}/n\mathbb{Z})$, and $H^1(k_{\mathfrak{p}}, \mu_n) = k_{\mathfrak{p}}^{\times}/(k_{\mathfrak{p}}^{\times})^n$ (Kummer theory). Thus Tate local duality induces an isomorphism

$$k_{\mathfrak{p}}^{\times}/(k_{\mathfrak{p}}^{\times})^n \simeq \mathrm{Gal}(k_{\mathfrak{p}}^{\mathrm{ab}}/k_{\mathfrak{p}})/n\mathrm{Gal}(k_{\mathfrak{p}}^{\mathrm{ab}}/k_{\mathfrak{p}}).$$

By taking the projective limit $\varprojlim_n$, we obtain the *reciprocity homomorphism* of local class field theory

$$\rho_{k_{\mathfrak{p}}} : k_{\mathfrak{p}}^{\times} \longrightarrow \mathrm{Gal}(k_{\mathfrak{p}}^{\mathrm{ab}}/k_{\mathfrak{p}}),$$

which is injective and has the dense image. Further, by taking the pull-back by $\rho_{k_{\mathfrak{p}}}$, one has a bijection between the set of open subgroups of $\mathrm{Gal}(k_{\mathfrak{p}}^{\mathrm{ab}}/k)$ and the set of finite-index open subgroups of $k_{\mathfrak{p}}^{\times}$. Let $k_{\mathfrak{p}}^{\mathrm{ur}}$ be the maximal unramified extension of $k_{\mathfrak{p}}$. Then we have the following commutative exact diagram:

$$
\begin{array}{ccccccccc}
0 & \longrightarrow & \mathcal{O}_{\mathfrak{p}}^{\times} & \longrightarrow & k_{\mathfrak{p}}^{\times} & \xrightarrow{v_{\mathfrak{p}}} & \mathbb{Z} & \longrightarrow & 0 \\
& & \downarrow & & \downarrow \rho_{k_{\mathfrak{p}}} & & \cap\downarrow & & \\
0 & \longrightarrow & \mathrm{Gal}(k_{\mathfrak{p}}^{\mathrm{ab}}/k_{\mathfrak{p}}^{\mathrm{ur}}) & \longrightarrow & \mathrm{Gal}(k_{\mathfrak{p}}^{\mathrm{ab}}/k) & \longrightarrow & \mathrm{Gal}(k_{\mathfrak{p}}^{\mathrm{ur}}/k) = \hat{\mathbb{Z}} & \longrightarrow & 0.
\end{array}
$$

Here the left vertical isomorphism is the restriction of $\rho_{k_{\mathfrak{p}}}$ to $\mathcal{O}_{\mathfrak{p}}^{\times}$ and the right vertical injection is the map sending 1 to the Frobenius automorphism $\sigma_{\mathfrak{p}}$. Therefore $\rho_{k_{\mathfrak{p}}}(\varpi) = \sigma_{\mathfrak{p}}$.

For a finite Abelian extension $K_{\mathfrak{P}}/k_{\mathfrak{p}}$, we define the *reciprocity homomorphism*

$$\rho_{K_{\mathfrak{P}}/k_{\mathfrak{p}}} : k_{\mathfrak{p}}^{\times} \longrightarrow \mathrm{Gal}(K_{\mathfrak{P}}/k_{\mathfrak{p}}) \tag{2.3}$$

by composing $\rho_{k_{\mathfrak{p}}}$ with the natural projection $\mathrm{Gal}(k_{\mathfrak{p}}^{\mathrm{ab}}/k_{\mathfrak{p}}) \to \mathrm{Gal}(K_{\mathfrak{P}}/k_{\mathfrak{p}})$. Then $\rho_{K_{\mathfrak{P}}/k_{\mathfrak{p}}}$ induces the isomorphism

$$k_{\mathfrak{p}}^{\times}/\mathrm{N}_{K_{\mathfrak{P}}/k_{\mathfrak{p}}}(K_{\mathfrak{P}}^{\times}) \simeq \mathrm{Gal}(K_{\mathfrak{P}}/k_{\mathfrak{p}}),$$

and it follows that any open subgroup of $k_{\mathfrak{p}}^{\times}$ with finite index is obtained as the norm group of the multiplicative group of a finite Abelian extension of $k_{\mathfrak{p}}$. Further, one has

$$\begin{aligned} K_{\mathfrak{P}}/k_{\mathfrak{p}} \text{ is unramified} &\Longleftrightarrow \rho_{K_{\mathfrak{P}}/k_{\mathfrak{p}}}(\mathcal{O}_{\mathfrak{p}}^{\times}) = \mathrm{id}_{K_{\mathfrak{P}}} \\ &\Longleftrightarrow \mathrm{N}_{K_{\mathfrak{P}}/k_{\mathfrak{p}}}(\mathcal{O}_{\mathfrak{P}}^{\times}) = \mathcal{O}_{\mathfrak{p}}^{\times}, \end{aligned} \tag{2.4}$$

and, in this case,

$$\rho_{k_{\mathfrak{p}}}(x) = \sigma^{v_{\mathfrak{p}}(x)},$$

where $\sigma \in \mathrm{Gal}(K_{\mathfrak{P}}/k_{\mathfrak{p}})$ is the Frobenius automorphism. On the other hand, if $K_{\mathfrak{P}}/k_{\mathfrak{p}}$ is totally ramified, the restriction of $\rho_{k_{\mathfrak{p}}}$ to $\mathcal{O}_{\mathfrak{p}}^{\times}$ induces the isomorphism

$$\mathcal{O}_{\mathfrak{p}}^{\times}/\mathrm{N}_{K_{\mathfrak{P}}/k_{\mathfrak{p}}}(\mathcal{O}_{\mathfrak{P}}^{\times}) \simeq \mathrm{Gal}(K_{\mathfrak{P}}/k_{\mathfrak{p}}).$$

We now assume that $k_{\mathfrak{p}}$ contains a primitive n-th root of unity for some integer $n \geq 2$. Then the *Hilbert symbol*

$$\left(\frac{\cdot,\cdot}{\mathfrak{p}}\right)_n : k_{\mathfrak{p}}^{\times}/(k_{\mathfrak{p}}^{\times})^n \times k_{\mathfrak{p}}^{\times}/(k_{\mathfrak{p}}^{\times})^n \longrightarrow \mu_n$$

is defined by

$$\left(\frac{a,b}{\mathfrak{p}}\right)_n := \frac{\rho_{k_{\mathfrak{p}}}(b)(\sqrt[n]{a})}{\sqrt[n]{a}}.$$

The Hilbert symbol is bi-multiplicative and skew symmetric and satisfies the following property:

$$\left(\frac{a, b}{\mathfrak{p}}\right)_n = 1 \iff b \in N_{k_\mathfrak{p}(\sqrt[n]{a})/k_\mathfrak{p}}(k_\mathfrak{p}(\sqrt[n]{a})^\times) \tag{2.5}$$
$$\iff a \in N_{k_\mathfrak{p}(\sqrt[n]{b})/k_\mathfrak{p}}(k_\mathfrak{p}(\sqrt[n]{b})^\times).$$

When $k_\mathfrak{p}(\sqrt[n]{a})/k_\mathfrak{p}$ $(a \in k_\mathfrak{p}^\times)$ is an unramified extension (this is the case if $a \in \mathcal{O}_\mathfrak{p}^\times$), the *n-th power residue symbol* is defined by

$$\left(\frac{a}{\mathfrak{p}}\right)_n := \left(\frac{a, \varpi}{\mathfrak{p}}\right)_n = \frac{\sigma(\sqrt[n]{a})}{\sqrt[n]{a}}, \tag{2.6}$$

where $\sigma = \rho_{k_\mathfrak{p}(\sqrt[n]{a})/k_\mathfrak{p}}(\pi) \in \mathrm{Gal}(k_\mathfrak{p}(\sqrt[n]{a})/k_\mathfrak{p})$ is the Frobenius automorphism. Then one has

$$\left(\frac{a}{\mathfrak{p}}\right)_n = 1 \iff a \in (k_\mathfrak{p}^\times)^n$$
$$\iff a \bmod \mathfrak{p} \in (\mathbb{F}_\mathfrak{p}^\times)^n \ (\text{if } a \in U_\mathfrak{p}).$$

Let $k_\mathfrak{p} = \mathbb{Q}_p$ for an odd prime number p and let a be an integer prime to p. Then the power residue symbol $\left(\frac{a}{p}\right)_2$ coincides with the Legendre symbol $\left(\frac{a}{p}\right)$.

As for the field $\mathbb{R}$ of real numbers, we also have the duality theorem by using Tate's modified cohomology groups ([205, Ch.VIII]). Let M be a finite $\mathrm{Gal}(\mathbb{C}/\mathbb{R})$-module and let $M' = \mathrm{Hom}(M, \mathbb{C}^\times)$. The action of $\mathrm{Gal}(\mathbb{C}/\mathbb{R})$ on M' is defined by $(g\varphi)(x) = g\varphi(g^{-1}x)$ $(g \in \mathrm{Gal}(\mathbb{C}/\mathbb{R}), \varphi \in M', x \in M)$. Then the cup product

$$\hat{H}^i(\mathbb{R}, M') \times \hat{H}^{2-i}(\mathbb{R}, M) \longrightarrow H^2(\mathbb{R}, \mathbb{C}^\times) \simeq \mathbb{F}_2 \ (i \in \mathbb{Z})$$

gives a non-degenerate pairing of finite Abelian groups. Letting $i = 1$ and $M = \mu_2$, we have the isomorphism

$$\rho_{\mathbb{C}/\mathbb{R}} : \ \mathbb{R}^\times/(\mathbb{R}^\times)^2 = H^1(\mathbb{R}, \mu_2) \simeq H^1(\mathbb{R}, \mathbb{F}_2)^* = \mathrm{Gal}(\mathbb{C}/\mathbb{R}).$$

The *reciprocity homomorphism* $\rho_\mathbb{R} : \ \mathbb{R}^\times \to \mathrm{Gal}(\mathbb{C}/\mathbb{R})$ is then defined by composing the natural projection $\mathbb{R}^\times \to \mathbb{R}^\times/(\mathbb{R}^\times)^2$ with $\rho_{\mathbb{C}/\mathbb{R}}$. Hence $\rho_\mathbb{R}$ is surjective and $\mathrm{Ker}(\rho_\mathbb{R}) = (\mathbb{R}^\times)^2$ (the connected component of 1 consisting of positive real numbers).

2.3.3 Number Rings

Let k be a number field of finite degree over $\mathbb{Q}$. Let $\mathcal{O}_k$ be the ring of integers of k and set $X = \mathrm{Spec}(\mathcal{O}_k)$, $\overline{X} := X \sqcup S_k^\infty$. An étale sheaf M of Abelian groups on X is said to be *constructible* if all stalks of M are finite and there is an open subset $U \subset X$ such that $M|_U$ is locally constant [248]. For a constructible sheaf M on X, the modified étale cohomology groups $\hat{H}^i(X, M)$ ($i \in \mathbb{Z}$), which take the infinite primes into account, are defined. For the definition of modified cohomology, we refer to [248]. (See [15], [99, §3], [147] for some other treatments on the modified étale cohomology taking the infinite primes into account.) We let $M' := \underline{\mathrm{Hom}}(M, \mathbb{G}_{m,X})$, where $\mathbb{G}_{m,X}$ is the étale sheaf on X defined by associating to a connected finite étale covering $\mathrm{Spec}(B) \to X$ the multiplicative group $\mathbb{G}_{m,X}(Y) = B^\times$.

Artin–Verdier Duality 2.3.2 *Let M be a constructible sheaf on X. There is a canonical isomorphism* $\mathrm{inv}_X : \hat{H}^3(X, \mathbb{G}_{m,X}) \xrightarrow{\sim} \mathbb{Q}/\mathbb{Z}$, *called the invariant isomorphism for X, and the natural pairing*

$$\hat{H}^i(X, M') \times \mathrm{Ext}_X^{3-i}(M, \mathbb{G}_{m,X}) \longrightarrow \hat{H}^3(X, \mathbb{G}_{m,X}) \simeq \mathbb{Q}/\mathbb{Z}$$

gives a non-degenerate pairing of finite Abelian groups. The étale cohomological dimension of $X = \mathrm{Spec}(\mathcal{O}_k)$ is 3, up to 2-torsion in the case that k has a real prime.

Let U be an open subset of X. In the following, we shall use the notations $X_0 := \mathrm{Max}(\mathcal{O}_k)$, $U_0 := U \cap \mathrm{Max}(\mathcal{O}_k)$. Let S_k^∞ be the set of infinite primes of k, and set $S = X \setminus U$, $\overline{S} = S \cup S_k^\infty$ so that $\pi_1(U) = G_S(k) = \mathrm{Gal}(k_S/k)$ (Example 2.2.20). Let M be a finite $G_S(k)$-module and we use the same notation M to denote the corresponding locally constant, finite étale sheaf on U. Assume $\#M \in \mathcal{O}(U)^\times$ (S-unit). Let $j : U \hookrightarrow X$ be the inclusion map and define the constructible sheaf $j_!M$ on X as follows: For a finite étale covering $h : Y \to X$, $j_!M(Y) := M$ if $h(Y) \subset U$, and $j_!M(Y) = 0$ otherwise. Then we have $\mathrm{Ext}_X^i(j_!M, \mathbb{G}_{m,X}) = H^i(U, M')$ and the pairing of Artin–Verdier duality becomes the cup product

$$\hat{H}^i(X, j_!M) \times H^{3-i}(U, M') \longrightarrow \hat{H}^3(X, \mathbb{G}_{m,X}) \simeq \mathbb{Q}/\mathbb{Z} \quad (i \in \mathbb{Z}).$$

Let V be an open subset of X so that $V \subset U$. Applying the excision

$$H_v^{i+1}(X, j_!M) = \begin{cases} \hat{H}^i(k_v, M) & (v \in S_k^\infty) \\ H^i(k_\mathfrak{p}, M) & (v = \mathfrak{p} \in S) \\ H_\mathfrak{p}^{i+1}(U, M) & (v = \mathfrak{p} \in U \setminus V) \end{cases}$$

to the relative étale cohomology sequence for the pair $V \subset X$ and taking the inductive limit $\varinjlim_{V:\,\text{smaller}}$, we obtain the following long exact sequence:

$$\cdots \longrightarrow H_c^i(U, M) \longrightarrow H^i(k, M) \longrightarrow \bigoplus_{v \in \overline{S}} H^i(k_v, M) \oplus \bigoplus_{\mathfrak{p} \in U_0} H_\mathfrak{p}^{i+1}(U, M)$$

$$\longrightarrow H_c^{i+1}(U, M) \longrightarrow \cdots.$$

Next we take the inductive limit $\varinjlim_U$ making U smaller (i.e., S larger) in the above exact sequence. Noting $H_{\mathfrak{p}}^{i+1}(U, M) = \mathrm{Coker}(H^i(\mathbb{F}_{\mathfrak{p}}, M) \to H^i(k_{\mathfrak{p}}, M))$, we obtain the Tate–Poitou exact sequence:

Tate–Poitou Exact Sequence 2.3.3 *Let M be a finite $\mathrm{Gal}(\overline{k}/k)$-module and set $M' = \mathrm{Hom}(M, \overline{k}^{\times})$. The action of $\mathrm{Gal}(\overline{k}/k)$ on M' is given by $(g\varphi)(x) = g\varphi(g^{-1}x)$ $(g \in \mathrm{Gal}(\overline{k}/k), \varphi \in M', x \in M)$. Then we have the following exact sequence of locally compact Abelian groups:*

$$0 \longrightarrow H^0(k, M) \longrightarrow P^0(k, M) \longrightarrow H^2(k, M')^* \longrightarrow H^1(k, M)$$
$$\downarrow$$
$$P^1(k, M)$$
$$\downarrow$$
$$0 \longleftarrow H^0(k, M')^* \longleftarrow P^2(k, M) \longleftarrow H^2(k, M) \longleftarrow H^1(k, M')^*$$

Here the cohomology groups $H^i(k, -)$, $H^i(k_v, -)$ are endowed with the discrete topology, and $P^i(k, M)$ is defined by

$$P^i(k, M) := \prod_{\mathfrak{p} \in X_0} H^i(k_{\mathfrak{p}}, M) \times \prod_{v \in S_k^{\infty}} \hat{H}^i(k_v, M),$$

where $\prod_{\mathfrak{p} \in X_0} H^i(k_{\mathfrak{p}}, M)$ means the restricted direct product of $H^i(k_{\mathfrak{p}}, M)$'s with respect to the subgroups

$$H_{\mathrm{ur}}^i(k_{\mathfrak{p}}, M) := \mathrm{Im}(H^i(\mathbb{F}_{\mathfrak{p}}, M) \to H^i(k_{\mathfrak{p}}, M)),$$

namely,

$$\prod_{\mathfrak{p} \in X_0} H^i(k_{\mathfrak{p}}, M) := \{(c_{\mathfrak{p}}) \mid c_{\mathfrak{p}} \in H_{\mathrm{ur}}^i(k_{\mathfrak{p}}, M) \text{ for all but finitely many of } \mathfrak{p}\text{'s }\}.$$

The topology of $P^i(k, M)$ is given as the restricted direct product topology, namely, the basis of neighborhoods of the identity is given by the compact groups

$$\prod_{v \in S_k^{\infty}} \hat{H}^i(k_v, M) \times \prod_{\mathfrak{p} \in S} H^i(k_{\mathfrak{p}}, M) \times \prod_{\mathfrak{p} \in U_0} H_{\mathrm{ur}}^i(k_{\mathfrak{p}}, M),$$

where U ranges over open subsets of X.

We define the *idèle group* J_k and the *idèle class group* C_k of k respectively by

$$J_k := \prod_{\mathfrak{p} \in X_0} k_{\mathfrak{p}}^{\times} \times \prod_{v \in S_k^{\infty}} k_v^{\times}, \quad C_k := J_k / P_k, \tag{2.7}$$

where $\prod_{\mathfrak{p} \in X_0} k_{\mathfrak{p}}^{\times}$ means the restricted direct product of $k_{\mathfrak{p}}^{\times}$'s with respect to $\mathcal{O}_{\mathfrak{p}}^{\times}$'s and P_k is the closed subgroup of J_k defined by the image of $k^{\times}$ embedded diagonally in J_k.

Now let us specialize M to be the group μ_n of n-th roots of unity in the Tate–Poitou exact sequence 2.3.3. Then

$$H^1(k, \mu_n) = k^{\times}/(k^{\times})^n, \quad P^1(k, \mu_n) = J_k/J_k^n,$$
$$H^2(k, \mu_n) = {}_n\mathrm{Br}(k), \quad P^2(k, M) = \bigoplus_v {}_n\mathrm{Br}(k_v).$$

Here $\mathrm{Br}(R)$ stands for the *Brauer group* of R ([176, Ch.VI,§3]) and ${}_nA := \{x \in A \mid nx = 0\}$ for an Abelian (additive) group A. Since $H^1(k, \mathbb{Z}/n\mathbb{Z})^* = \mathrm{Gal}(k^{\mathrm{ab}}/k)/n\mathrm{Gal}(k^{\mathrm{ab}}/k)$ and the localization map $\mathrm{Br}(k) \to \bigoplus_{v \in X_0 \cup S_k^{\infty}} \mathrm{Br}(k_v)$ is injective (Hasse principle for the Brauer group ([ibid., Ch.VIII,§1])), the Tate–Poitou exact sequence yields the following isomorphism:

$$C_k/C_k^n \simeq \mathrm{Gal}(k^{\mathrm{ab}}/k)/n\mathrm{Gal}(k^{\mathrm{ab}}/k).$$

Taking the projective limit $\varprojlim_n$, we obtain the *reciprocity homomorphism* of class field theory,

$$\rho_k : C_k \longrightarrow \mathrm{Gal}(k^{\mathrm{ab}}/k). \tag{2.8}$$

The map ρ_k is surjective and $\mathrm{Ker}(\rho_k)$ coincides with the connected component of 1 in C_k. Further, taking the pull-back by ρ_k, one has a bijection between the set of open subgroups of $\mathrm{Gal}(k^{\mathrm{ab}}/k)$ and the set of open subgroups of C_k. The relation with local class field theory is given as follows: Let $\iota_v : k_v^{\times} \to C_k$ be the map defined by $\iota_v(a_v) = [(1, \ldots, 1, a_v, 1, \ldots)]$. Then one has the following commutative diagram:

$$
\begin{array}{ccc}
k_v^{\times} & \xrightarrow{\rho_{k_v}} & \mathrm{Gal}(k_v^{\mathrm{ab}}/k_v) \\
\iota_v \downarrow & & \downarrow \\
C_k & \xrightarrow{\rho_k} & \mathrm{Gal}(k^{\mathrm{ab}}/k).
\end{array}
\tag{2.9}
$$

For a finite Abelian extension K/k, the *reciprocity homomorphism*

$$\rho_{K/k} : C_k \longrightarrow \mathrm{Gal}(K/k) \tag{2.10}$$

is defined by the composing ρ_k with the natural projection $\mathrm{Gal}(k^{\mathrm{ab}}/k) \to \mathrm{Gal}(K/k)$. Then $\rho_{K/k}$ induces the isomorphism

$$C_k/N_{K/k}(C_K) \simeq \mathrm{Gal}(K/k)$$

and it follows that any open subgroup of C_k is obtained as the norm group of the idèle class group of a finite Abelian extension of k. Further, one has

$$
\begin{aligned}
\mathfrak{p} \text{ is completely decomposed in } K/k &\iff \rho_{K/k} \circ \iota_{\mathfrak{p}}(k_{\mathfrak{p}}^{\times}) = \mathrm{id}, \\
v \text{ is unramified in } K/k &\iff \rho_{K/k} \circ \iota_{\mathfrak{p}}(\mathcal{O}_v^{\times}) = \mathrm{id},
\end{aligned}
\tag{2.11}
$$

where we set $\mathcal{O}_v^{\times} := k_v^{\times}$ if $v \in S_k^{\infty}$.

Example 2.3.4 (Unramified Class Field Theory) Let $\tilde{k}_+^{\mathrm{ab}}$ be the maximal Abelian extension of k such that any $\mathfrak{p} \in X_0$ is unramified. Then we have

$$
\pi_1^{\mathrm{ab}}(\mathrm{Spec}(\mathcal{O}_k)) = \mathrm{Gal}(\tilde{k}_+^{\mathrm{ab}}/k).
$$

By (2.11), the fundamental map ρ_k induces the isomorphism

$$
J_k/k^{\times}\left(\prod_{v \in S_k^{\infty}} (k_v^{\times})^2 \times \prod_{\mathfrak{p} \in X_0} \mathcal{O}_{\mathfrak{p}}^{\times} \right) \simeq \mathrm{Gal}(\tilde{k}_+^{\mathrm{ab}}/k).
$$

Note that the left-hand side is isomorphic to the narrow ideal class group $H^+(k)$ by the correspondence $J_k \ni (a_v) \mapsto \prod_{\mathfrak{p} \in X_0} \mathfrak{p}^{v_{\mathfrak{p}}(a_{\mathfrak{p}})} \in I_k$. Therefore we have the following canonical isomorphism:

$$
H^+(k) \simeq \mathrm{Gal}(\tilde{k}_+^{\mathrm{ab}}/k),
$$

Let $\tilde{k}^{ab}$ be the maximal Abelian extension such that any prime of k is unramified, called the *Hilbert class field of k*. Then the Galois group $\mathrm{Gal}(\tilde{k}^{ab}/k)$ is then canonically isomorphic to the ideal class group $H(k)$ of k:

$$
H(k) \simeq \mathrm{Gal}(\tilde{k}^{\mathrm{ab}}/k).
$$

The above two isomorphisms are regarded as arithmetic analogues of the Hurewicz isomorphism in Example 2.1.13.

Example 2.3.5 Let S be a finite subset of $\mathrm{Max}(\mathcal{O}_k)$ and $\overline{S} = S \cup S_k^{\infty}$. Let $k_{\overline{S}}^{\mathrm{ab}}$ be the maximal Abelian extension of k unramified outside $\overline{S}$ so that $G_S(k)^{\mathrm{ab}} = \pi_1^{\mathrm{ab}}(\mathrm{Spec}(\mathcal{O}_k) \setminus S) = \mathrm{Gal}(k_{\overline{S}}^{\mathrm{ab}}/k)$ (Example 2.2.20). By (2.11), the reciprocity homomorphism ρ_k induces the isomorphism

$$
\overline{J_k/k^{\times}\left(\prod_{v \in S_k^{\infty}} (k_v^{\times})^2 \times \prod_{\mathfrak{p} \in X \setminus S} \mathcal{O}_{\mathfrak{p}}^{\times} \right)} \simeq \mathrm{Gal}(k_{\overline{S}}^{\mathrm{ab}}/k),
$$

where $\overline{k^\times(\cdots)}$ means the topological closure. By Example 2.3.4, $\mathrm{Gal}(\tilde{k}^{ab}_+/k) \simeq H^+(k) = J_k/k^\times(\prod_{v \in S_k^\infty}(k_v^\times)^2 \times \prod_{\mathfrak{p} \in X_0} \mathcal{O}_\mathfrak{p}^\times)$ and

$$
k^\times\left(\prod_{v \in S_k^\infty}(k_v^\times)^2 \times \prod_{\mathfrak{p} \in X_0}\mathcal{O}_\mathfrak{p}^\times\right)\Big/\overline{k^\times\left(\prod_{v \in S_k^\infty}(k_v^\times)^2 \times \prod_{\mathfrak{p} \in X \backslash S}\mathcal{O}_\mathfrak{p}^\times\right)}
$$
$$
\simeq \prod_{\mathfrak{p} \in S}\mathcal{O}_\mathfrak{p}^\times\Big/\left(\prod_{\mathfrak{p} \in S}\mathcal{O}_\mathfrak{p}^\times \cap \overline{k^\times\left(\prod_{v \in S_k^\infty}(k_v^\times)^2 \times \prod_{\mathfrak{p} \in X \backslash S}\mathcal{O}_\mathfrak{p}^\times\right)}\right)
$$
$$
\simeq \prod_{\mathfrak{p} \in S}\mathcal{O}_\mathfrak{p}^\times/\overline{\mathcal{O}_k^+},
$$

where $\mathcal{O}_k^+ := \{a \in \mathcal{O}_k^\times \mid a \text{ is totally positive}\}$ and $\overline{\mathcal{O}_k^+}$ denotes the topological closure of the diagonal image of $\mathcal{O}_k^+$ in $\prod_{\mathfrak{p} \in S}\mathcal{O}_\mathfrak{p}^\times$. Hence we have the following exact sequence:

$$
0 \longrightarrow \prod_{\mathfrak{p} \in S}U_\mathfrak{p}/\overline{\mathcal{O}_k^+} \longrightarrow G_S(k)^{ab} \longrightarrow H^+(k) \longrightarrow 0
$$

As $\mathcal{O}_\mathfrak{p}^\times = \mathbb{F}_\mathfrak{p}^\times \times (1 + \mathfrak{p})$, this exact sequence gives some restrictions on ramified primes in S. For example, if $\mathfrak{p}$ is ramified in a pro-l extension for some prime number l, one must have $N\mathfrak{p} \equiv 1$ or $0 \bmod l$.

Example 2.3.6 Let $k = \mathbb{Q}$ and $S = \{(p_1), \ldots, (p_r)\}$ in Example 2.3.3.9. For this case, we have $H^+(\mathbb{Q}) = 1$ and $\mathbb{Z}^+ = \{1\}$ and hence

$$
G_S^{ab} \simeq \prod_{i=1}^r \mathbb{Z}_{p_i}^\times.
$$

It follows that $\mathbb{Q}_S^{ab} = \mathbb{Q}(\mu_{p_i^\infty} \mid 1 \le i \le r)$, where $\mu_{p_i^\infty} := \bigcup_{d \ge 1} \mu_{p_i^d}$, $\mu_{p_i^d}$ being the group of p_i^d-th roots of unity.

Suppose that $p_i \equiv 1 \bmod n$ $(1 \le i \le r)$ for some integer $n (\ge 2)$. Fix a primitive root $\alpha_i \bmod p_i$, $\mathbb{F}_{p_i}^\times = \langle \alpha_i \rangle$. Let

$$
\psi : \prod_{i=1}^r \mathbb{Z}_{p_i}^\times = \prod_{i=1}^r \mathbb{F}_{p_i}^\times \times (1 + p_i\mathbb{Z}_{p_i}) \longrightarrow \mathbb{Z}/n\mathbb{Z}
$$

be the homomorphism defined by $\psi(\alpha_i) = 1$, $\psi(1 + p_i\mathbb{Z}_{p_i}) = 0$. Let k be the subfield of $\mathbb{Q}_S^{ab}$ corresponding to $\mathrm{Ker}(\psi)$ which is independent of the choice of α_i. The field k is the cyclic extension of $\mathbb{Q}$ of degree n such that any prime outside $\overline{S} = S \cup \{\infty\}$ is unramified and each prime in S is totally ramified in $k/\mathbb{Q}$.

Next, let $S = \{(p)\}$. Then one has $\mathbb{Q}_{\overline{S}} = \mathbb{Q}(\mu_{p^\infty})$, $G_{\{p\}} = \mathrm{Gal}(\mathbb{Q}(\mu_{p^\infty})/\mathbb{Q}) \simeq \mathbb{Z}_p^\times$. We set

$$q := \begin{cases} p \ (p \text{ is a odd prime number}) \\ 4 \ (p = 2), \end{cases}$$

and let

$$\psi : \mathbb{Z}_p^\times = \mathbb{F}_p^\times \times (1 + p\mathbb{Z}_p) \longrightarrow 1 + q\mathbb{Z}_p \simeq \mathbb{Z}_p$$

be the projection on $1 + q\mathbb{Z}_p$. Let $\mathbb{Q}_\infty$ denote the subfield of $\mathbb{Q}(\mu_{p^\infty})$ corresponding to $\mathrm{Ker}(\psi)$. The field $\mathbb{Q}_\infty$ is then the unique Galois extension of $\mathbb{Q}$ whose Galois group $\mathrm{Gal}(\mathbb{Q}_\infty/\mathbb{Q})$ is isomorphic to the additive group $\mathbb{Z}_p$. Note that only (p) is ramified in the extension $\mathbb{Q}_\infty/\mathbb{Q}$ and it is totally ramified.

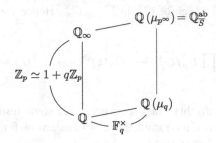

In general, for a number field F of finite degree over $\mathbb{Q}$, $F_\infty := F\mathbb{Q}_\infty$ is a Galois extension with $\mathrm{Gal}(F_\infty/F)$ being isomorphic to $\mathbb{Z}_p$ such that only primes over p are ramified in F_∞/F. The extension F_∞ is called the *cyclotomic $\mathbb{Z}_p$-extension* of F.

As is seen above, the Artin–Verdier duality and the Tate–Poitou exact sequence, which contain the main content of class field theory, are arithmetic analogues of the 3-dimensional Poincaré duality and the relative cohomology sequence (+excision) in topology respectively. Readers may find similar features between Example 2.3.6 and Examples 2.1.12, 2.1.15, Example 2.3.4 and Example 2.1.13. We shall discuss these analogies more precisely in the subsequent chapters.

Chapter 3
Knots and Primes, 3-Manifolds and Number Rings

In this chapter we explain the basic analogies between knots and primes, 3-manifolds and number rings, which will be fundamental in subsequent chapters.

By Examples 2.1.1 and 2.2.9, we find the following analogy between the fundamental groups of a circle S^1 and of a finite field $\mathbb{F}_q$.

$$
\begin{array}{|c|c|}
\hline
\pi_1(S^1) = \mathrm{Gal}(\mathbb{R}/S^1) & \pi_1(\mathrm{Spec}(\mathbb{F}_q)) = \mathrm{Gal}(\overline{\mathbb{F}}_q/\mathbb{F}_q) \\
= \langle [l] \rangle & = \langle \sigma \rangle \\
\simeq \mathbb{Z} & \simeq \hat{\mathbb{Z}} \\
\hline
\end{array}
\tag{3.1}
$$

Here, the loop l and the Frobenius automorphism σ, the universal covering $\mathbb{R}$ and the separable closure $\overline{\mathbb{F}}_q$, and the cyclic covering $\mathbb{R}/n\mathbb{Z} \to S^1$ and the cyclic extension $\mathbb{F}_{q^n}/\mathbb{F}_q$ are corresponding, respectively. Since, in fact, $\pi_i(S^1)$ and $\pi_i(\mathrm{Spec}(\mathbb{F}_q))$ are trivial for $i \geq 2$, S^1 is the Eilenberg–MacLane space $K(\mathbb{Z}, 1)$ homotopically and $\mathrm{Spec}(\mathbb{F}_q)$ is regarded as an étale homotopical analogue $K(\hat{\mathbb{Z}}, 1)$.

$$
\begin{array}{|c|c|}
\hline
\text{circle} & \text{finite field} \\
S^1 = K(\mathbb{Z}, 1) & \mathrm{Spec}(\mathbb{F}_q) = K(\hat{\mathbb{Z}}, 1) \\
\hline
\end{array}
\tag{3.2}
$$

A tubular neighborhood $V = S^1 \times D^2$ of a circle S^1 is homotopically equivalent to the core S^1 and $V \setminus S^1$ is homotopically equivalent to the 2-dimensional torus ∂V. On the other hand, for a $\mathfrak{p}$-adic integer ring $\mathcal{O}_{\mathfrak{p}}$ whose residue field is $\mathbb{F}_q$ and quotient field is $k_{\mathfrak{p}}$ ($\mathfrak{p}$-adic field), $\mathrm{Spec}(\mathcal{O}_{\mathfrak{p}})$ is étale homotopically equivalent to $\mathrm{Spec}(\mathbb{F}_q)$ and $\mathrm{Spec}(\mathcal{O}_{\mathfrak{p}}) \setminus \mathrm{Spec}(\mathbb{F}_q) = \mathrm{Spec}(k_{\mathfrak{p}})$. Hence $\mathrm{Spec}(\mathcal{O}_{\mathfrak{p}})$ and $\mathfrak{p}$-adic field $\mathrm{Spec}(k_{\mathfrak{p}})$ are analogous to V and ∂V. In fact, we find the following analogy between $\pi_1(\partial V)$ and $\pi_1(\mathrm{Spec}(k_{\mathfrak{p}}))$. In the natural homomorphism $\pi_1(\partial V) \to \pi_1(V) = \pi_1(S^1)$, the image of a longitude $\beta = S^1 \times \{b\}$ ($b \in \partial D^2$) is a generator $l \in \pi_1(S^1)$ and the kernel is the infinite cyclic group generated by a meridian $\alpha = \{a\} \times \partial D^2$ ($a \in S^1$).

© The Author(s), under exclusive license to Springer Nature Singapore Pte Ltd. 2024
M. Morishita, *Knots and Primes*, Universitext,
https://doi.org/10.1007/978-981-99-9255-3_3

Thus $\pi_1(\partial V)$ is a free abelian group generated by α and β which are subject to the relation $[\alpha, \beta] := \alpha\beta\alpha^{-1}\beta^{-1} = 1$ (Example 2.1.2). On the other hand, in the natural homomorphism $\pi_1(\mathrm{Spec}(k_\mathfrak{p})) \to \pi_1(\mathrm{Spec}(\mathcal{O}_\mathfrak{p})) = \pi_1(\mathrm{Spec}(\mathbb{F}_q))$, the image of an extension of the Frobenius automorphism (denoted by σ as well) in $\pi_1(\mathrm{Spec}(k_\mathfrak{p})) = \mathrm{Gal}(\overline{k}_\mathfrak{p}/k_\mathfrak{p})$ is $\sigma \in \pi_1(\mathrm{Spec}(\mathbb{F}_q)) = \mathrm{Gal}(\overline{\mathbb{F}}_q/\mathbb{F}_q)$ and the kernel is the inertia group $I_{k_\mathfrak{p}}$. We call an element of $I_{k_\mathfrak{p}}$ *monodromy*. Although which monodromy we take as an analogue of a meridian depends on the situation we are considering, there is a canonical generator τ, which corresponds to an meridian, in the maximal tame quotient $I^{\mathrm{t}}_{k_\mathfrak{p}}$ of $I_{k_\mathfrak{p}}$. In fact, the tame fundamental group $\pi^{\mathrm{t}}_1(\mathrm{Spec}(k_\mathfrak{p}))$ is a profinite group generated by τ and σ which are subject to the relation $\tau^{q-1}[\tau, \sigma] = 1$ (Example 2.2.23).

tubular neighborhood V	$\mathfrak{p}$-adic integer ring $\mathrm{Spec}(\mathcal{O}_\mathfrak{p})$
boundary ∂V	$\mathfrak{p}$-adic field $\mathrm{Spec}(k_\mathfrak{p})$
$1 \to \langle\alpha\rangle \to \pi_1(\partial V) \to \langle\beta\rangle \to 1$	$1 \to I_{k_\mathfrak{p}} \to \pi_1(\mathrm{Spec}(k_\mathfrak{p})) \to \langle\sigma\rangle \to 1$
β : longitude	σ : Frobenius automorphism
α : meridian	τ : monodromy ($\in I_{k_\mathfrak{p}}$)
$\pi_1(\partial V) = \langle\alpha, \beta \,|\, [\alpha, \beta] = 1\rangle$	$\pi^{\mathrm{t}}_1(\mathrm{Spec}(k_\mathfrak{p})) = \langle\tau, \sigma \,|\, \tau^{q-1}[\tau, \sigma] = 1\rangle$

$$\tag{3.3}$$

A knot is an embedding of S^1 into a 3-manifold M. On the other hand, as seen in Sect. 2.3, for the ring of integers $\mathcal{O}_k$ of a finite algebraic number field k, $\mathrm{Spec}(\mathcal{O}_k)$ is seen as an analogue of a 3-dimensional manifold, since $\mathrm{Spec}(\mathcal{O}_k)$ has the étale cohomological dimension 3 (except for 2-torsion, when k has a real place) and satisfies the Artin–Verdier duality 2.3.2 which is an arithmetic analogue of 3-dimensional Poincaré duality.

3-manifold M	number ring $\mathrm{Spec}(\mathcal{O}_k)$

$$\tag{3.4}$$

For a non-zero prime ideal $\mathfrak{p}$ of $\mathcal{O}_k$, by (1.2) and (1.4), the natural embedding $\mathrm{Spec}(\mathbb{F}_\mathfrak{p}) \hookrightarrow \mathrm{Spec}(\mathcal{O}_k)$ is regarded as an analogue of a knot.

knot	prime ideal
$S^1 \hookrightarrow M$	$\mathrm{Spec}(\mathbb{F}_\mathfrak{p}) \longhookrightarrow \mathrm{Spec}(\mathcal{O}_k)$

$$\tag{3.5}$$

In particular, since $\pi_1(\mathrm{Spec}(\mathbb{Z})) = 1$ (Example 2.2.19), $\mathrm{Spec}(\mathbb{Z})$ with the infinite prime ∞ is seen as an analogue of $S^3 = \mathbb{R} \cup \{\infty\}$, and the prime ideal (p) (p being a prime number) is seen as an analogue of a knot in $\mathbb{R}^3$.

knot	rational prime
$S^1 \hookrightarrow \mathbb{R}^3 \cup \{\infty\} = S^3$	$\mathrm{Spec}(\mathbb{F}_p) \longhookrightarrow \mathrm{Spec}(\mathbb{Z}) \cup \{\infty\}$

$$\tag{3.6}$$

Here we regard S^3 as the end-compactification of $\mathbb{R}^3$ and see the infinite prime as an analogue of the end. In general, the set of ends E_M of a non-compact 3-manifold M corresponds to the set of infinite primes S_k^∞ of a number field k [41, 190].

$$\boxed{\text{ends } E_M \,\big|\, \text{infinite primes } S_k^\infty} \tag{3.7}$$

We should also note the analogy that any connected oriented 3-manifold is a finite covering of S^3 branched along a link (Alexander's theorem), just as any finite algebraic number field is a finite extension of $\mathbb{Q}$ ramified over a finite set of primes.

For a knot K in a 3-manifold M, let V_K be a tubular neighborhood of K, $X_K :=$ $M \setminus \text{int}(V_K)$ the knot complement, $G_K = G_K(M) = \pi_1(X_K) = \pi_1(M \setminus K)$ the knot group (Example 0.1.6). The knot group G_K is the object which reflects how K is knotted in M. In fact, it is known that a prime knot K in S^3 is determined by the knot group G_K. Namely, for prime knots $K, L \subset S^3$, we have the following [69, 243]:

$$G_K \simeq G_L \iff K \simeq L. \tag{3.8}$$

Here $K \simeq L$ means that there is an auto-homeomorphism h of S^3 such that $h(K) = L$. Let α and β be a meridian and a longitude of K respectively. The image of the homomorphism $\pi_1(\partial X_K) = \langle \alpha, \beta \mid [\alpha, \beta] = 1 \rangle \to G_K$ induced by the inclusion $\partial X_K \hookrightarrow X_K$ is called the *peripheral group*, denoted by D_K. We also denote by I_K the image of the subgroup of generated by α in G_K. When $M = S^3$, we see by the Wirtinger presentation of G_K that G_K is generated by conjugates of I_K (Example 2.1.6).

On the other hand, by (3.3), for a prime ideal $\mathfrak{p}$ ($\neq (0)$) of $\mathcal{O}_k$, the $\mathfrak{p}$-adic field $\text{Spec}(k_\mathfrak{p})$ plays a role of the "boundary" of $X_{\{\mathfrak{p}\}} := \text{Spec}(\mathcal{O}_k) \setminus \{\mathfrak{p}\}$. How $\mathfrak{p}$ is knotted in $\text{Spec}(\mathcal{O}_k)$ is reflected in the structure of the étale fundamental group $G_{\{\mathfrak{p}\}} := \pi_1(X_{\{\mathfrak{p}\}})$. Following the model of the knot group, we call $G_{\{\mathfrak{p}\}}$ the *prime group*. For prime numbers p, q, we have the following analogy of (3.8):

$$G_{\{(p)\}} \simeq G_{\{(q)\}} \iff p = q. \tag{3.9}$$

This is because the abelianization $G_{\{\mathfrak{p}\}}^{\text{ab}}$ of $G_{\{\mathfrak{p}\}}$ is isomorphic to $\mathbb{Z}_p^\times$ (Example 2.3.6). The arithmetic analogue of the peripheral group is the image of the natural homomorphism $\pi_1(\text{Spec}(k_\mathfrak{p})) \to G_{\{\mathfrak{p}\}}$ induced by the inclusion $\text{Spec}(k_\mathfrak{p}) \to X_{\{\mathfrak{p}\}}$, namely, the decomposition group $D_\mathfrak{p}$ over $\mathfrak{p}$, and the analogue of I_K is the inertia group $I_\mathfrak{p}$ over $\mathfrak{p}$ (Example 2.2.24).[1] When $k = \mathbb{Q}$, $G_{\{\mathfrak{p}\}}$ is generated by conjugates of $I_{\{p\}}$. In fact, if F denotes the extension of $\mathbb{Q}$ corresponding to the subgroup H of $G_{\{p\}}$ generated by conjugates of $I_{\{p\}}$, $F/\mathbb{Q}$ is an unramified

---

[1] Although we should write $G_{\{\mathfrak{p}\}}$, $D_{\{\mathfrak{p}\}}$ and $I_{\{\mathfrak{p}\}}$ as analogues of G_K, D_K and I_K respectively, we often write $G_\mathfrak{p}$, $D_\mathfrak{p}$ and $I_\mathfrak{p}$ for simplicity. For a prime number p, we also write $G_{\{p\}}$, $D_{\{p\}}$ and $I_{\{p\}}$ or simply G_p, D_p and I_p for $G_{\{(p)\}}$, $D_{\{(p)\}}$ and $I_{\{(p)\}}$ respectively.

extension in the narrow sense, and thus $F = \mathbb{Q}$ by $\pi_1(\mathrm{Spec}(\mathbb{Z})) = 1$. Hence $H = G$. This may be regarded as a weak analogue of the Wirtinger presentation.

boundary	p-adic field	
$\partial V_K \subset M \setminus \mathrm{int}(V_K)$	$\mathrm{Spec}(k_\mathfrak{p}) \subset \mathrm{Spec}(\mathcal{O}_k) \setminus \{\mathfrak{p}\}$	(3.10)
peripheral group D_K	decomposition group $D_{\{\mathfrak{p}\}}$	

In general, for a number field k of finite degree over $\mathbb{Q}$ and a finite subset S of $\mathrm{Max}(\mathcal{O}_k)$, the étale fundamental group $\pi_1(\mathrm{Spec}(\mathcal{O}_k) \setminus S)$, namely, the Galois group $G_S(k) = \mathrm{Gal}(k_{\overline{S}}/k)$ of the maximal Galois extension $k_{\overline{S}}$ over k unramified outside $\overline{S} := S \cup S_k^\infty$ (Example 2.2.20) is regarded as an analogue of a link group G_L.

link group	Galois group with restricted ramification	
$G_L(M) = \pi_1(M \setminus L)$	$G_S(k) = \pi_1(\mathrm{Spec}(\mathcal{O}_k) \setminus S)$	(3.11)

The pro-finite group $G_S(k)$ is huge in general, and it is unknown if even the prime group $G_{\{p\}}$ for a prime number p is finitely generated or not. It is a fundamental problem in algebraic number theory to understand the structure of $G_{\{p\}}$, in other words, to understand in the sense of étale homotopy theory

"how a prime number $\mathrm{Spec}(\mathbb{F}_p)$ is embedded in $\mathrm{Spec}(\mathbb{Z})$".

We note that this is the central problem in knot theory. The fact that the Galois group G_S is huge and complicated shows that the set S of primes is embedded in $\mathrm{Spec}(\mathcal{O}_k)$ in a very complicated, invisible manner. As we see in the subsequent chapters, however, it is possible to understand a glimpse of the "shape" of a prime number and how prime numbers are "linked", by taking the various quotients of G_S and comparing them with a link group G_L.

Remark 3.0.1

(1) We note that the analogies (3.4), (3.5) are conceptual, and it is not meant that there is a one-to-one correspondence between 3-manifolds and number rings, knots and primes. For instance, it is known that one has $\pi_1(\mathrm{Spec}(\mathcal{O}_k)) = 1$ for all imaginary quadratic fields of class number 1. Nevertheless, we have the following analogue of the Poincaré conjecture:

$$\hat{H}^i(\mathrm{Spec}(\mathcal{O}_k), \mathbb{Z}/n\mathbb{Z}) = \hat{H}^i(\mathrm{Spec}(\mathbb{Z}), \mathbb{Z}/n\mathbb{Z}) \ (i \in \mathbb{Z}, n \geq 2) \Longleftrightarrow \mathcal{O}_k = \mathbb{Z}.$$

This follows from the Artin–Verdier duality 2.3.2 and Dirichlet's unit theorem 2.2.15. It should be noted that Dirichlet's unit theorem is analytic in nature.

(2) It has been commonly considered that a number ring $\mathrm{Spec}(\mathcal{O}_k)$ is an analogue of an algebraic curve C over a finite field $\mathbb{F}_q$ and so a prime ideal of a number ring corresponds to a point on a curve. Let $\overline{C} := C \otimes_{\mathbb{F}_q} \overline{\mathbb{F}}_q$. Since $\overline{C}$ is an algebraic curve over an algebraically closed field, $\overline{C}$ is of dimension 2 étale homotopically. (Note that a curve over the field of complex numbers is a Riemann surface.) In

view of the fibration $\overline{C} \to C \to \text{Spec}(\mathbb{F}_q)$, C may be viewed as a surface-bundle over S^1.

| algebraic curve over $\mathbb{F}_q$ | surface-bundle over S^1 |

Since a number ring has no constant field, the structure of $\pi_1(\text{Spec}(\mathcal{O}_k))$ has no analogy with that of $\pi_1(C)$ and is quite random (it can be both finite and infinite). We may also note that the analogy of (3.8), (3.9) for 2 points on an affine line does not hold. Thus it is appropriate to compare $\pi_1(\text{Spec}(\mathcal{O}_k))$ with 3-manifold groups in general. One might regard the set of roots of unity in a number ring $\mathcal{O}_k$ as like the constant field of $\mathcal{O}_k$. In fact, from this viewpoint, Iwasawa considered the extension k_∞/k obtained by adjoining all p-th power of roots of unity to k as an analogue of the extension $\overline{\mathbb{F}}_q(\overline{C})/\mathbb{F}_q(C)$ of constant fields and showed some analogies between the number field k_∞ and the algebraic function field $\overline{\mathbb{F}}_q(\overline{C})$ [92]. See also [176, Ch.XI]. However it should be noted that the extension k_∞/k is ramified over p unlike the unramified extension $\overline{\mathbb{F}}_q(\overline{C})/\mathbb{F}_q(C)$. So, again, it is more natural to regard k_∞/k as an analogue of the tower of cyclic coverings of a 3-manifold ramified over a knot. It is our viewpoint throughout this book to regard a prime of a number field as an analogue of a knot in a 3-manifold rather than a point on an algebraic curve (Riemann surface classically), and the Galois group $G_S(k)$ as an analogue of a link group rather than the fundamental group of an algebraic curve (surface group). Although the idea of regarding a prime as an analogue of a circle has already been known in the study of closed geodesics in a Riemannian manifold or periodic orbits of a dynamical system ([188], [203], [216] etc.), it is essential in our analogy to see a number ring as being 3-dimensional, together with the local analogies such as (3.3) and (3.10). For a dynamical view in arithmetic topology, we also refer to [140], [144] (cf. Sect. 7.1).

(3) In a series of works on the dynamical study of number-theoretical zeta functions, C. Deninger pointed out that there are close analogies between arithmetic schemes and *foliated dynamical systems* of odd dimension, namely, smooth manifolds equipped with 1-codimensional (complex) foliation and transversal flow satisfying certain conditions (cf. [40]–[45]. See also [117], [118], [153]). In particular, a 3-dimensional foliated dynamical system may be regarded as a geometric analogue of a number ring, where closed orbits (knots) correspond to finite primes. So his theory fits and refines the analogies in arithmetic topology. Moreover, Deninger proposed a conjecture that to any finite algebraic number field k one could associate a certain 3-dimensional space M_k, which is no longer a manifold, with 2-dimensional foliation $\mathcal{F}$ and a dynamical system ϕ^t ($t \in \mathbb{R}$) on M_k. Here the dynamical $\mathbb{R}$-action on the foliation cohomology $H^*_{\mathcal{F}}(M_k)$ would play a similar role to the $\text{Gal}(\overline{\mathbb{F}}_q/\mathbb{F}_q)$-action on the geometric l-adic cohomology $H^*(\overline{C})$ of an algebraic curve C over $\mathbb{F}_q$, and the conjecture also asserts that there would be a precise one-to-one correspondence between the set of maximal ideals $\mathfrak{p}$ of $\mathcal{O}_k$ and the set of closed $\mathbb{R}$-orbits (knots) γ in M_k under the equality $N\mathfrak{p} = $ length of γ. See also the recent work [46].

Chapter 4
Linking Numbers and Legendre Symbols

In this chapter, we shall discuss the analogy between the linking number and the Legendre symbol, based on the analogies between knots and primes in Chap. 3.

4.1 Linking Numbers

Let $K \cup L$ be a 2-component link in S^3. The *linking number* $\mathrm{lk}(L, K)$ is described in terms of the monodromy as follows. Let $X_L = S^3 \setminus \mathrm{int}(V_L)$ be the exterior of L and let $G_L = \pi_1(X_L)$ be the knot group of L. For a meridian α of L, let $\psi_\infty : G_L \to \mathbb{Z}$ be the surjective homomorphism sending α to 1. Let X_∞ be the infinite cyclic cover of X_L corresponding to $\mathrm{Ker}(\psi)$, and let τ denote the generator of $\mathrm{Gal}(X_\infty/X_L)$ corresponding to $1 \in \mathbb{Z}$ (Example 2.1.12). Let $\rho_\infty : G_L \to \mathrm{Gal}(X_\infty/X_L)$ be the natural homomorphism (monodromy permutation representation).

Proposition 4.1.1 $\rho_\infty([K]) = \tau^{\mathrm{lk}(L,K)}$.

Proof We construct X_∞ as in Example 2. Namely, let Y be the space obtained by cutting X_L along the Seifert surface Σ_L of L, and we construct X_∞ by gluing the copies Y_i ($i \in \mathbb{Z}$) of Y as follows (Fig. 4.1):

Let $\tilde{K}$ be a lift of K in X_∞. According to where K crosses Σ_L with intersection number $+1$ (resp. -1), $\tilde{K}$ crosses from Y_i to Y_{i+1} (resp. from Y_{i+1} to Y_i) for some i. Therefore, if the starting point y_0 of $\tilde{K}$ is in Y_0, the terminus of $\tilde{K}$ is in Y_l, $l = \mathrm{lk}(L, K)$. Hence we have $\rho_\infty([K])(y_0) \in Y_l$. Since τ is the map sending Y_i to Y_{i+1}, this means $\rho_\infty([K]) = \tau^l$. □

Let $\psi_2 : G_L \to \mathbb{Z}/2\mathbb{Z}$ be the composite map of ψ_∞ with the natural homomorphism $\mathbb{Z} \to \mathbb{Z}/2\mathbb{Z}$ and let $h_2 : X_2 \to X_L$ be the double covering of X_L corresponding to $\mathrm{Ker}(\psi_2)$. Let $\rho_2 : G_L \to \mathrm{Gal}(X_2/X_L)$ be the natural homomorphism. By Proposition 2.1.1, we have the following:

Fig. 4.1 A lift of a knot in
the infinite cyclic cover

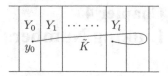

Fig. 4.2 A 2-componet link
$K \cup L$ with even linking
number

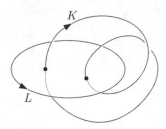

Corollary 4.1.2 *The image of* $[K]$ *under the composite map* $G_L \xrightarrow{\rho_2} \mathrm{Gal}(X_2/X_L) \simeq$
$\mathbb{Z}/2\mathbb{Z}$ *is given by* $\mathrm{lk}(L, K)$ *mod* 2:

$$G_L \xrightarrow{\rho_2} \mathrm{Gal}(X_2/X_L) \simeq \mathbb{Z}/2\mathbb{Z},$$
$$[K] \longmapsto \quad \mathrm{lk}(L, K) \bmod 2.$$

For $y \in h_2^{-1}(x)$ $(x \in K)$, we see that

$$\rho_2([K])(y) = y.[K]$$
$$= \text{the terminus of the lift of } K \text{ with starting point } y.$$

This implies

$$\rho_2([K]) = \mathrm{id}_{X_2} \iff h_2^{-1}(K) = K_1 \cup K_2 \text{ (2-component link in } X_2),$$
$$\rho_2([K]) = \tau \quad \iff h_2^{-1}(K) = \mathfrak{K} \text{ (knot in } X_2).$$

Hence, by Corollary 4.1.2, we have the following:

$$h_2^{-1}(K) = \begin{cases} K_1 \cup K_2 & \mathrm{lk}(L, K) \equiv 0 \bmod 2, \\ \mathfrak{K} & \mathrm{lk}(L, K) \equiv 1 \bmod 2. \end{cases} \tag{4.1}$$

Example 4.1.3 Let $K \cup L$ be the following link (Fig. 4.2).

Since $\mathrm{lk}(L, K) = 2$, K is decomposed in X_2 as $h_2^{-1}(K) = K_1 \cup K_2$. In fact,
$h_2^{-1}(K)$ is drawn in X_2 as follows (Fig. 4.3).

Let $K \cup L$ be the following link (Fig. 4.4).

Since $\mathrm{lk}(L, K) = 3$, K is lifted to a knot $h_2^{-1}(K) = \mathfrak{K}$ in X_2. In fact, $h_2^{-1}(K)$ is
drawn in X_2 as follows (Fig. 4.5).

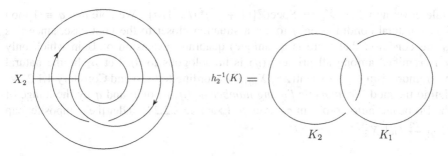

Fig. 4.3 The double covering of S^3 ramified along L and the inverse image of K

Fig. 4.4 A 2-component link
$K \cup L$ with odd linking
number

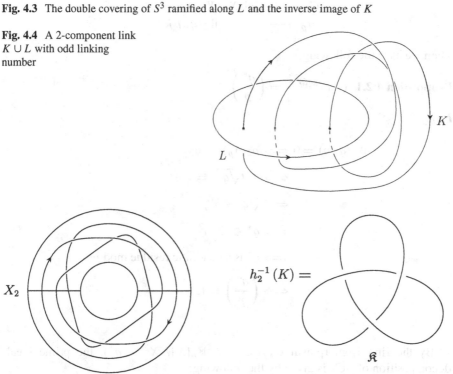

Fig. 4.5 The double covering of S^3 ramified along L and the inverse image of K

4.2 Legendre Symbols

Let p and q be odd prime numbers. Let $X_{\{q\}} := \mathrm{Spec}(\mathbb{Z}) \setminus \{q\} = \mathrm{Spec}(\mathbb{Z}[1/q])$ and let $G_{\{q\}} = \pi_1(X_{\{q\}})$ be the prime group of q. Let α be a primitive root mod q and let $\psi_2 : G_{\{q\}} = \mathbb{Z}_q^\times = \mathbb{F}_q^\times \times (1 + q\mathbb{Z}_q) \to \mathbb{Z}/2\mathbb{Z}$ be the surjective homomorphism defined by $\psi_2(\alpha) = 1$, $\psi_2(1 + q\mathbb{Z}_q) = 0$. Let k be the quadratic extension of $\mathbb{Q}$ corresponding to $\mathrm{Ker}(\psi_2)$ (Example 2.3.6). Namely, $k = \mathbb{Q}(\sqrt{q^*})$, $q^* := (-1)^{\frac{q-1}{2}} q$. The field k is the unique quadratic extension of $\mathbb{Q}$ such that only q is ramified among all finite primes. Let $h_2 : X_2 \to X_{\{q\}}$ be the associated double étale covering where $X_2 := \text{Spec}(\mathbb{Z}[(1 + \sqrt{q^*})/2, 1/q])$. We note that $q \equiv 1 \mod 4$ is a natural condition on q to get a situation closer to the knot case, since this is the condition that there is a (unique) quadratic extension of $\mathbb{Q}$ in which only q is ramified among all primes ((q) is homologous to 0). Let ρ_2 be the natural homomorphism $G_{\{q\}} \to \text{Gal}(X_2/X_{\{q\}})$. According to (3.6) and Corollary 4.1.2, we define the mod 2 *arithmetic linking number* $\text{lk}_2(q, p)$ of p and q by the image of the Frobenius automorphism σ_p over p (Example 2.2.24) under the composite map $G_{\{q\}} \xrightarrow{\rho_2} \text{Gal}(X_2/X_{\{q\}}) \simeq \mathbb{Z}/2\mathbb{Z}$:

$$G_{\{q\}} \xrightarrow{\rho_2} \text{Gal}(X_2/X_{\{q\}}) \simeq \mathbb{Z}/2\mathbb{Z}$$
$$\sigma_p \longmapsto \qquad\qquad \text{lk}_2(q, p).$$

Then we have the following:

Proposition 4.2.1 $(-1)^{\text{lk}_2(q,p)} = \left(\dfrac{q^*}{p}\right)$.

Proof

$$\text{lk}_2(q, p) = 0 \Longleftrightarrow \rho_2(\sigma_p) = \text{id}_{X_2}$$
$$\Longleftrightarrow \sigma_p(\sqrt{q^*}) = \sqrt{q^*}$$
$$\Longleftrightarrow \sqrt{q^*} \in \mathbb{F}_p^\times$$
$$\Longleftrightarrow q^* \in (\mathbb{F}_p^\times)^2$$
$$\Longleftrightarrow q^* \text{ is a quadratic residue mod } p$$
$$\Longleftrightarrow \left(\frac{q^*}{p}\right) = 1.$$

$\square$

By the ring isomorphism $\mathcal{O}_k/q\mathcal{O}_k \simeq \mathbb{F}_p[X]/(X^2 - q^*)$, the prime ideal decomposition of $p\mathcal{O}_k$ is given by the following:

$$p\mathcal{O}_k = \begin{cases} \mathfrak{p}_1\mathfrak{p}_2, & q^* \text{ is a quadratic residue mod } p \\ \mathfrak{p}, & q^* \text{ is a quadratic non-residue mod } p. \end{cases}$$

By Proposition 4.2.1, we have

$$h_2^{-1}(\{p\}) = \begin{cases} \{\mathfrak{p}_1, \mathfrak{p}_2\}, & \text{lk}_2(p, q) = 0 \\ \mathfrak{p}, & \text{lk}_2(p, q) = 1. \end{cases} \qquad (4.2)$$

This is the arithmetic analogue of (4.1). For example, the 5 prime ideals $\{(5), (13), (17), (29), (149)\}$ are linked mod 2 as in Fig. 4.6 (*Olympic primes*):

By Proposition 4.2.1, we see that the Legendre symbol is an arithmetic analogue of the mod 2 linking number, and the symmetry of the linking number corresponds

Fig. 4.6 An Olympic link

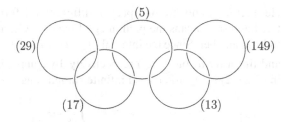

to the symmetry of the Legendre symbol, namely, the quadratic reciprocity law, for prime numbers $p, q \equiv 1 \bmod 4$.

Remark 4.2.2 The linking number is also given by the cup product. For a link $L = K_1 \cup K_2$ in S^3, let $\alpha_i^* \in H^1(X)$ (X being the exterior of L in S^3) be the dual of a meridian α_i of K_i. Under the isomorphism $H^2(X) \simeq \mathbb{Z}$ given by a generator of $H^2(X)$ which is the Lefschetz dual to a path from K_1 to K_2, the cup product $\alpha_1 \cup \alpha_2$ is sent to $\mathrm{lk}(K_1, K_2)$. Similarly, the Legendre symbol is also interpreted as a cup product in the étale (or Galois) cohomology group (cf. [111, 8.11], [165, 2], [240]).

Remark 4.2.3 What is an arithmetic analogue of Gauss' integral expression for the linking number in the Introduction ? In order to answer this question, we first reformulate Gauss' formula as follows:

$$\int_{x \in K} \int_{y \in L} \omega(x - y) = \mathrm{lk}(K, L), \tag{4.3}$$

where ω is the differential 1-form on $\mathbb{R}^3$ given by

$$\omega = \frac{1}{4\pi \|x\|^3} (x_1 \, dx_2 \wedge dx_3 + x_2 \, dx_3 \wedge dx_1 + x_3 \, dx_1 \wedge dx_2).$$

Furthermore, using the infinite-dimensional integral, we rewrite the left hand side of (4.3) in the following gauge-theoretic manner ([115, 3.3], [201]): For a framed link $K_1 \cup K_2$, one has

$$\int_{A(\mathbb{R}^3)} \exp\left(\frac{\sqrt{-1}}{4\pi} \int_{\mathbb{R}^3} a \wedge da + \sqrt{-1} \int_{K_1} a + \sqrt{-1} \int_{K_2} a \right) \mathcal{D}a$$

$$= \exp\left(\sum_{1 \le i,j \le 2} \int_{x \in K_i} \int_{y \in K_j} \omega(x - y) \right) \tag{4.4}$$

$$= \exp\left(\frac{\sqrt{-1}}{4} \sum_{1 \le i,j \le 2} \mathrm{lk}(K_i, K_j) \right).$$

Here $A(\mathbb{R}^3)$ stands for the space of differential 1-forms on $\mathbb{R}^3$ and the integral in the left-hand side means the path integral over $A(\mathbb{R}^3)$, and $\mathrm{lk}(K_i, K_i)$ denotes the self-linking number. Since the integrals $\int_{\mathbb{R}^3} a \wedge da$ and $\int_{K_i} a$ are regarded as the quadratic and linear forms on $A(\mathbb{R}^3)$ respectively, by completing the square, the path integral in (2.2.4.2) is regarded as an infinite-dimensional analogue of the Gaussian integral

$$\int_{\mathbb{R}} e^{-x^2} \, dx. \tag{4.5}$$

On the other hand, the arithmetic analogue of the Gaussian integral (4.5) on the finite field $\mathbb{F}_q$ is nothing but the Gaussian sum

$$\sum_{x \in \mathbb{F}_q} \zeta_q^{x^2},$$

where ζ_q is a q-th root of unity in an algebraic closure of $\mathbb{F}_p$. As Gauss showed, the Legendre symbol is expressed by the Gaussian sum as follows [204, 3.3, Lemma 2]:

$$\left(\sum_{x \in \mathbb{F}_q} \zeta_q^{x^2} \right)^{p-1} = \left(\frac{q^*}{p} \right) = (-1)^{\mathrm{lk}_2(q,p)}. \tag{4.6}$$

Comparing (4.4) and (4.6), one may find that the Gauss' integral formula for the linking number is analogous to the formula expressing the Legendre symbol by the Gaussian sum.

Summary

Linking number $\mathrm{lk}(L, K)$	Legendre symbol $\left(\dfrac{q^*}{p} \right)$
$\mathrm{lk}(L, K) = \mathrm{lk}(K, L)$	$\left(\dfrac{q}{p} \right) = \left(\dfrac{p}{q} \right)$ $(p, q \equiv 1 \bmod 4)$
Gauss' linking integral	Gaussian sum

Chapter 5
Decompositions of Knots and Primes

As we have seen in Sect. 4.2, the Legendre symbol describes how a prime number is decomposed in a quadratic extension. More generally, the Hilbert theory deals with, in a group-theoretic manner, the decomposition of a prime in a finite Galois extension of number fields, and further the Artin reciprocity of class field theory describes, in an arithmetic manner, the decomposition law of a prime in a finite Abelian extension. Based on the analogies in Chap. 3, in this chapter, we shall give a topological analogue of the Hilbert theory for coverings of 3-manifolds.

5.1 Decomposition of a Knot

Let $h : M \rightarrow S^3$ be a finite Galois covering of oriented, connected, closed 3-manifolds ramified over a link $L \subset S^3$, and let $X := S^3 \setminus L$, $Y := M \setminus h^{-1}(L)$, $G := \mathrm{Gal}(Y/X) = \mathrm{Gal}(M/S^3)$ and $n := \#G \, (\geq 1)$. Let K be a knot in S^3 which is a component of L or disjoint from L, and suppose $h^{-1}(K) = K_1 \cup \cdots \cup K_r$ ($r = r_K$-component link). For a tubular neighborhood V_K of K, let V_{K_i} be the connected component of $h^{-1}(V_K)$ containing K_i. Fix a base point $x \in \partial V_K$. Suppose $h^{-1}(x) = \{y_1, \cdots, y_n\}$. Let $\rho : G_L = \pi_1(X, x) \rightarrow \mathrm{Aut}(h^{-1}(x))$ be the monodromy permutation representation which induces an isomorphism $\pi_1(X, x)/h_*(\pi_1(Y, y_i)) \simeq \mathrm{Im}(\rho) \simeq G$. Note that $\pi_1(X, x)$ and hence G acts transitively on the set of knots $S_K := \{K_1, \ldots, K_r\}$ lying over K. We call the stabilizer D_{K_i} of K_i the *decomposition group* of K_i:

$$D_{K_i} := \{g \in G \mid g(K_i) = K_i\}.$$

Since we have the bijection $G/D_{K_i} \simeq S_K$ for each i, $\#D_{K_i} = n/r$ is independent of K_i. In fact, if $g(K_i) = K_j \, (g \in G)$, $D_{K_j} = g D_{K_i} g^{-1}$. Since each $g \in G$ induces a homeomorphism $g|_{\partial V_{K_i}} : \partial V_{K_i} \xrightarrow{\approx} \partial V_{g(K_i)}$, $g|_{\partial V_{K_i}}$ is a covering transformation

M. Morishita, *Knots and Primes*, Universitext,
https://doi.org/10.1007/978-981-99-9255-3_5

of ∂V_{K_i} over ∂V_K for $g \in D_{K_i}$, and the correspondence $g \mapsto g|_{\partial V_{K_i}}$ gives an isomorphism

$$D_{K_i} \simeq \mathrm{Gal}(\partial V_{K_i}/\partial V_K).$$

The Fox completion of the subcovering space of Y over X corresponding to D_{K_i} is called the *decomposition (covering) space* of K_i and is denoted by Z_{K_i}. The map $g \mapsto \overline{g} := g|_{K_i}$ induces the homomorphism

$$D_{K_i} \longrightarrow \mathrm{Gal}(K_i/K)$$

whose kernel is called the *inertia group* of K_i and is denoted by I_{K_i}:

$$I_{K_i} := \{g \in D_{K_i} \mid \overline{g} = \mathrm{id}_{K_i}\}.$$

If $K_j = g(K_i)$ $(g \in G)$, one has $I_{K_j} = gI_{K_i}g^{-1}$ and hence $\#I_{K_i}$ is independent of K_i. Set $e = e_K := \#I_{K_i}$. The Fox completion of the subcovering space of Y over X corresponding to I_{K_i} is called the *inertia (covering) space* of K_i and is denoted by T_{K_i}:

$$M \longrightarrow T_{K_i} \longrightarrow Z_{K_i} \longrightarrow S^3.$$

Since the Galois group G acts on $h^{-1}(x)$ simply-transitively, we can understand D_{K_i} and I_{K_i} by looking at their actions on $h^{-1}(x)$. Let α be a meridian of K and β be a longitude of K, and take $y_{i_1} \in \partial V_i$. The orbits of y_{i_1} under the action of D_{K_i} and I_{K_i} coincide with the orbits of y_{i_1} under the action of $\rho(\langle \alpha, \beta \rangle)$ and $\rho(\langle \alpha \rangle)$ respectively. Let $\rho(\langle \alpha \rangle)(y_{i_1}) = \{y_{i_1}, \ldots, y_{i_e}\}$. If we write $\rho(\alpha)$ as the product of mutually disjoint cyclic permutations, the cyclic permutation $(y_{i_1} \cdots y_{i_e})$ is the factor containing y_{i_1} and e is the ramification index of K_i over K. On the other hand, $\rho(\langle \alpha, \beta \rangle)(y_{i_1})$ is nothing but the set of $y_j \in h^{-1}(x)$ so that $y_j \in \partial V_{K_i}$. Let $f = f_K$ be the minimum $m \in \mathbb{N}$ such that $y_{i_1}^m := \rho(\beta^m)(y_{i_1}) \in \{y_{i_1}, \ldots, y_{i_e}\}$. Then f is the covering degree of K_i over K and we have (Fig. 5.1)

$$\rho(\langle \alpha, \beta \rangle)(y_{i_1}) = \{y_j \in h^{-1}(x) \mid y_j \in \partial V_i\}$$
$$= \{y_{i_1}^m, \ldots, y_{i_e}^m \mid 0 \le m < f\}.$$

Hence we have the equalities

$$\#D_{K_i} = \#\rho(\langle \alpha, \beta \rangle)(y_{i_1}) = ef,$$
$$\#I_{K_i} = \#\rho(\langle \alpha \rangle)(y_{i_1}) = e, \#\mathrm{Gal}(K_i/K) = f.$$

By comparing the orders, we see that the homomorphism $D_{K_i} \ni g \mapsto \overline{g} \in \mathrm{Gal}(K_i/K)$ is surjective:

$$1 \longrightarrow I_{K_i} \longrightarrow D_{K_i} \longrightarrow \mathrm{Gal}(K_i/K) \longrightarrow 1 \quad \text{(exact)}.$$

Fig. 5.1 The visualized action of the decomposition group
$D_{K_i} \simeq \mathrm{Gal}(\partial V_{K_i} / \partial V_K)$ on ∂V_{K_i}

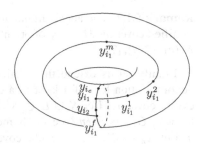

In particular, we have the equality $efr = n$, and therefore

$$D_{K_i} = 1 \iff Z_{K_i} = M \iff e = f = 1, r = n,$$
$$D_{K_i} = G \iff Z_{K_i} = S^3 \iff ef = n, r = 1,$$
$$I_{K_i} = 1 \iff T_{K_i} = M \iff e = 1, fr = n,$$
$$I_{K_i} = G \iff T_{K_i} = S^3 \iff e = n, f = r = 1.$$

In general one has the following theorem. Let $K_{i,T}$ be the image of K_i under $M \to T_{K_i}$ and let $K_{i,Z}$ be the image of $K_{i,T}$ under $T_{K_i} \to Z_{K_i}$.

Theorem 5.1.1 *The map $M \to T_{K_i}$ is a ramified covering of degree e such that the ramification index of K_i over $K_{i,T}$ is e. The map $T_{K_i} \to Z_{K_i}$ is a cyclic covering of degree f such that the covering degree of $K_{i,T}$ over $K_{i,Z}$ is f. The map $Z_{K_i} \to S^3$ is a covering of degree r such that K is completely decomposed into an r-component link containing $K_{i,Z}$ as a component.*

Now assume that G is an Abelian group. Then D_{K_i}, I_{K_i} and Z_{K_i}, T_{K_i} are independent of K_i and so we write respectively D_K, I_K and Z_K, T_K for them. In the covering $h : M \to S^3$, K is decomposed, covered and ramified as follows:

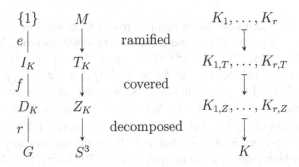

Assume further that K is unramified, namely, K is disjoint from L. Since $I_K = 1$, we have an isomorphism $D_K \simeq \mathrm{Gal}(K_i/K)$. Let σ_K be the generator of D_K which is defined by the inverse image under this isomorphism of the generator of $\mathrm{Gal}(K_i/K)$ corresponding the loop going once around K counterclockwise (Example 2.1.9). By the equality $fr = n$, the decomposition law of K in the covering $h : M \to S^3$ is determined by f, namely, the order of σ_K in G.

Remark 5.1.2 The above argument and results hold similarly for any finite Galois ramified covering $M \to N$ of oriented, connected, closed 3-manifolds and a knot in N (cf. [227], [228]).

Finally, let us extend the relation between the mod 2 linking number and the decomposition law of a knot in a double covering in Sect. 4.1 to the case of any cyclic covering. Let $K \cup L \subset S^3$ be a 2-component link. For an integer $n \geq 2$, let $\psi : G_L \to \mathbb{Z}/n\mathbb{Z}$ be the homomorphism sending a meridian of L to 1 mod n. Let $X_n \to X_L$ be the cyclic covering of degree n corresponding to Ker(ψ) and let $h_n : M \to S^3$ be its Fox completion. Let τ be the generator of Gal(X_n/X_L) corresponding to 1 mod n (Example 2.1.12). Finally, let $\rho : G_L \to \text{Gal}(X_n/X_L)$ be the natural homomorphism. Then we have, by the definition of σ_K above,

$$\sigma_K = \rho([K]).$$

Therefore, by Proposition 4.1.1,

$$\sigma_K = \tau^{\text{lk}(L,K)}. \tag{5.1}$$

Hence one has, for a positive divisor m of n,

$$\text{g.c.d}(\text{lk}(L, K), n) = m \Longleftrightarrow h_n^{-1}(K) = K_1 \cup \cdots \cup K_m.$$

In particular, K is decomposed completely in X_n (i.e., decomposed into an n-component link) if and only if $\text{lk}(L, K) \equiv 0 \bmod n$.

5.2 Decomposition of a Prime

Let $k/\mathbb{Q}$ be a finite Galois extension ramified over a finite set S of prime numbers. Let $h : \text{Spec}(\mathcal{O}_k) \to \text{Spec}(\mathbb{Z})$ be the associated ramified covering. We set $X := \text{Spec}(\mathbb{Z}) \setminus S$, $Y := \text{Spec}(\mathcal{O}_k) \setminus h^{-1}(S)$, $G := \text{Gal}(Y/X) = \text{Gal}(k/\mathbb{Q})$, $n := \#G$ (≥ 1). Let p be a prime number and let $S_p := h^{-1}(\{p\}) = \{\mathfrak{p}_1, \ldots, \mathfrak{p}_r\}$ ($r = r_p$). One has $\mathcal{O}_k \otimes_{\mathbb{Z}} \mathbb{Z}_p = \prod_{i=1}^{r} \mathcal{O}_{\mathfrak{p}_i}$ where $\mathcal{O}_{\mathfrak{p}_i}$ is the $\mathfrak{p}_i$-adic completion of $\mathcal{O}_k$. Fix an algebraic closure $\overline{\mathbb{Q}}_p$ of $\mathbb{Q}_p$ and let $\overline{x} : \text{Spec}(\overline{\mathbb{Q}}_p) \to X$ be the base point induced by the inclusion $\mathbb{Z}[1/S] \subset \overline{\mathbb{Q}}_p$. Let $F_{\overline{x}}(Y) = \text{Hom}_X(\text{Spec}(\overline{\mathbb{Q}}_p), Y) = \{y_1, \ldots, y_n\}$ and let $\rho : G_S := \pi_1(X, \overline{x}) \to \text{Aut}(F_{\overline{x}}(Y))$ be the monodromy permutation representation which induces an isomorphism $\pi_1(X, \overline{x})/h_*(\pi_1(Y, y_i)) \simeq \text{Im}(\rho) \simeq G$. Note that $\pi_1(X, \overline{x})$ and hence G acts on S_p transitively. We call the stabilizer $D_{\mathfrak{p}_i}$ of $\mathfrak{p}_i$ the *decomposition group* of $\mathfrak{p}_i$:

$$D_{\mathfrak{p}_i} := \{g \in G \mid g(\mathfrak{p}_i) = \mathfrak{p}_i\}.$$

Since we have the bijection $G/D_{\mathfrak{p}_i} \simeq S_p$, $\#D_{\mathfrak{p}_i} = n/r$ is independent of $\mathfrak{p}_i$. In fact, if $\mathfrak{p}_j = g(\mathfrak{p}_i)$ ($g \in G$), then $D_{\mathfrak{p}_j} = g D_{\mathfrak{p}_i} g^{-1}$. Since $g \in G$ induces an isomorphism

$\hat{g} : k_{\mathfrak{p}_i} \xrightarrow{\sim} k_{g(\mathfrak{p}_i)}$, $\hat{g}$ gives an isomorphism of $k_{\mathfrak{p}_i}$ over $\mathbb{Q}_p$ if $g \in D_{\mathfrak{p}_i}$, and the correspondence $g \mapsto \hat{g}$ gives the isomorphism

$$D_{\mathfrak{p}_i} \simeq \mathrm{Gal}(k_{\mathfrak{p}_i}/\mathbb{Q}_p).$$

The subfield of k corresponding to $D_{\mathfrak{p}_i}$ is called the *decomposition field* of $\mathfrak{p}_i$ and is denoted by $Z_{\mathfrak{p}_i}$. Furthermore, $g \in D_{\mathfrak{p}_i}$ induces the isomorphism $\bar{g}$ of $\mathbb{F}_{\mathfrak{p}_i} = \mathcal{O}_k/\mathfrak{p}_i$ over $\mathbb{F}_p$ defined by $\bar{g}(\alpha \bmod \mathfrak{p}_i) := g(\alpha) \bmod \mathfrak{p}_i$ ($\alpha \in \mathcal{O}_k$). The map $g \mapsto \bar{g}$ gives the homomorphism

$$D_{\mathfrak{p}_i} \longrightarrow \mathrm{Gal}(\mathbb{F}_{\mathfrak{p}_i}/\mathbb{F}_p),$$

whose kernel is called the *inertia group* of $\mathfrak{p}_i$ and is denoted by $I_{\mathfrak{p}_i}$:

$$I_{\mathfrak{p}_i} := \{g \in D_{\mathfrak{p}_i} \mid \bar{g} = \mathrm{id}_{\mathbb{F}_{\mathfrak{p}_i}}\}.$$

If $g(\mathfrak{p}_i) = \mathfrak{p}_j$ ($g \in G$), one has $I_{\mathfrak{p}_j} = g I_{\mathfrak{p}_i} g^{-1}$ and hence $\#I_{\mathfrak{p}_i}$ is independent of $\mathfrak{p}_i$. Set $e = e_p := \#I_{\mathfrak{p}_i}$. The subfield of k corresponding to $I_{\mathfrak{p}_i}$ is called the *inertia field* of $\mathfrak{p}_i$ and is denoted by $T_{\mathfrak{p}_i}$:

$$k \supset T_{\mathfrak{p}_i} \supset Z_{\mathfrak{p}_i} \supset \mathbb{Q}.$$

Lemma 5.2.1 *The homomorphism $D_{\mathfrak{p}_i} \ni g \mapsto \bar{g} \in \mathrm{Gal}(\mathbb{F}_{\mathfrak{p}_i}/\mathbb{F}_p)$ is surjective:*

$$1 \longrightarrow I_{\mathfrak{p}_i} \longrightarrow D_{\mathfrak{p}_i} \longrightarrow \mathrm{Gal}(\mathbb{F}_{\mathfrak{p}_i}/\mathbb{F}_p) \longrightarrow 1 \quad \text{(exact)}.$$

Proof Since $\mathrm{Gal}(\mathbb{F}_{\mathfrak{p}_i}/\mathbb{F}_p)$ is generated by the Frobenius automorphism σ, it suffices to show that there is $g \in D_{\mathfrak{p}_i}$ such that $\bar{g} = \sigma$. Take $\theta \in \mathcal{O}_k$ so that $\mathbb{F}_{\mathfrak{p}_i} = \mathbb{F}_p(\bar{\theta})$, $\bar{\theta} = \theta \bmod \mathfrak{p}_i$. By the Chinese remainder theorem, we have an $\alpha \in \mathcal{O}_k$ such that $\alpha \equiv \theta \bmod \mathfrak{p}_i$ and $\alpha \not\equiv 0 \bmod g(\mathfrak{p}_i), \forall g \notin D_{\mathfrak{p}_i}$. Using such an α, consider the polynomial

$$f(X) := \prod_{g \in G} (X - g(\alpha)).$$

Then we see easily $f(X) \in \mathbb{Z}[X]$, and $\bar{f}(X) := f(X) \bmod p \in \mathbb{F}_p[X]$ is decomposed as

$$\bar{f}(X) = X^m \bar{f}_1(X) \ (m \geq 1), \quad \bar{f}_1(X) := \prod_{g \in D_{\mathfrak{p}_i}} (X - g(\alpha)).$$

Since $0 = \bar{f}(\bar{\alpha}) = \bar{f}(\bar{\theta}) = \bar{\theta}^m \bar{f}_1(\bar{\theta})$ ($\bar{\alpha} := \alpha \bmod \mathfrak{p}_i$) and $\bar{\theta} \neq 0$, we have $\bar{f}_1(\bar{\theta}) = 0$. Let $h(X) \in \mathbb{F}_p[X]$ be the minimal polynomial of $\bar{\theta}$ over $\mathbb{F}_p$. Then $h(X) | \bar{f}_1(X)$. Since $\sigma(\bar{\theta})$ is a root of $h(X) = 0$, so is $\bar{f}_1(X) = 0$. Hence there is $g \in D_{\mathfrak{p}_i}$ such that $\sigma(\bar{\theta}) = \bar{g}(\bar{\alpha}) = \bar{g}(\bar{\theta})$.

Let $G = \sqcup_{i=1}^{r} g_i D_{\mathfrak{p}_i}$ be the coset decomposition such that $\mathfrak{p}_i = g_i(\mathfrak{p}_1)$ ($g_1 := 1$). If we denote by e_i the ramification index of $\mathfrak{p}_i$, we have $p\mathcal{O}_k = \mathfrak{p}_1^{e_1} \cdots \mathfrak{p}_r^{e_r} = g_1(\mathfrak{p}_1)^{e_1} \cdots g_r(\mathfrak{p}_1)^{e_r}$. Letting $g \in G$ act on both sides, we see $e_1 = \cdots = e_r = e$ by the uniqueness of the prime decomposition: $p\mathcal{O}_k = (\mathfrak{p}_1 \cdots \mathfrak{p}_r)^e$. Further, since $\mathbb{F}_{\mathfrak{p}_i} = \mathcal{O}_k/\mathfrak{p}_1^g \simeq \mathcal{O}_k/\mathfrak{p}_1 = \mathbb{F}_{\mathfrak{p}_1}$, $[\mathbb{F}_{\mathfrak{p}_i} : \mathbb{F}_p] = f_p = f$ is also independent of $\mathfrak{p}_i$: $N\mathfrak{p}_i = p^f$. Taking the norm of both sides of $p\mathcal{O}_k = (\mathfrak{p}_1 \cdots \mathfrak{p}_r)^e$, we have $p^n = p^{efr}$. Hence $n = efr$ and $\#D_{\mathfrak{p}_i} = n/r = ef$. By Lemma 3.2.1, $\#I_{\mathfrak{p}_i} = e$. Therefore

$$
\begin{aligned}
D_{\mathfrak{p}_i} = 1 &\Longleftrightarrow Z_{\mathfrak{p}_i} = k \Longleftrightarrow e = f = 1, r = n, \\
D_{\mathfrak{p}_i} = G &\Longleftrightarrow Z_{\mathfrak{p}_i} = \mathbb{Q} \Longleftrightarrow ef = n, r = 1, \\
I_{\mathfrak{p}_i} = 1 &\Longleftrightarrow T_{\mathfrak{p}_i} = k \Longleftrightarrow e = 1, fr = n, \\
I_{\mathfrak{p}_i} = G &\Longleftrightarrow T_{\mathfrak{p}_i} = \mathbb{Q} \Longleftrightarrow e = n, f = r = 1.
\end{aligned}
$$

In general, one has the following theorem. Let $\mathfrak{p}_{i,T} := \mathfrak{p}_i \cap \mathcal{O}_{T_{\mathfrak{p}_i}}$ and $\mathfrak{p}_{i,Z} := \mathfrak{p}_i \cap \mathcal{O}_{Z_{\mathfrak{p}_i}}$.

Theorem 5.2.2 *The extension $k/T_{\mathfrak{p}_i}$ is a ramified extension of degree e such that the ramification index of $\mathfrak{p}_i$ over $\mathfrak{p}_{i,T}$ is e. The extension $T_{\mathfrak{p}_i}/Z_{\mathfrak{p}_i}$ is a cyclic extension of degree f such that the covering (inert) degree of $\mathfrak{p}_{i,T}$ over $\mathfrak{p}_{i,Z}$ is f. The extension $Z_{\mathfrak{p}_i}/\mathbb{Q}$ is an extension of degree r such that p is completely decomposed into r prime ideals containing $\mathfrak{p}_{i,Z}$ as one prime factor.*

Now assume that G is an Abelian group. Then $D_{\mathfrak{p}_i}$, $I_{\mathfrak{p}_i}$ and $Z_{\mathfrak{p}_i}$, $T_{\mathfrak{p}_i}$ are independent of $\mathfrak{p}_i$ and so we write respectively D_p, I_p and Z_p, T_p for them. In the extension $k/\mathbb{Q}$, p is decomposed, covered and ramified as follows:

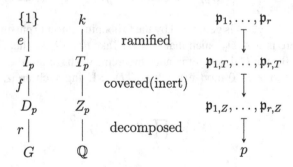

Assume further that $\mathfrak{p}$ is unramified. Since $I_{\mathfrak{p}} = 1$, we have the isomorphism $D_p \simeq \text{Gal}(\mathbb{F}_{\mathfrak{p}_i}/\mathbb{F}_p)$. Let σ_p be the generator of D_p which is defined by the inverse image under this isomorphism of the Frobenius automorphism of $\text{Gal}(\mathbb{F}_{\mathfrak{p}_i}/\mathbb{F}_p)$ (Example 0.2.9). By the equality $fr = n$, the decomposition law of p in the extension $k/\mathbb{Q}$ is determined by f, namely, the order of σ_p in G.

Remark 5.2.3 The above argument and results hold similarly for any finite Galois extension of number fields k/F and a prime ideal of F (cf. [186, Chap. 1, §27]).

Finally, we extend the relation between the Legendre symbol and the decomposition law of a prime number in a quadratic extension in Sect. 4.2 to the case of any cyclic extension. Let $n \geq 2$ be an integer and let p and q be distinct prime numbers such that $p, q \equiv 1 \bmod n$. We fix a primitive root $\alpha \bmod q$ and let $\psi : G_{\{q\}}^{ab} \simeq \mathbb{Z}_q^\times \to \mathbb{Z}/n\mathbb{Z}$ be the homomorphism defined by $\psi(\alpha) = 1, \psi(1 + q\mathbb{Z}_q) = 1$. Let $k/\mathbb{Q}$ be the cyclic extension of degree n corresponding to $\mathrm{Ker}(\psi)$ and let $h_n : \mathrm{Spec}(\mathcal{O}_k) \to \mathrm{Spec}(\mathbb{Z})$ be the associated ramified covering. Let τ be the generator of $\mathrm{Gal}(k/\mathbb{Q})$ corresponding to $1 \in \mathbb{Z}/n\mathbb{Z}$ (Example 2.3.3). According to (5.1), we define the *mod n linking number* $\mathrm{lk}_n(q, p) \in \mathbb{Z}/n\mathbb{Z}$ of p and q by

$$\sigma_p = \tau^{\mathrm{lk}_n(q,p)}.$$

(This depends on the choice of α.) Let $\mathfrak{q}$ be the unique prime ideal of $\mathcal{O}_k$ lying over q. Then $k_\mathfrak{q} = \mathbb{Q}_q(\sqrt[n]{q})$. Note that the image of τ under the canonical isomorphism $\mathrm{Gal}(k/\mathbb{Q}) \overset{\sim}{\to} \mathrm{Gal}(k_\mathfrak{q}/\mathbb{Q}_q)$ is given by $\rho_{k_\mathfrak{q}/\mathbb{Q}_q}(\alpha)$, where $\rho_{k_\mathfrak{q}/\mathbb{Q}_q}$ is the reciprocity homomorphism of local class field theory (2.3). Define the primitive n-th root of unity $\zeta \in \mathbb{Q}_q$ by

$$\zeta = \frac{\rho_{k_\mathfrak{q}/\mathbb{Q}_q}(\alpha)(\sqrt[n]{q})}{\sqrt[n]{q}}. \tag{5.2}$$

Proposition 5.2.4 *We have* $\zeta^{\mathrm{lk}_n(q,p)} = \left(\dfrac{p}{q}\right)_n$, *where* $\left(\dfrac{*}{q}\right)_n$ *stands for the n-th power residue symbol in* $\mathbb{Q}_q$.

Proof Let l be an integer $(\bmod\ q - 1)$ so that $p^{-1} \equiv \alpha^l \bmod q$. Then we shall show that

$$\mathrm{lk}_n(q, p) = l \bmod n.$$

Let **a** be the idèle of $\mathbb{Q}$ whose p-component is p and other components are all 1, **b** the idèle of $\mathbb{Q}$ whose q-component is p and other components are all 1 and **c** the idèle of $\mathbb{Q}$ whose p, q-components are 1 and other components are all p (0.3.3.3). Then $p = \mathbf{abc}$ in the idèle group $J_\mathbb{Q}$ of $\mathbb{Q}$. Let $\rho_{k/\mathbb{Q}} : C_\mathbb{Q} = J_\mathbb{Q}/\mathbb{Q}^\times \to \mathrm{Gal}(k/\mathbb{Q})$ be the reciprocity homomorphism in class field theory (2.10). Then $\rho_{k/\mathbb{Q}}(q) = \mathrm{id}$ and $\rho_{k/\mathbb{Q}}(\mathbf{a}) = \sigma_p$. Since $k/\mathbb{Q}$ is unramified outside $\{q, \infty\}$, by (2.11), $\rho_{k/\mathbb{Q}}(\mathbf{c}) = \mathrm{id}$. Therefore $\sigma_p = \rho_{k/\mathbb{Q}}(\mathbf{b}^{-1})$. By the following commutative diagram (cf. (2.9))

$$
\begin{array}{ccc}
\mathbb{Q}_q^\times & \overset{\rho_{k_\mathfrak{q}/\mathbb{Q}_q}}{\longrightarrow} & \mathrm{Gal}(k_\mathfrak{q}/\mathbb{Q}_q) \\
\downarrow & & \wr\downarrow \iota_\mathfrak{q} \\
C_\mathbb{Q} & \overset{\rho_{k/\mathbb{Q}}}{\longrightarrow} & \mathrm{Gal}(k/\mathbb{Q}),
\end{array}
$$

we have

$$
\begin{aligned}
\tau^l &= \iota_\mathfrak{q}(\rho_{k_\mathfrak{q}/\mathbb{Q}_q}(\alpha^l)) \\
&= \rho_{k/\mathbb{Q}}(\mathbf{b}^{-1}) \quad (p^{-1} = \alpha^l \bmod q) \\
&= \sigma_p.
\end{aligned}
$$

Hence $\mathrm{lk}_n(q, p) = l \bmod n$. Also

$$
\begin{aligned}
\left(\frac{p}{q}\right)_n &= \left(\frac{p, q}{q}\right)_n \quad \text{(by (2.6))} \\
&= \left(\frac{q, p^{-1}}{q}\right)_n \\
&= \frac{\rho_{k_q/\mathbb{Q}_q}(p^{-1})(\sqrt[n]{q})}{\sqrt[n]{q}} \\
&= \frac{\tau^l(\sqrt[n]{q})}{\sqrt[n]{q}} \\
&= \zeta^l \quad \text{(by (5.2))}.
\end{aligned}
$$

This yields the assertion.

By the definition of $\mathrm{lk}_n(q, p)$, one has, for a positive divisor m of n,

$$
\mathrm{g.c.d}(\mathrm{lk}(q, p), n) = m \iff h_n^{-1}(\{p\}) = \{\mathfrak{p}_1, \dots, \mathfrak{p}_m\},
$$

where $\mathrm{lk}(q, p) \in \mathbb{Z}$, $\mathrm{lk}(q, p) \bmod n = \mathrm{lk}_n(q, p)$. In particular, p is decomposed completely in k if and only if $\mathrm{lk}_n(q, p) = 0 \bmod n$.

Remark 5.2.5

(1) When $n = 2$, we have $(p/q)_2 = (q^*/p)_2$ and hence Proposition 5.2.4 is regarded as an extension of Proposition 4.2.1.
(2) When $n > 2$, one can not obtain the reciprocity law for $(q/p)_n$ since $\mathbb{Q}$ has no primitive n-th root of unity. In order to obtain the reciprocity law, one should consider the n-th power residue symbol for two principal prime ideal of a number field containing a primitive n-th root of unity [82, §154], as an analogue of the linking number of two null-homologous knots in a 3-manifold.

Summary

Galois covering of degree n	Galois covering of degree n
$M \xrightarrow{h} S^3$	$\mathrm{Spec}(\mathcal{O}_k) \xrightarrow{h} \mathrm{Spec}(\mathbb{Z})$
$K_i \longmapsto K$	$\mathfrak{p}_i \longmapsto (p)$
I_{K_i}: inertia group of K_i	$I_{\mathfrak{p}_i}$: inertia group of $\mathfrak{p}_i$
T_{K_i}: inertia space of K_i	$T_{\mathfrak{p}_i}$ inertia field of $\mathfrak{p}_i$
$\#I_{K_i} = [M : I_{K_i}] = e$	$\#I_{\mathfrak{p}_i} = [k : T_{\mathfrak{p}_i}] = e$
D_{K_i}: decomposition group of K_i	$D_{\mathfrak{p}_i}$: decomposition group of $\mathfrak{p}_i$
Z_{K_i}: decomposition space of K_i	$Z_{\mathfrak{p}_i}$: decomposition field of $\mathfrak{p}_i$
$\#D_{K_i} = [M : Z_{K_i}] = ef$	$\#D_{\mathfrak{p}_i} = [k : Z_{\mathfrak{p}_i}] = ef$
$efr = n$	$efr = n$
$(h^{-1}(K) = K_1 \cup \cdots \cup K_r)$	$(h^{-1}(\{p\}) = \{\mathfrak{p}_1, \dots, \mathfrak{p}_r\})$

Chapter 6
Homology Groups and Ideal Class Groups I: Genus Theory

In this chapter, we review Gauss' genus theory from the link-theoretic point of view. We shall see that the notion of genera is defined by using the idea analogous to the linking number. We also present, vice versa, a topological analogue of the genus theory. For this, firstly, in Sect. 6.1, we make clear the analogies between homology groups and ideal class groups, Hurewicz theorem and Artin reciprocity law.

6.1 Homology Groups and Ideal Class Groups

Firstly, let us see the analogy between the 1st homology group of a 3-manifold and the ideal class group of a number field ([98], [194]). Let M be a connected oriented 3-manifold. Recall that knots in M generate the group $Z_1(M)$ of 1-cycles of M. The boundaries ∂D of 2-chains $D \in C_2(M)$ generate the subgroup $B_1(M)$ of $Z_1(M)$, and the 1st homology group $H_1(M)$ is defined by the quotient group:

$$H_1(M) = Z_1(M)/B_1(M).$$

On the other hand, let k be a number field of finite degree over $\mathbb{Q}$. Recall that prime ideals ($\neq 0$) of the ring $\mathcal{O}_k$ of integers of k generate the ideal group $I(k)$ of k. The principal ideals (a) generated by numbers $a \in k^\times$ (resp. totally positive $a \in k^\times$) form the subgroup $P(k)$ (resp. $P^+(k)$) of $I(k)$, and the ideal class group (resp. the narrow ideal class group) is defined by the quotient group:

$$H(k) = I(k)/P(k) \text{ (resp. } H^+(k) = I(k)/P^+(k)).$$

Note that 2-chains D with $\partial D = 0$ form the 2nd homology group of M, while numbers $a \in k^\times$ with $(a) = \mathcal{O}_k$ form the unit group $\mathcal{O}_k^\times$.

Since the (narrow) ideal class group of a number field is finite (cf. 2.2.14), 3-manifolds M with finite $H_1(M)$, namely, *rational homology 3-spheres* are closer analogues of number rings in the above analogy.

Summing up, we have the following analogies:

$C_2(M) \longrightarrow Z_1(M)$	$k^\times \longrightarrow I(k)$
$D \longmapsto \partial D$	$a \longmapsto (a)$
$B_1(M)$	$P(k) \ (P^+(k))$
1st homology group	(narrow) ideal class group
$H_1(M) = Z_1(M)/B_1(M)$	$H(k) = I(k)/P(k) \ (H^+(k) = I(k)/P^+(k))$
2nd homology group	Unit group
$H_2(M)$	$\mathcal{O}_k^\times$

Finally we summarize the analogy between the Hurewicz theorem (for 3-manifolds) and the Artin reciprocity.

Hurewicz Theorem Let $f : M \to S^3$ be a finite Abelian covering ramified over a link L. Set $X := S^3 \setminus L, Y := f^{-1}(X), G := \mathrm{Gal}(Y/X)$. For a knot $K \subset X$, let σ_K be the generator of the decomposition group D_K defined in Sect. 5.1. Defining $\sigma_c := \prod_K \sigma_K^{n_K}$ for a 1-cycle $c = \sum_K n_K K \in Z_1(X)$, we get the homomorphism, which we call the *Hurewicz map*,

$$\sigma_{M/S^3} : Z_1(X) \longrightarrow G; \ c \longmapsto \sigma_c.$$

Then σ_{M/S^3} is surjective and $\mathrm{Ker}(\sigma_{M/S^3}) = f_*(Z_1(Y)) + B_1(X)$. Hence one has the isomorphism, which we call the *Hurewicz isomorphism*,

$$\sigma_{M/S^3} : H_1(X)/f_*(H_1(Y)) \simeq G.$$

For a knot K in X, let f be the order of $[K] \in H_1(X)$ in $H_1(X)/f_*(H_1(Y))$. Then K is decomposed in F into $r := d/f$ different knots $f^{-1}(K) = \{\mathfrak{K}_1, \ldots, \mathfrak{K}_r\}$.

Let L have r-components, $L = K_1 \cup \cdots \cup K_r$. Then the homology groups $H_i(X)$ can be computed by the Mayer–Vietoris sequence. For the decomposition $S^3 = X_L \cup V_L$ with $X_L \cap V_L = \partial V_L$, together with $H_i(S^3) = H_i(V_L) = 0 \ (i = 1, 2)$ and $H_3(\partial V_L) = 0$, the Mayer–Vietoris exact sequence reads

$$0 \longrightarrow H_3(X_L) \longrightarrow H_3(S^3) \overset{\Delta}{\longrightarrow} H_2(\partial V_L) \longrightarrow H_2(X_L) \longrightarrow 0,$$
$$0 \longrightarrow H_1(\partial V_L) \overset{i_1}{\longrightarrow} H_1(V_L) \oplus H_1(X_L) \longrightarrow H_0(\partial V_L) \overset{i_0}{\longrightarrow} H_0(V_L) \oplus H_0(X_L).$$

By the dimension reason, $H_i(X_L) = 0$ for $i > 3$. Since the generator $[S^3]$ of $H_3(S^3)$ is sent to $([\partial V_{K_i}])$ in $\bigoplus_{i=1}^r H_2(\partial V_K) = H_2(\partial V_L)$, Δ is injective and hence $H_3(X_L) = 0$, $H_2(X_L) \simeq \mathbb{Z}^{r-1}$. Since i_0 is injective, $H_1(X_L)$ is the cokernel of the injective map $i_1 : H_1(\partial V_L) \to H_1(V_L) \oplus H_1(X_L)$. Since $H_1(\partial V_L) = H_1(V_L) \oplus \bigoplus_{i=1}^r \mathbb{Z}[\alpha_{K_i}]$, $H_1(X_L) = \bigoplus_{i=1}^r \mathbb{Z}[\alpha_{K_i}] \simeq \mathbb{Z}^r$. Finally, $H_0(X_L) = \mathbb{Z}$ for X_L is connected.

Artin Reciprocity Law   Let $k/\mathbb{Q}$ be a finite Abelian extension ramified over a finite set S of prime numbers, and let $f : \mathrm{Spec}(\mathcal{O}_k) \to \mathrm{Spec}(\mathbb{Z})$ be the associated covering of the rings of integers. Set $X := \mathrm{Spec}(\mathbb{Z}) \setminus S$, $X_0 := \mathrm{Max}(\mathbb{Z}) \setminus S$, $S_k := f^{-1}(S)$, $Y := \mathrm{Spec}(\mathcal{O}_k) \setminus S_k$, $Y_0 := \mathrm{Max}(\mathcal{O}_k) \setminus S_k$, $G = \mathrm{Gal}(k/\mathbb{Q})$. Further, we set

$I(X) := \bigoplus_{p \in X_0} \mathbb{Z}$, $P(X) := \{(a) \in P^+(\mathbb{Q}) \mid a \equiv 1 \bmod q \ (\forall q \in S)\}$,
$H(X) := I(X)/P(X)$,
$I(Y) := \bigoplus_{p \in Y_0} \mathbb{Z}$, $P(Y) := \{(\alpha) \in P^+(k) \mid \alpha \equiv 1 \bmod \mathfrak{q} \ (\forall \mathfrak{q} \in S_k)\}$,
$H(Y) := I(Y)/P(Y)$.

For a prime number $p \in X_0$, we have $\sigma_p \in G$ defined as in Section 5.2. Defining $\sigma_{\mathfrak{a}} := \prod_{p \in X_0} \sigma_p^{n_p}$ for $\mathfrak{a} = \prod_{p \in X_0} p^{n_p} \in I(X)$, we get the homomorphism, which we call the *Artin map*,

$$\sigma_{k/\mathbb{Q}} : I(X) \longrightarrow G; \ \mathfrak{a} \longmapsto \sigma_{\mathfrak{a}}.$$

Then $\sigma_{k/\mathbb{Q}}$ is surjective and $\mathrm{Ker}(\sigma_{k/\mathbb{Q}}) = \mathrm{N}_{k/\mathbb{Q}}(I(Y))P(X)$. Hence we have the isomorphism, called the (ideal theoretic) *Artin reciprocity law*,

$$\sigma_{k/\mathbb{Q}} : H(X)/\mathrm{N}_{k/\mathbb{Q}}(H(Y)) \simeq G.$$

For a prime ideal $(p) \in I(X)$, let f be the order of $[(p)] \in H(X)$ in $H(X)/\mathrm{N}_{F/k}(H(Y))$. Then (p) is decomposed in F into a product $(p) = \mathfrak{p}_1 \cdots \mathfrak{p}_r$ of $r := d/f$ different maximal ideals.

Let $m := \prod_{p \in S} p$. Then the *ray class group* $H(X)$ mod m can be computed as follows. Define the homomorphism $I(X) \to (\mathbb{Z}/m\mathbb{Z})^\times$ by $(a) \mapsto a \bmod m$ for $a \in \mathbb{N}$. It is obviously surjective and the kernel is $P(X)$. Thus we have $H(X) \simeq (\mathbb{Z}/m\mathbb{Z})^\times$.

6.2   Genus Theory for a Link

Let $L = K_1 \cup \cdots \cup K_r \subset S^3$ be an r-component link and let $X_L := S^3 \setminus \mathrm{int}(V_L)$ the link exterior and $G_L := \pi_1(X_L)$. For an integer $n \geq 2$, let $\psi : G_L \to \mathbb{Z}/n\mathbb{Z}$ be the surjective homomorphism sending each meridian α_i of K_i to $1 \in \mathbb{Z}/n\mathbb{Z}$. Let $h : Y \to X_L$ be the cyclic covering of degree n corresponding to $\mathrm{Ker}(\psi)$. The Fox completion $f : M \to S^3$ (Example 2.1.15) is a cyclic covering of degree n over S^3 ramified along L. Let τ be the generator of $\mathrm{Gal}(M/S^3)$ corresponding to $1 \in \mathbb{Z}/n\mathbb{Z}$. In the following, a 1-cycle representing a homology class of $H_1(M)$ will be taken to be disjoint from $f^{-1}(L)$. Now we say that $[a], [b] \in H_1(M)$ belong to the same *genus*, written as $[a] \approx [b]$, if the following holds:

$$\mathrm{lk}(f_*(a), K_i) \equiv \mathrm{lk}(f_*(b), K_i) \mod n \quad (1 \le i \le r).$$

This definition is shown to be independent of the choice of 1-cycles representing homology classes as follows. Suppose that $[a] = 0 \in H_1(M)$. It suffices to show $\mathrm{lk}(f_*(a), K_i) \equiv 0 \mod n$. The relative homology exact sequence $H_2(M, Y) \overset{\partial}{\to} H_1(Y) \to H_1(M)$ yields $[a] \in \mathrm{Im}(\partial)$. By the excision, $H_2(M, Y)$ is generated by 2-cycles whose boundaries are meridians $\tilde{\alpha}_i$ of components of $f^{-1}(L)$, and so $\mathrm{Im}(\partial)$ is generated by $[\tilde{\alpha}_i]$ $(1 \le i \le r)$. Since $f_*([\tilde{\alpha}_i]) = n[\alpha_i] \in H_1(X_L)$, $\mathrm{lk}(f_*(a), K_i) \equiv 0 \mod n$.

Theorem 6.2.1 ([156]) *Let* $\chi : H_1(M) \to (\mathbb{Z}/n\mathbb{Z})^r$ *be the homomorphism defined by* $\chi([a]) := (\mathrm{lk}(f_*(a), K_i) \mod n)$. *Then one has the following:*

$$\mathrm{Im}(\chi) = \left\{ (\varepsilon_i) \in (\mathbb{Z}/n\mathbb{Z})^r \;\middle|\; \sum_{i=1}^r \varepsilon_i = 0 \right\}, \quad \mathrm{Ker}(\chi) = (\tau - 1)H_1(M)$$

and hence

$$H_1(M)/\approx \; \simeq H_1(M)/(\tau - 1)H_1(M) \simeq (\mathbb{Z}/n\mathbb{Z})^{r-1}.$$

Proof Let j denote the inclusion $Y \hookrightarrow M$. Then $j_* : H_1(Y) \to H_1(M)$ is surjective and, as explained before the Theorem, we have $B := \mathrm{Ker}(j_*) = \mathbb{Z}([\tilde{\alpha}_1]) \oplus \cdots \oplus \mathbb{Z}([\tilde{\alpha}_r])$, where $\tilde{\alpha}_i$ is a meridian of a component of $f^{-1}(L)$ lying over K_i. Hence $f_*(B) = h_*(B) = \mathbb{Z}(n[\alpha_1]) \oplus \cdots \oplus \mathbb{Z}(n[\alpha_r]) \subset H_1(X_L) = \mathbb{Z}[\alpha_1] \oplus \cdots \oplus \mathbb{Z}[\alpha_r]$. By the lower terms obtained from the spectral sequence associated with the finite cyclic covering $Y \to X_L$, we have the exact sequence

$$H_0(\mathrm{Gal}(Y/X_L), H_1(Y)) \longrightarrow H_1(X_L) \longrightarrow H_1(\mathrm{Gal}(Y/X_L), \mathbb{Z}) \longrightarrow 0,$$

which gives the exact sequence

$$H_1(Y) \overset{\tau-1}{\longrightarrow} H_1(Y) \overset{h_*}{\longrightarrow} H_1(X_L) \longrightarrow \mathrm{Gal}(Y/X_L) \to 0.$$

Therefore we obtain the following commutative exact diagram:

$$
\begin{array}{ccccc}
 & 0 & & 0 & \\
 & \downarrow & & \downarrow & \\
 & B & \xrightarrow{f_*} & f_*(B) & \longrightarrow 0 \\
 & \downarrow & & \downarrow & \\
0 \longrightarrow (\tau-1)H_1(Y) \longrightarrow & H_1(Y) & \xrightarrow{f_*} & f_*(H_1(Y)) & \longrightarrow 0 \\
\downarrow{j_*} & \downarrow{j_*} & & & \\
0 \longrightarrow (\tau-1)H_1(M) \longrightarrow & H_1(M) & & & \\
\downarrow & \downarrow & & & \\
0 & 0 & & &
\end{array}
$$

This yields the exact sequence

$$0 \to (\tau-1)H_1(M) \to H_1(M) \to f_*(H_1(Y))/f_*(B) \to 0. \qquad (6.1)$$

Define $\varphi : H_1(X_L) \to (\mathbb{Z}/n\mathbb{Z})^r$ by $\varphi(c) := (\mathrm{lk}(c, K_l) \bmod n)$. It is easy to see that φ is surjective and $\mathrm{Ker}(\varphi) = \mathbb{Z}(n[\alpha_1]) \oplus \cdots \oplus \mathbb{Z}(n[\alpha_r]) = f_*(B)$. So we have the following commutative exact diagram for a covering $Y \to X_L$:

$$
\begin{array}{ccccc}
0 & & 0 & & \\
\downarrow & & \downarrow & & \\
f_*(B) & = & f_*(B) & & \\
\downarrow & & \downarrow & & \\
0 \longrightarrow f_*(H_1(Y)) & \longrightarrow & H_1(X_L) & \longrightarrow \mathrm{Gal}(Y/X_L) & \longrightarrow 0 \\
\downarrow & & \downarrow{\varphi} & \updownarrow{\wr} & \\
f_*(H_1(Y))/f_*(B) & & (\mathbb{Z}/n\mathbb{Z})^r & \xrightarrow{\Sigma} \mathbb{Z}/n\mathbb{Z} & \\
\downarrow & & \downarrow & & \\
0 & & 0 & &
\end{array}
$$

Here $\Sigma : (\mathbb{Z}/n\mathbb{Z})^r \to \mathbb{Z}/n\mathbb{Z}$ is the homomorphism defined by $\Sigma((\varepsilon_i)) := \sum_{i=1}^{r} \varepsilon_i$, and $\mathrm{Gal}(Y/X) \simeq \mathbb{Z}/n\mathbb{Z}$ is the isomorphism sending τ to 1 mod n. Hence

$$f_*(H_1(Y))/f_*(B) \overset{\varphi}{\simeq} \mathrm{Ker}(\Sigma) \simeq (\mathbb{Z}/n\mathbb{Z})^{r-1}. \qquad (6.2)$$

By (6.1), (6.2) and $\chi = \varphi \circ f_*$, we obtain the exact sequence

$$0 \longrightarrow (\tau-1)H_1(M) \longrightarrow H_1(M) \xrightarrow{\chi} (\mathbb{Z}/n\mathbb{Z})^{r-1} \longrightarrow 0$$

which yields our assertion.   □

For the case of $n = 2$, we have the following topological analogue of Gauss' theorem:

Corollary 6.2.2 *Let* $f : M \to S^3$ *be a double covering of connected oriented closed 3-manifolds ramified over an* r-*component link* $L = K_1 \cup \cdots \cup K_r$. *Then the homomorphism* $\chi : H_1(M) \to (\mathbb{Z}/2\mathbb{Z})^r$ *defined by* $\chi([a]) := (\mathrm{lk}(f_*(a), K_i) \bmod 2)$ *induces the following isomorphism:*

$$H_1(M)/2H_1(M) \simeq \left\{ (\varepsilon_i) \in (\mathbb{Z}/2\mathbb{Z})^r \ \bigg| \ \sum_{i=1}^r \varepsilon_i = 0 \right\} \simeq (\mathbb{Z}/2\mathbb{Z})^{r-1}.$$

Proof By Theorem 6.2.1, it suffices to show $(\tau - 1)H_1(M) = 2H_1(M)$. Let $tr : H_1(S^3) \to H_1(M)$ denote the transfer map. Since $tr \circ f_* : H_1(M) \to H_1(M)$ is $1 + \tau$, one has $\tau = -1$ as $H_1(S^3) = 0$. Hence $(\tau - 1)H_1(M) = 2H_1(M)$. □

Example 6.2.3 Let $L = B(a, b)$ be a 2-bridge link where a is an even integer (≥ 2) and $0 < b < a, (a, b) = 1$. The double covering M of S^3 ramified over L is the lens space $L(a, b)$ (Example 2.1.16). Then one has $H_1(M) \simeq \mathbb{Z}/a\mathbb{Z}$, and $H_1(M)/\approx = H_1(M)/2H_1(M) \simeq \mathbb{Z}/2\mathbb{Z}$.

6.3 Genus Theory for Primes

Let $n \geq 2$ be an integer and let $S = \{p_1, \ldots, p_r\}$ be the set of r distinct prime numbers such that $p_i \equiv 1 \bmod n$ $(1 \leq i \leq r)$. Let $G_S := \pi_1(\mathrm{Spec}(\mathbb{Z}[1/(p_1 \cdots p_r)])) = \mathrm{Gal}(\mathbb{Q}_S/\mathbb{Q})$, where $\mathbb{Q}_S$ is the maximal Galois extension of $\mathbb{Q}$ unramified outside $S \cup \{\infty\}$ (Example 2.2.20). For each p_i, we fix a primitive root $\alpha_i \bmod p_i$. Let $\psi : G_S^{\mathrm{ab}} = \prod_{i=1}^r \mathbb{Z}_{p_i}^\times = \prod_{i=1}^r \mathbb{F}_{p_i}^\times \times (1 + p_i\mathbb{Z}_{p_i}) \to \mathbb{Z}/n\mathbb{Z}$ be the homomorphism defined by $\psi(\alpha_i) = 1, \psi(1 + p_i\mathbb{Z}_{p_i}) = 0$ $(1 \leq i \leq r)$. Let k be the cyclic extension of $\mathbb{Q}$ of degree n corresponding to $\mathrm{Ker}(\psi)$. Let τ be the generator of $\mathrm{Gal}(k/\mathbb{Q})$ corresponding to $1 \in \mathbb{Z}/n\mathbb{Z}$ (Example 2.3.6). Let $\mu_n \subset \overline{\mathbb{Q}}$ denote the group of n-th roots of unity and we fix an embedding $\mathbb{Q}(\mu_n) \subset \mathbb{Q}_{p_i}$ $(1 \leq i \leq r)$. In the following, an ideal representing an ideal class of the narrow ideal class group $H^+(k)$ will be taken to an ideal of $\mathcal{O}_k$ disjoint from S. Now we say that $[\mathfrak{a}], [\mathfrak{b}] \in H^+(k)$ belong to the same *genus* – written as $[\mathfrak{a}] \approx [\mathfrak{b}]$ –, if the following holds:

$$\left(\frac{N\mathfrak{a}}{p_i} \right)_n = \left(\frac{N\mathfrak{b}}{p_i} \right)_n \quad (1 \leq i \leq r),$$

where $\left(\frac{*}{p_i} \right)_n$ denotes the n-th power residue symbol in $\mathbb{Q}_{p_i}$ taking the value in μ_n. This definition is shown to be independent of the choice of ideals representing ideal classes as follows. Suppose $[\mathfrak{a}] = 0 \in H^+(k)$. It suffices to show $(\frac{N\mathfrak{a}}{p_i})_n = 1$. There is a totally positive $\alpha \in k^\times$ such that $\mathfrak{a} = (\alpha)$. So $N\mathfrak{a} = N_{k/\mathbb{Q}}(\alpha) = N_{k_{\mathfrak{p}_i}/\mathbb{Q}_{p_i}}(\alpha)$

(here $\mathfrak{p}_i$ is a prime ideal of k lying over p_i and $k_{\mathfrak{p}_i} = \mathbb{Q}_{p_i}(\sqrt[n]{p_i})$). By (2.5), (2.6), we have $\left(\frac{N\mathfrak{a}}{p_i}\right)_n = 1$.

Theorem 6.3.1 ([94]) *Let* $\chi : H^+(k) \to \mu_n^r$ *be the homomorphism defined by* $\chi([\mathfrak{a}]) := \left(\left(\frac{N\mathfrak{a}}{p_i}\right)_n\right)$. *Then one has the following:*

$$\mathrm{Im}(\chi) = \left\{ (\zeta_i) \in \mu_n^r \;\middle|\; \prod_{i=1}^r \zeta_i = 1 \right\}, \quad \mathrm{Ker}(\chi) = H^+(k)^{\tau-1},$$

and hence

$$H^+(k)/\approx \; \simeq \; H^+(k)/H^+(k)^{\tau-1} \simeq (\mathbb{Z}/n\mathbb{Z})^{r-1}.$$

Proof Let $J_{\mathbb{Q}}$ and J_k be the idèle group of $\mathbb{Q}$ and k respectively. We set $U_k := \prod_{\mathfrak{p}\in\mathrm{Max}(\mathcal{O}_k)} \mathcal{O}_{\mathfrak{p}}^\times \times \prod_{v\in S_k^\infty} (k_v^\times)^2$ and we then have the isomorphism $J_k/U_k k^\times \simeq H^+(k)$ (Example 2.3.4). Next, we shall show that the kernel of the norm map $N_{k/\mathbb{Q}} : J_k/k^\times \to N_{k/\mathbb{Q}}(J_k)\mathbb{Q}^\times/\mathbb{Q}^\times$ induced on the idèle class groups is given by $(J_k/k^\times)^{\tau-1}$. First, it is obvious that $(J_k/k^\times)^{\tau-1} \subset \mathrm{Ker}(N_{k/\mathbb{Q}})$. For $\mathbf{a} \in J_k$, assume $N_{k/\mathbb{Q}}(\mathbf{a}) \in \mathbb{Q}^\times$. By the Hasse norm theorem one has

$$N_{k/\mathbb{Q}}(J_k) \cap \mathbb{Q}^\times = N_{k/\mathbb{Q}}(k^\times)$$

and hence there is $\alpha \in k^\times$ such that $N_{k/\mathbb{Q}}(\mathbf{a}) = N_{k/\mathbb{Q}}(\alpha)$. The Hilbert theorem 90 asserts that for $\mathbf{b} \in J_k$,

$$N_{k/\mathbb{Q}}(\mathbf{b}) = 1 \implies \exists \mathbf{c} \in J_k, \; \mathbf{b} = \mathbf{c}^{\tau-1}$$

and so there is $\mathbf{c} \in J_k$ such that $\mathbf{a} = \alpha\mathbf{c}^{\tau-1}$. Therefore $\mathbf{a}k^\times = \mathbf{c}^{\tau-1}k^\times \in (J_k/k^\times)^{\tau-1}$. Hence we have the following commutative exact diagram:

$$
\begin{array}{ccc}
& 0 & \quad 0 \\
& \downarrow & \quad \downarrow \\
& U_k k^\times/k^\times \xrightarrow{N_{k/\mathbb{Q}}} N_{k/\mathbb{Q}}(U_k)\mathbb{Q}^\times/\mathbb{Q}^\times \longrightarrow 0 \\
& \downarrow & \quad \downarrow \\
0 \longrightarrow (J_k/k^\times)^{\tau-1} \longrightarrow J_k/k^\times \xrightarrow{N_{k/\mathbb{Q}}} N_{k/\mathbb{Q}}(J_k)\mathbb{Q}^\times/\mathbb{Q}^\times \longrightarrow 0 \\
& \downarrow & \quad \downarrow \\
0 \longrightarrow H^+(k)^{\tau-1} \longrightarrow H^+(k) \\
& \downarrow & \quad \downarrow \\
& 0 & \quad 0
\end{array}
$$

This yields the exact sequence

$$0 \longrightarrow H^+(k)^{\tau-1} \longrightarrow H^+(k) \longrightarrow \mathrm{N}_{k/\mathbb{Q}}(J_k)\mathbb{Q}^\times/\mathrm{N}_{k/\mathbb{Q}}(U_k)\mathbb{Q}^\times \longrightarrow 0. \qquad (6.3)$$

Since $H_\mathbb{Q}^+ = 1$, we have $J_\mathbb{Q} = \mathbb{Q}^\times((\mathbb{R}^\times)^2 \times \prod_p \mathbb{Z}_p^\times)$. Therefore we can choose uniquely an idèle of the form $\mathbf{a} = ((a_p), a_\infty)$ with $a_p \in \mathbb{Z}_p^\times$ ($\forall p \in \mathrm{Max}(\mathbb{Z})$) and $a_\infty > 0$ as a representative of each idèle class in $J_\mathbb{Q}/\mathbb{Q}^\times$. We then define the homomorphism $\varphi : J_\mathbb{Q}/\mathbb{Q}^\times \to \mu_n^r$ by

$$\varphi(\mathbf{a}\mathbb{Q}^\times) := \left(\left(\frac{a_{p_i}}{p_i}\right)_n\right).$$

We first note that φ is surjective, since the map $\mathbb{Z}_{p_i}^\times \ni u \mapsto \left(\frac{u}{p_i}\right)_n \in \mu_n$ is surjective ($1 \le i \le r$) as $\left(\frac{\alpha_i}{p_i}\right)_n$ is a primitive n-th root of unity. Next, we will show that $\mathrm{Ker}(\varphi) = \mathrm{N}_{k/\mathbb{Q}}(U_k)\mathbb{Q}^\times/\mathbb{Q}^\times$. By (2.5) we have $\left(\frac{a_{p_i}}{p_i}\right)_n = 1 \Leftrightarrow a_{p_i} \in \mathrm{N}_{k_{\mathfrak{p}_i}/\mathbb{Q}_{p_i}}(\mathcal{O}_{\mathfrak{p}_i}^\times)$ for any prime ideal $\mathfrak{p}_i$ of k over p_i. If $p \notin S$, then $\mathrm{N}_{k_\mathfrak{p}/\mathbb{Q}_p}(\mathcal{O}_\mathfrak{p}^\times) = \mathbb{Z}_p^\times$ for any prime ideal $\mathfrak{p}$ of k, since p is unramified in $k/\mathbb{Q}$ (cf. 2.4), and $\mathrm{N}_{k_v/\mathbb{R}}((k_v^\times)^2) = (\mathbb{R}^\times)^2$ for $v \in S_k^\infty$. Therefore $\mathrm{Ker}(\varphi) = \mathrm{N}_{k/\mathbb{Q}}(U_k)\mathbb{Q}^\times/\mathbb{Q}^\times$. Finally, noting the isomorphism $\rho_{k/\mathbb{Q}} : J_\mathbb{Q}/\mathrm{N}_{k/\mathbb{Q}}(J_k)\mathbb{Q}^\times \simeq G$ in class field theory (2.10), we have the following commutative exact diagram:

$$
\begin{array}{ccccccccc}
 & & 0 & & 0 & & & & \\
 & & \downarrow & & \downarrow & & & & \\
 & & \mathrm{N}_{k/\mathbb{Q}}(U_k)\mathbb{Q}^\times/\mathbb{Q}^\times & = & \mathrm{N}_{k/\mathbb{Q}}(U_k)\mathbb{Q}^\times/\mathbb{Q}^\times & & & \\
 & & \downarrow & & \downarrow & & & & \\
0 & \longrightarrow & \mathrm{N}_{k/\mathbb{Q}}(J_k)\mathbb{Q}^\times/\mathbb{Q}^\times & \longrightarrow & J_\mathbb{Q}/\mathbb{Q}^\times & \stackrel{\rho_{k/\mathbb{Q}}}{\longrightarrow} & \mathrm{Gal}(k/\mathbb{Q}) & \longrightarrow & 0 \\
 & & \downarrow & & \downarrow\varphi & & \downarrow & & \\
 & & \mathrm{N}_{k/\mathbb{Q}}(J_k)\mathbb{Q}^\times/N(U_k)\mathbb{Q}^\times & & \mu_n^r & \stackrel{\Sigma}{\longrightarrow} & \mathbb{Z}/n\mathbb{Z} & & \\
 & & \downarrow & & \downarrow & & & & \\
 & & 0 & & 0 & & & &
\end{array}
$$

Here $\Sigma : \mu_n^r \to \mathbb{Z}/n\mathbb{Z}$ is defined as follows: Let $\xi_i : \mu_n \stackrel{\sim}{\to} \mathbb{Z}/n\mathbb{Z}$ be the isomorphism defined by sending $(\frac{\alpha_i}{p_i})_n$ to 1 mod n. Then we set $\Sigma((\zeta_i)) := \sum_{i=1}^r \xi_i(\zeta_i)$. The isomorphism $\mathrm{Gal}(k/\mathbb{Q}) \stackrel{\sim}{\to} \mathbb{Z}/n\mathbb{Z}$ is defined by sending τ to 1 mod n. From the diagram above,

$$\mathrm{N}_{k/\mathbb{Q}}(J_k)\mathbb{Q}^\times/\mathrm{N}_{k/\mathbb{Q}}(U_k)\mathbb{Q}^\times \stackrel{\varphi}{\simeq} \mathrm{Ker}(\Sigma) \simeq \mu_n^{r-1}. \qquad (6.4)$$

Noting (6.3), (6.4) and $\chi = \varphi \circ N_{k/\mathbb{Q}}$, we obtain the exact sequence

$$0 \longrightarrow H^+(k)^{\tau-1} \longrightarrow H^+(k) \xrightarrow{\chi} (\mathbb{Z}/2\mathbb{Z})^{r-1} \longrightarrow 0,$$

which yields our assertion.   □

For the case that $n = 2$, we have the following Gauss' theorem:

Corollary 6.3.2 *Let $k/\mathbb{Q}$ be a quadratic extension ramified over r odd primes $p_1, \ldots, p_r$ (the infinite prime of $\mathbb{Q}$ is possibly ramified). Then the homomorphism $\chi : H^+(k) \rightarrow \{\pm 1\}^r$ defined by $\chi([\mathfrak{a}]) := \left(\left(\frac{N\mathfrak{a}}{p_i}\right)\right)$ induces the following isomorphism:*

$$H^+(k)/H^+(k)^2 \simeq \left\{ (\zeta_i) \in \{\pm 1\}^r \,\middle|\, \prod_{i=1}^{r} \zeta_i = 1 \right\} \simeq (\mathbb{Z}/2\mathbb{Z})^{r-1}.$$

Proof By Theorem 6.3.1, it suffices to show that $H^+(k)^{\tau-1} = H^+(k)^2$. Since $H^+(\mathbb{Q}) = 1$, we have $N_{k/\mathbb{Q}}([\mathfrak{a}]) = [\mathfrak{a}][\mathfrak{a}]^\tau = 1$ for $[\mathfrak{a}] \in H^+(k)$ and so $\tau = -1$. Hence $H^+(k)^{\tau-1} = H^+(k)^2$.   □

Example 6.3.3 Let $k = \mathbb{Q}(\sqrt{145})$, which is a quadratic extension of $\mathbb{Q}$ ramified over $\{5, 29\}$. Let $\mathfrak{p} := (2, (1+\sqrt{145})/2)$. Then one has $H^+(k)(= H(k)) = \langle[\mathfrak{p}]\rangle \simeq \mathbb{Z}/4\mathbb{Z}$ and $H^+(k)/\approx\, = H^+(k)/H^+(k)^2 \simeq \mathbb{Z}/2\mathbb{Z}$.

Remark 6.3.4

(1) Genus theory in this chapter is generalized for relative extensions (resp. coverings) of number fields (resp. 3-manifolds). We refer to [60] for number fields and [229] for 3-manifolds.

   In [209], A. Sikora studied the analogies between a group action on a 3-manifold M and a number field k and showed some analogous formulas relating the number of ramified knots in a cyclic covering $M \rightarrow M/G$ (resp. ramified primes in a cyclic covering k/k^G) to the cyclic group G-action on $H_1(M)$ (resp. $H(k)$). In [152], B. Morin gave a unified proof of Sikora's results in the arithmetic and topological cases introducing the equivariant étale cohomology.

(2) Besides the Hilbert theory and genus theory, some analogies for 3-manifolds of the capitulation problem and class tower problem have also been investigated (see [59], [159], [193], [195], [227]).

Summary

1st homology group $H_1(M)$	(narrow) ideal class group $H^+(k)$
Classification of homology classes by the linking numbers	Classification of ideal classes by the Legendre symbols
$H_1(M)/2H_1(M) \simeq (\mathbb{Z}/2\mathbb{Z})^{r-1}$ ($M \to S^3$: double ramified covering)	$H^+(k)/H^+(k)^2 \simeq (\mathbb{Z}/2\mathbb{Z})^{r-1}$ ($k/\mathbb{Q}$: quadratic extension)

In Chaps. 4–6, we re-examined Gauss' theory on linking numbers, quadratic residues and genus theory from the viewpoint of the analogies between knots and primes, 3-manifolds and number rings in Chap. 3. In the rest of this book, we shall try to bridge knot theory and algebraic number theory, which branched out after the works of Gauss and have evolved in their separate ways, from the viewpoint of this analogy.

Chapter 7
Idelic Class Field Theory for 3-Manifolds and Number Rings

Idelic class field theory describes Abelian extensions of a number field k in terms of the idèles of k. Firstly, we construct local class field theory for the local field at each prime of k. Getting local theories together over all primes of k, we obtain idelic class field theory of k. Both local and global class field theory are derived from the arithmetic duality theorems, as we explained in Sect. 2.3. In Sect. 7.3, we state the main results of local and global class field theory again as in Sect. 2.3. In order to develop a topological analogue of idelic class field theory for a 3-manifold M, it is a fundamental problem to find out the good countable set of knots in M, which plays a role similar to the set of all primes of a number field. A candidate of such a set of knots was firstly introduced by Niibo and Ueki as the notion of a *very admissible link*, which was refined by Mihara later as the notion of a *stably generic link*, for the purpose of constructing idelic class field theory for a 3-manifold M [145], [179], [181]. As in the case of number theory, we firstly construct local class field theory for the boundary torus of a tubular neighborhood of each component knot in a stably generic link $\mathfrak{L}$, and then, getting these local theories together over all knots in $\mathfrak{L}$, we construct idelic class field theory for M. We describe these topological counterparts in Sects. 7.1 and 7.2. As examples of stably generic links, we consider links satisfying Chebotarev type density law, called *Chebotarev links*, following Mazur and McMullen [140], [144].

7.1 Stably Generic Links and Chebotarev Links

In this section, we introduce the notion of a stably generic link in a 3-manifold, which plays a role similar to the set of all finite primes of a finite algebraic number field, for the purpose of constructing idelic class field theory for 3-manifolds.

© The Author(s), under exclusive license to Springer Nature Singapore Pte Ltd. 2024   81
M. Morishita, *Knots and Primes*, Universitext,
https://doi.org/10.1007/978-981-99-9255-3_7

Let M be an oriented, connected closed 3-manifold. In Sects. 7.1 and 7.2, we consider a link consisting of disjoint, countably many knots in M and we call a usual link of finitely many components (Example 2.1.6) a *finite link*.

Definition 7.1.1 Let $\mathcal{L}$ be a link consisting of disjoint countably many knots in M. We call $\mathcal{L}$ an *admissible link* if the homology classes of all components of $\mathcal{L}$ generates $H_1(M)$. We call $\mathcal{L}$ a *generic link* if for any finite sublink L of $\mathcal{L}$, the homology classes of all components of $\mathcal{L} \setminus L$ generates $H_1(M \setminus L)$.

We call $\mathcal{L}$ a *stably admissible link* if for any finite covering $f : N \to M$ of oriented, connected, closed 3-manifolds, ramified over a finite sublink of $\mathcal{L}$, $f^{-1}(\mathcal{L})$ is an admissible link in N. We call $\mathcal{L}$ a *stably generic link* if for any finite covering $f : N \to M$ of oriented, connected, closed 3-manifolds, ramified over a finite sublink of $\mathcal{L}$, $f^{-1}(\mathcal{L})$ is a generic link in N.

Remark 7.1.2

(1) Obviously, a generic link (resp. a stably generic link) is an admissible link (resp. a stably admissible link). If $\mathcal{L}$ is an admissible link, the link $\mathcal{L} \sqcup K$ obtained by adding any knot K is also an admissible link. On the other hand, if $\mathcal{L}$ is a generic link and K is a knot which is separated from $\mathcal{L}$, then $\mathcal{L} \sqcup K$ is not a generic link, because $H_1(M \setminus K)$ is not generated by the homology classes of components of $\mathcal{L} \setminus K$. In this sense, a generic link has no superfluous component.

(2) By the analogy between the 1st homology group of a 3-manifold and the ideal class group of a number field in Sect. 6.1, the notions of a stably admissible link and a stable generic link are given in a way imitating a weak form of the Chebotarev density theorem for a finite Abelian extension of number fields. See Definition 7.1.5 below.

The existence of a stably admissible link was shown by Niibo–Ueki in [181] and then Mihara [145] showed the existence of a stably generic link. In Theorem 7.1.4 below, we prove the existence of a stably generic link, following [145]. For this, we prepare the following:

Lemma 7.1.3 *For a finite link L in an oriented, connected, compact 3-manifold M, we denote by* $\mathrm{FCov}(M, L)$ *the set of finite coverings $f : N \to M$ of oriented, connected closed 3-manifolds ramified over a sublink of L. Then* $\mathrm{FCov}(M, L)$ *is a countable set. So we can give an order on the set* $\mathrm{FCov}(M, L)$ *written as*

$$\mathrm{FCov}(M, L) = \{f_m : N_m \longrightarrow M\}_{m \in \mathbb{N} \cup \{0\}},$$

where $f_0 = \mathrm{id}_M, N_0 = M$.

Proof By covering theory in Sect. 2.1, elements of $\mathrm{FCov}(M, L)$ correspond bijectively to subgroups of $\pi_1(M \setminus L)$ of finite index, in other words, finite sets on which $\pi_1(M \setminus L)$ acts transitively. Since $\pi_1(M \setminus L)$ is a finitely generated group, the set of finite $\pi_1(M \setminus L)$-sets with transitive action is a countable set. $\square$

Theorem 7.1.4 ([145]) *A stably generic link exists in an oriented, connected, closed 3-manifold.*

Proof In this proof, for a link L in a 3-manifold X, we write $\langle L \rangle$ to denote the subgroup of $H_1(X)$ generated by the homology classes of components of L:

$$\langle L \rangle := \langle [K] \in H_1(X) \mid K \text{ is a component of } L \rangle,$$

to avoid the complexity of notation.

To prove the assertion, we show the following claim. For $l, m, n \in \mathbb{N} \cup \{0\}$, we set $a(l, m, n) := 2^n (2(2^m(2l + 1) - 1) + 1)$.

Claim We can construct the increasing sequence of finite links in M

$$L_0 \subset L_1 \subset \cdots \subset L_n \subset \cdots$$

which satisfies

$$\langle f_{n,m}^{-1}(L_{a(l,m,n)} \setminus L_{n+l}) \rangle = H_1(N_{n,m} \setminus f^{-1}(L_{n+l})), \qquad (\star)$$

where we write $\mathrm{FCov}(M, L_n) = \{f_{n,m} : N_{n,m} \to M\}_{m \in \mathbb{N} \cup \{0\}}$ by Lemma 7.1.3.

We prove the Claim by induction on n. First, we take any non-empty finite link L_0 such that $\langle L_0 \rangle = H_1(M)$. By Lemma 7.1.3, we have $\mathrm{FCov}(M, L_0) = \{f_{0m} : N_{0m} \to M\}_{m \in \mathbb{N} \cup \{0\}}$. Suppose $n > 0$ and we are given $L_{n'}$ and $\mathrm{FCov}(M, L_{n'})$ for $0 \leq n' < n$. We set

$$\begin{aligned} s &:= \max\{e \in \mathbb{N} \cup \{0\} \mid 2^{-e} n \in \mathbb{Z}\}, \\ t &:= \max\{e \in \mathbb{N} \cup \{0\} \mid 2^{-e}(2^{-1}(2^{-s}n - 1) + 1) \in \mathbb{Z}\}, \\ u &:= 2^{-1}(2^t(2^{-1}(2^{-s}n - 1) + 1) - 1). \end{aligned}$$

Then we see easily $s + u < n$. Consider $f_{s,t} : N_{s,t} \to M \in \mathrm{FCov}(M, L_s)$. Since $H_1(N_{s,t} \setminus f^{-1}(L_{n-1}))$ is a finitely generated Abelian group, there is a finite link (large enough) L_n satisfying

$$\langle f_{s,t}^{-1}(L_n \setminus L_{n-1}) \rangle = H_1(N_{s,t} \setminus f^{-1}(L_{n-1})).$$

Since $L_{s+u} \subset L_{n-1}$, $H_1(N_{s,t} \setminus f^{-1}(L_{s+u}))$ is a quotient of $H_1(N_{s,t} \setminus f^{-1}(L_{n-1}))$ by the Mayer–Vietoris exact sequence and so we have

$$\langle f_{s,t}^{-1}(L_n \setminus L_{s+u}) \rangle = H_1(N_{s,t} \setminus f^{-1}(L_{s+t})).$$

Thus we constructed L_n and so the countable set $\mathrm{FCov}(M, L_n)$ by Lemma 7.1.3. When we take $a(l, m, n)$ for n here, we have $s = n$, $t = m$ and $u = l$. Therefore we proved the Claim.

Now our assertion follows from the Claim as follows. Let $\mathfrak{L} := \bigsqcup_{n \geq 0} L_n$. Let $f : N \to M$ be a finite covering of oriented, connected, closed 3-manifold ramified over a finite sublink L of $\mathfrak{L}$. Then there is n (large enough) such that $L \subset L_n$ and so $f \in \mathrm{FCov}(M, L_n)$. Therefore we may let $f = f_{n,m} : N_{n,m} \to M$. It suffices to show that $f_{n,m}^{-1}(\mathfrak{L})$ is a generic link in $N_{n,m}$. Let $L' \subset f_{n,m}^{-1}(\mathfrak{L})$ be a finite sublink. Taking l large, we can let $L' \subset f_{n,m}^{-1}(L_{n+l})$. Combining $(\star)$, we obtain

$$\langle f_{n,m}^{-1}(\mathfrak{L}) \setminus L' \rangle = H_1(N_{n,m} \setminus L').$$

This completes the proof. $\hfill\square$

In [140] and [144], Mazur and McMullen introduced the notion of a *link obeying the Chebotarev law*, which we call simply a *Chebotarev link*, following the Chebtarev density theorem for number field extensions. Ueki [234] introduced the weak version of a Chebotarev link.

Definition 7.1.5 Let $(K_n)_{n \in \mathbb{N}}$ be a sequence of disjoint knots in an oriented, connected closed 3-manifold M. We say that $(K_n)_{n \in \mathbb{N}}$ obeys the *Chebotarev density law* if for any finite Galois covering $f : N \to M$ of oriented, connected, closed 3-manifolds, ramified over a finite link $\bigsqcup_{i=1}^{r} K_{n_i}$ with Galois group $G = \mathrm{Gal}(N/M)$ and for any conjugacy class C of G, we have

$$\lim_{x \to \infty} \frac{1}{x} \# \{ K_n \mid n \neq n_i \ (i = 1, \ldots, r), n \leq x, C([K_n]) = C \} = \frac{\#C}{\#G},$$

where $C([K_n])$ denotes the conjugacy class of G containing $\rho([K_n])$ for the natural monodromy homomorphism $\rho : \pi_1(M \setminus \bigsqcup_{i=1}^{r} K_{n_i}) \to G$. Then we call the link $\mathfrak{L} = \bigsqcup_{n=1}^{\infty} K_n$ of countable infinitely many components a *Chebotarev link*. Namely, a Chebotarev link is a link whose components have a total order satisfying the Chebotarev density law.

A link $\mathfrak{L} = \bigsqcup_{n=1}^{\infty} K_n$ consisting of countable infinitely many knots K_n's in a 3-manifold X is called a *weakly Chebotarev link* if for any finite Galois covering $h : Y \to X$ with Galois group $G = \mathrm{Gal}(Y/X)$ and for any conjugacy class C of G, there is a component K_n of $\mathfrak{L}$ such that $C(\rho([K_n])) = C$ holds for the natural homomorphism $\rho : \pi_(X) \to G$. Obviously, a Chebotarev link is a weakly Chebotarev link.

Among other things, McMullen provided examples of Chebotarev links obtained from the dynamical systems on complements of knots/links. Example 7.1.6 and Theorem 7.1.7 below are special cases of the more general theorem in [144] for a topologically mixing flow on a 3-manifold.

Example 7.1.6 ([144]) Let K be a hyperbolic fibered knot in S^3, for example, the *figure eight knot* $B(5, 3)$ (Fig. 7.1).

Fig. 7.1 The figure eight knot

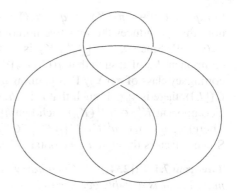

Then there is a fibration $\pi : S^3 \setminus \text{int}(V_K) \to S^1$ whose fibers are homeomorphic a surface Σ with one boundary component. Let $\varphi : \Sigma \overset{\approx}{\to} \Sigma$ be the monodromy of the fibration π. Since K is hyperbolic, φ is of pseudo-Anosov type. We then have the continuous dynamical system on $S^3 \setminus \text{int}(V_K)$ obtained as the suspension of φ, which has infinitely many countable closed orbits in $S^3 \setminus V_K$. We order all closed orbits by their periods to get the sequence $(K_n)_{n \in \mathbb{N}}$ of knots and let $\mathfrak{L} := \bigsqcup_{n=1}^{\infty} K_n$.

Theorem 7.1.7 ([144]) $\mathfrak{L}$ *is a Chebotarev link in* S^3.

The proof uses the method of symbolic dynamics and we refer to [144] for details. We note that this $\mathfrak{L}$ has the following remarkable property [65], [66]: for any finite link L' in S^3, there is a finite sublink L'' of $\mathfrak{L}$ such that L'' is equivalent to L'.

The construction of a Chebotarev link in Theorem 7.1.7 can be generalized for any orientable, connected, closed 3-manifold, owing to Soma's theorem [211] asserting that any orientable, connected, closed 3-manifold contains a hyperbolic, fibered link.

It is clear that a Chebotarev link in a 3-manifold may be regarded as a topological analogue of the set of all finite primes of a number field. Theorem 7.1.9 below, due to Ueki, gives the relation between Chebotarev and stably generic links. To show it, we firstly note the following:

Lemma 7.1.8 ([234]) *Let* $\mathfrak{L} = \bigsqcup_{n=1}^{\infty} K_n$ *be a weakly Chebotarev link in a 3-manifold* X. *Let* $h : Y \to X$ *be a finite covering. Then* $h^{-1}(\mathfrak{L})$ *is a weakly Chebotarev link in* Y.

Proof Let $Z \to Y$ be a finite Galois covering with Galois group G and let C be a conjugacy class of G. Let $\rho : \pi_1(Y) \twoheadrightarrow G$ be the natural homomorphism. We need to show that there is a component K_n' of $h^{-1}(\mathfrak{L})$ such that $C([K_n']) = C$, where $C([K_n'])$ denotes the conjugacy class of G containing $\rho([K_n'])$. Since ρ is surjective, we can take $[L'] \in \pi_1(N)$ such that $C([L']) = C$. We set $L := h(L')$. Here we may assume that $h|_{L'} : L' \to L$ has the covering degree one. Let Γ be the maximal normal subgroup of $\pi_1(X)$ contained in $\text{Ker}(\rho)$. Then Γ is of finite index in $\pi_1(X)$.

Let $q : \pi_1(X) \to \pi_1(X)/\Gamma, q' : \pi_1(Y) \to \pi_1(Y)/\Gamma$ be the quotient maps and we note that h_* induces the injective homomorphism $\pi_1(Y) \hookrightarrow \pi_1(X), \pi_1(Y)/\Gamma \hookrightarrow \pi_1(X)/\Gamma$. Since $\mathfrak{L} = \bigsqcup_{n=1}^{\infty} K_n$ is a weakly Chebotarev link in X, there is a component K_n of $\mathfrak{L}$ such that $q(C([K_n])) = q(C([L]))$, where $C(q(*))$ denotes the conjugacy class of $\pi_1(X)/\Gamma$ containing $q(*)$. Let $l' \in C([L'])$. Since $h_*(C([L']) \subset C([L])$, there is $\gamma \in \Gamma$ such that $k := h_*(l')\gamma \in C([K_n]) \cap h_*(\pi_1(Y))$. So there is a component $K'_m \in h^{-1}(K_n)$ such that $h|_{K'_m} : K'_m \to K_n$ has covering degree one. Therefore $q'(l') \in q'(C([L'])) \cap q'(C([K'_m]))$ Hence $q'(C[L']) = q'(C([K'_m]))$. Since ρ factors through q', we obtain $C = \rho([L']) = \rho([K'_m])$. □

Theorem 7.1.9 ([234]) *A Chebotarev link in an oriented, connected closed 3-manifold M is a stably generic link in M.*

Proof Let $\mathfrak{L} = \bigsqcup_{n=1}^{\infty} K_n$ be a Chebotarev link in M. For $m \in \mathbb{N}$, we set $L_m := \bigsqcup_{n=1}^{m} K_n$. Let $f : N \to M$ be a finite covering of oriented, connected, closed 3-manifolds ramified over a finite sublink L of $\mathfrak{L}$. We need to show that $f^{-1}(\mathfrak{L})$ is a generic link in N. For this, let L' be any finite sublink of $f^{-1}(\mathfrak{L})$. It suffices to prove that all homology classes of components of $f^{-1}(\mathfrak{L}) \setminus L'$ generate $H_1(N \setminus L')$.

Since $L \cup f(L')$ is a finite sublink of $\mathfrak{L}$, we can take large enough $m \in \mathbb{N}$ such that $L \cup f(L') \subset L_m$. We set $X := M \setminus L_m$. First, we show that $\mathfrak{L} \setminus L_m$ is a weakly Chebotarev link in X. For this, let $Z \to X$ be a finite Galois covering with Galois group $G := \mathrm{Gal}(Y/X)$, which corresponds to a finite Galois covering of M with Galois group G and ramification over L_m. Let C be any conjugacy class G. Since $\mathfrak{L}$ is a Chebotarev link in M, there is K_t with $t > m$ such that $C([K_t]) = C$. Therefore $\mathfrak{L} \setminus L_m$ is a weakly Chebotarev link in X.

Set $Y := N \setminus f^{-1}(L_m)$. Then $f|_Y : Y \to X$ is a finite covering. By Lemma 7.1.5, $f^{-1}(\mathfrak{L} \setminus L_m)$ is a weakly Chebotarev link in Y. Then the homology classes $c(K'_n)$'s of all components of $f^{-1}(\mathfrak{L} \setminus L_m)$ generate $H_1(Y)$. In fact, if not, $H_1(Y)/\langle c(K'_n); K'_n \subset f^{-1}(\mathfrak{L} \setminus L_m)\rangle$ is a non-trivial abelian group and so there is a surjective homomorphism $H_1(Y) \twoheadrightarrow H_1(Y)/\langle c(K'_n); K'_n \subset f^{-1}(\mathfrak{L} \setminus L_m)\rangle \twoheadrightarrow A$ for some non-trivial finite Abelian group A. It means that there is a finite Abelian covering of Y with Galois group A such that $C([K'_n]) = 0$ in A for all component K'_n of $f^{-1}(\mathfrak{L} \setminus L_m)$. This contradicts that $f^{-1}(\mathfrak{L} \setminus L_m)$ is a weakly Chebotarev link in Y.

Since $f^{-1}(\mathfrak{L} \setminus L_m) \subset f^{-1}(\mathfrak{L}) \setminus L'$ and $H_1(N \setminus L')$ is a quotient of $H_1(Y)$, the homology classes of $f^{-1}(\mathfrak{L}) \setminus L'$ generate $H_1(N \setminus L')$. □

Remark 7.1.10 Since M is supposed to be closed, it does not have an end, which may be seen as an analogue of an infinite prime of a number field k (cf. (3.7)). So a stably generic link or a Chebotarev link is an analogue for a 3-manifold of the set of finite primes $\mathrm{Max}(\mathcal{O}_k)$. For analogues of infinite primes from the viewpoint of 3-dimensional foliated dynamical systems, see Remark 7.2.7.

7.2 Idelic Class Field Theory for 3-Manifolds

Let M be an oriented, connected, closed 3-manifold and we fix, once and for all, a stably generic link $\mathfrak{L}$ in M. We present local and global class field theory for $(M, \mathfrak{L})$. In view of the analogies discussed in previous chapters, we use a homological approach, following [181]. For a cohomological approach, we refer to [145].

We start by constructing local class field theory for each component knot of $\mathfrak{L}$. Let K be a knot which is a component of $\mathfrak{L}$. Let V_K be a tubular neighborhood of K so that the boundary ∂V_K is a 2-dimensional torus. Let α_K and β_K be a meridian and longitude on the boundary ∂V_K of V_L as in Example 2.1.6. Then $H_1(\partial V_K)$ is a free Abelian group generated by the homology classes of a meridian α_K and a longitude β_K, $H_1(\partial V_K) = \mathbb{Z}[\alpha_K] \oplus \mathbb{Z}[\beta_K]$, so that any homology class $c \in H_1(\partial V_K)$ is written uniquely by $c = u_K(c)[\alpha_K] + v_K(c)[\beta_K]$. Here v_K is the homomorphism $v_K : H_1(\partial V_K) \to H_1(V_K) = \mathbb{Z}[\beta_K] \simeq \mathbb{Z}$ induced by the inclusion $\partial V_K \hookrightarrow V_K$. We set $U_K := \mathrm{Ker}(v_K) = \mathbb{Z}[\alpha_K]$, which coincides with $H_2(V_K, \partial V_K)$ by the relative homology exact sequence. Thus we have the exact sequence

$$0 \longrightarrow U_K \longrightarrow H_1(\partial V_K) \xrightarrow{v_K} \mathbb{Z} \to 0.$$

We regard $H_1(\partial V_K)$, $U_K = \mathbb{Z}[\alpha_K]$ and v_K as topological analogues for a knot K of the multiplicative group $k_\mathfrak{p}^\times$ of the $\mathfrak{p}$-adic field, the unit group $U_\mathfrak{p} := \mathcal{O}_\mathfrak{p}^\times$ and the $\mathfrak{p}$-adic additive valuation, respectively, for a finite prime $\mathfrak{p}$ of a number field k (cf. Sect. 7.3). We equip $H_1(\partial V_K)$ with the unique topology such that $H_1(\partial V_K)$ is a topological group and U_K is an open subgroup.

Let $\partial V_K^{\mathrm{ab}}$ be the maximal Abelian covering of ∂V_K (which is the universal covering of ∂V_K). Since $V_K \setminus K$ is homotopically equivalent to ∂V_K, coverings of ∂V_K correspond to ramified coverings of V_K ramified along K. Let $\partial V_K^{\mathrm{ur}}$ denote the maximal covering of ∂V_K, which comes from the maximal (unramified) covering of V_K by pulling back.

As in Example 2.1.13, we have the Hurewicz isomorphism for ∂V_K, which we call the *local reciprocity isomorphism*,

$$\rho_{\partial V_K} : H_1(\partial V_K) \xrightarrow{\sim} \mathrm{Gal}(\partial V_K^{\mathrm{ab}}/\partial V_K).$$

For a finite Abelian covering $T \to \partial V_K$, we obtain the *reciprocity homomorphism* for $T \to \partial V_K$ by composing $\rho_{\partial V_K}$ with the natural quotient map $\mathrm{Gal}(\partial V_K^{\mathrm{ab}}/\partial V_K) \to \mathrm{Gal}(T/\partial V_K)$

$$\rho_{T/\partial V_K} : H_1(\partial V_K) \longrightarrow \mathrm{Gal}(T/\partial V_K).$$

Then a topological analogue for ∂V_K of local class field theory is stated as follows. The proof follows simply from the covering theory over ∂V_K.

Theorem 7.2.1

(1) *We have the following commutative diagram:*

$$
\begin{array}{ccccccccc}
0 & \longrightarrow & U_K & \longrightarrow & H_1(\partial V_K) & \xrightarrow{v_K} & \mathbb{Z} & \longrightarrow & 0 \\
 & & \downarrow\wr & & \downarrow{\rho_{\partial V_K}} & & \downarrow\wr & & \\
0 & \longrightarrow & \mathrm{Gal}(\partial V_K^{\mathrm{ab}}/\partial V_K^{\mathrm{ur}}) & \longrightarrow & \mathrm{Gal}(\partial V_K^{\mathrm{ab}}/\partial V_K) & \longrightarrow & \mathrm{Gal}(\partial V_K^{\mathrm{ur}}/\partial V_K) & \longrightarrow & 0.
\end{array}
$$

(2) *For a finite Abelian covering $h : T \to \partial V_K$, the reciprocity homomorphism $\rho_{\partial V_K}$ induces the isomorphism*

$$
\rho_{T/\partial V_K} : H_1(\partial V_K)/h_*(H_1(T)) \xrightarrow{\sim} \mathrm{Gal}(T/\partial V_K).
$$

(3) *The map $T \to H_1(\partial V_K)/h_*(H_1(T))$ gives the bijection between the set of finite Abelian coverings of ∂V_K and the set of open subgroups of $H_1(\partial V_K)$ of finite index.*

Remark 7.2.2 By the Poincaré duality $H^1(\partial V_K) \simeq H_1(\partial V_K)$, the local reciprocity isomorphism yields the isomorphism

$$
H^1(\partial V_K) \xrightarrow{\sim} \mathrm{Gal}(\partial V_K^{\mathrm{ab}}/\partial V_K).
$$

This cohomological version of the local reciprocity isomorphism [145] may be seen as being closer to the isomorphism $H^1(k_{\mathfrak{p}}, \hat{\mathbb{Z}}(1)) \simeq \mathrm{Gal}(k_{\mathfrak{p}}^{\mathrm{ab}}/k_{\mathfrak{p}})$, where $\hat{\mathbb{Z}}(1) = \varprojlim_n \mu_n$, obtained by the Tate local duality 2.3.1.

Now we construct idelic class field theory for $(M, \mathfrak{L})$ by getting local theories together over all components of $\mathfrak{L}$. In the following, for a link $\mathcal{L}$ of countably many or finitely many components, we use the notations:

$S_{\mathcal{L}} :=$ the set of components of $\mathcal{L}$,
$F(\mathcal{L}) :=$ the set of finite sublinks of $\mathcal{L}$.

We define the *idèle group $J_{(M,\mathfrak{L})}$* by

$$
J_{(M,\mathfrak{L})} := \prod_{K \in S_{\mathfrak{L}}} H_1(\partial V_K),
$$

where the right-hand side means the restricted direct product of $H_1(\partial V_K)$ with respect to $U_K = \mathbb{Z}[\alpha_K]$, namely,

$$J_{(M,\mathfrak{L})} := \left\{ (a_K)_{K \in S_{\mathfrak{L}}} \in \prod_{K \in S_{\mathfrak{L}}} H_1(\partial V_K) \;\middle|\; a_K \in U_K \text{ except for } \right.$$
$$\left. \text{finitely many } K\text{'s in } S_{\mathfrak{L}} \right\}. \tag{7.1}$$

An element of $J_{(M,\mathfrak{L})}$ is called an *idèle* of $(M, \mathfrak{L})$. The idèle group $J_{(M,\mathfrak{L})}$ is equipped with the restricted product topology. Namely, a basis of neighborhood of the identity $0 \in J_{(M,\mathfrak{L})}$ is given by sets

$$\prod_{K \in F(L)} \mathcal{U}_K \times \prod_{K \notin F(L)} U_K,$$

where L runs over $F(\mathfrak{L})$ and $\mathcal{U}_K$ runs over a basis of neighborhood of $0 \in H_1(\partial V_K)$.

For $L \in F(\mathfrak{L})$, let X_L be the exterior of L, $X_L := M \setminus \mathrm{int}(V_L) \simeq M \setminus L$. When $L \subset L'$ ($L, L' \in F(\mathfrak{L})$), let $\varphi_{L,L'} : H_2(M, L) \to H_2(M, L')$ be the induced natural homomorphism. Noting that $\{H_2(M, L); \varphi_{L,L'}\}_{L \in F(\mathfrak{L})}$ forms a direct system of Abelian groups, we set

$$H_2(M, \mathfrak{L}) := \varinjlim_{L \in F(\mathfrak{L})} H_2(M, L).$$

By the excision and the relative homology sequence, we obtain the following composite map:

$$\partial_L : H_2(M, L) \simeq H_2(M, V_L) \simeq H_2(X_L, \partial V_L) \longrightarrow H_1(\partial V_L).$$

Noting $H_1(\partial V_L) = \prod_{K \in S_{\mathfrak{L}}} H_1(\partial V_K)$, let $p_{L',L} : H_1(\partial V_{L'}) \to H_1(\partial V_L)$ be the projection for $L \subset L'$ and so we have a projective system $\{H_1(\partial V_L); p_{L',L}\}_{L \in F(\mathfrak{L})}$ of Abelian groups. For $L \subset L'$, we notice the following commutative diagram, namely, $p_{L',L} \circ \partial_{L'} \circ \varphi_{L,L'} = \partial_L$:

$$
\begin{array}{ccc}
H_2(M, L) & \xrightarrow{\partial_L} & H_1(\partial V_L) \\
\varphi_{L,L'} \downarrow & & \uparrow p_{L',L} \\
H_2(M, L') & \xrightarrow{\partial_{L'}} & H_1(\partial V_{L'}).
\end{array}
$$

By this diagram and $\varprojlim_{L \in F(\mathfrak{L})} H_1(\partial V_L) \subset \prod_{K \in S_\mathfrak{L}} H_1(\partial V_K)$, we obtain the following (well-defined) homomorphism, denoted by $\Delta_{(M,\mathfrak{L})}$,

$$\Delta_{(M,\mathfrak{L})} : H_2(M, \mathfrak{L}) \longrightarrow \prod_{K \in S_\mathfrak{L}} H_1(\partial V_K),$$

where $\Delta_{(M,\mathfrak{L})}$ on $H_2(M, L)$ is ∂_L above. If $K \in S_{L' \backslash L}$, the K-component of the image of $\partial_{L'} \circ \varphi_{L,L'}$ is in $\mathbb{Z}[\alpha_K]$. Therefore we see $\mathrm{Im}(\Delta_{(M,\mathfrak{L})}) \subset J_{(M,\mathfrak{L})}$. Hence we obtain the homomorphism

$$\Delta_{(M,\mathfrak{L})} : H_2(M, \mathfrak{L}) \longrightarrow J_{(M,\mathfrak{L})}.$$

We define the *principal idèle group* $P_{(M,\mathfrak{L})}$ of $(M, \mathfrak{L})$ by $P_{(M,\mathfrak{L})} := \mathrm{Im}(\Delta_{(M,\mathfrak{L})})$ and call an element of $P_{(M,\mathfrak{L})}$ a *principal idèle* of $(M, \mathfrak{L})$. We then define the *idèle class group* $C_{(M,\mathfrak{L})}$ of $(M, \mathfrak{L})$ by

$$C_{(M,\mathfrak{L})} := J_{(M,\mathfrak{L})}/P_{(M,\mathfrak{L})}.$$

Let $H_1(X_L) \xrightarrow{\sim} \mathrm{Gal}(X_L^{\mathrm{ab}}/X_L)$ be the Hurewicz isomorphism for $L \in F(\mathfrak{L})$. When $L \subset L'$, the inclusion $X_{L'} \hookrightarrow X_L$ induces the homomorphism $\psi_{L',L} : \mathrm{Gal}(X_{L'}^{\mathrm{ab}}/X_{L'}) \simeq H_1(X_{L'}) \rightarrow H_1(X_L) \simeq \mathrm{Gal}(X_L^{\mathrm{ab}}/X_L)$. So $\{\mathrm{Gal}(X_L^{\mathrm{ab}}/X_L); \psi_{L',L}\}_{L \in F(\mathfrak{L})}$ forms a projective system of Abelian group and we set

$$\mathrm{Gal}(M, \mathfrak{L})^{\mathrm{ab}} := \varprojlim_{L \in F(\mathfrak{L})} \mathrm{Gal}(X_L^{\mathrm{ab}}/X_L),$$

which may be regarded as a topological analogue for $(M, \mathfrak{L})$ of the Galois group of the maximal Abelian extension of a number filed (cf. Sect. 7.3). For each $K \in S_\mathfrak{L}$ and $L \in F(\mathfrak{L})$, the inclusion $\partial V_K \hookrightarrow X_L$ induces the homomorphism $\rho_{K,L} : H_1(V_K) \rightarrow H_1(X_L) \simeq \mathrm{Gal}(X_L^{\mathrm{ab}}/X_L)$. Since $\rho_{K,L}(\alpha_K) = 0$ if $K \notin S_L$, by the definition (7.1) of $J_{(M,\mathfrak{L})}$, $\sum_{K \in S_\mathfrak{L}} \rho_{K,L}(a_K)$ is well defined for $(a_K)_{K \in S_\mathfrak{L}} \in J_{(M,\mathfrak{L})}$. Thus we have the homomorphism

$$\rho_L := \sum_{K \in S_\mathfrak{L}} \rho_{K,L} : J_{(M,\mathfrak{L})} \longrightarrow \mathrm{Gal}(X_L^{\mathrm{ab}}/X_L).$$

Since $(\rho_L(\mathbf{a}))_{L \in F(\mathfrak{L})}$ $(\mathbf{a} \in J_{(M,\mathfrak{L})})$ is an element of the projective limit of $\mathrm{Gal}(X_L^{\mathrm{ab}}/X_L)$'s, we obtain the homomorphism

$$\rho_{(M,\mathfrak{L})} : J_{(M,\mathfrak{L})} \longrightarrow \mathrm{Gal}(M, \mathfrak{L})^{\mathrm{ab}}.$$

Let $f : N \to M$ be a finite Abelian covering of oriented, connected, closed 3-manifolds, ramified over $L \in F(\mathfrak{L})$ and set $Y_L := N \setminus f^{-1}(L)$. Composing the natural homomorphism $H_1(X) \to \mathrm{Gal}(Y_{f^{-1}(L)}/X_L)$ with the projection $\varprojlim_{L \in S_{\mathfrak{L}}} H_1(X_L) \to H_1(X_L)$, we have the homomorphism $q_L : \mathrm{Gal}(M, \mathfrak{L})^{\mathrm{ab}} \to \mathrm{Gal}(Y_L/X_L) = \mathrm{Gal}(N/M)$. Thus we obtain the homomorphism, denoted by $\rho_{(N, f^{-1}(\mathfrak{L}))/(M, \mathfrak{L})}$, obtained as the composite of $\rho_{(M, \mathfrak{L})}$ and q_L:

$$\rho_{(N, f^{-1}(\mathfrak{L}))/(M, \mathfrak{L})} : J_{(M, \mathfrak{L})} \longrightarrow \mathrm{Gal}(N/M).$$

The main results of idelic class field theory for $(M, \mathfrak{L})$ are stated as follows.

Theorem 7.2.3 ([181]) *Let the notations be as above.*

(1) *We have $\rho_{(M, \mathfrak{L})}(P_{(M, \mathfrak{L})}) = \{\mathrm{id}\}$ and $\rho_{(M, \mathfrak{L})}$ induces the isomorphism, called the reciprocity isomorphism*

$$\rho_{(M, \mathfrak{L})} : C_{(M, \mathfrak{L})} \xrightarrow{\sim} \mathrm{Gal}(M, \mathfrak{L})^{\mathrm{ab}}.$$

(2) *For a finite Abelian covering $f : N \to M$ ramified over a finite sublink of $\mathfrak{L}$, $\rho_{N/M}$ induces the isomorphism*

$$\rho_{(N, f^{-1}(\mathfrak{L}))/(M, \mathfrak{L})} : C_{(M, \mathfrak{L})}/f_*(C_{(N, f^{-1}(\mathfrak{L}))}) \xrightarrow{\sim} \mathrm{Gal}(N/M).$$

(3) *The relation with local class field theory is given as follows: Let $\iota_K : H_1(\partial V_K) \to C_{(M, \mathfrak{L})}$ be the map defined by $\iota_K(a_K) = [(0, \ldots, 1, a_K, 0, \ldots)]$, and let $r_K : \mathrm{Gal}(\partial V_K^{\mathrm{ab}}/\partial V_K) \to \mathrm{Gal}(M, \mathfrak{L})^{\mathrm{ab}}$ be the map defined by $r_K(a_K) := (\rho_{K, L}(a_K))_{L \in S_{\mathfrak{L}}}$. Then one has the following commutative diagram:*

$$
\begin{array}{ccc}
H_1(\partial V_K) & \xrightarrow{\rho_{\partial V_K}} & \mathrm{Gal}(\partial V_K^{\mathrm{ab}}/\partial V_K) \\
\iota_K \downarrow & & \downarrow r_K \\
C_{(M, \mathfrak{L})} & \xrightarrow{\rho_{(M, \mathfrak{L})}} & \mathrm{Gal}(M, \mathfrak{L})^{\mathrm{ab}}.
\end{array}
$$

For $K \in S_{\mathfrak{L}}$,

> *K is completely decomposed in $N \to M$ $\Leftrightarrow$ $\rho_{(N, f^{-1}(\mathfrak{L}))/(M, \mathfrak{L})} \circ \iota_K$*
> *$(H_1(\partial V_K)) = \{\mathrm{id}\}$,*
> *K is unramified in $N \to M$ $\Leftrightarrow$ $\rho_{(N, f^{-1}(\mathfrak{L}))/(M, \mathfrak{L})} \circ \iota_K(U_K) = \{\mathrm{id}\}$.*

(4) *The map $C_{(N, f^{-1}(\mathfrak{L}))} \mapsto f_*(C_{(N, f^{-1}(\mathfrak{L}))})$ gives the bijection between the set of finite Abelian coverings of M ramified over a finite sublink of $\mathfrak{L}$ and the set of open subgroups of $C_{(M, \mathfrak{L})}$ of finite index.*

Proof

(1) The surjectivity of $\rho_{(M,\mathfrak{L})} : J_{(M,\mathfrak{L})} \to \mathrm{Gal}(M, \mathfrak{L})^{\mathrm{ab}} = \varprojlim_{L \in S_{\mathfrak{L}}} H_1(X_L)$ follows

from that $\mathfrak{L}$ is a generic link in M. So it suffices to prove $\mathrm{Ker}(\rho_{(M,\mathfrak{L})}) = P_{(M,\mathfrak{L})}$.

$P_{(M,\mathfrak{L})} \subset \mathrm{Ker}(\rho_{(M,\mathfrak{L})})$: Take any $c \in H_2(M, \mathfrak{L})$ represented by $c_L \in H_2(M, L)$ ($L \in S_{\mathfrak{L}}$). By the excision and the relative homology exact sequence for the pair $(X_L, \partial V_L)$, we have the exact sequence

$$H_2(M, L) \simeq H_2(X_L, \partial V_L) \xrightarrow{\partial_L} H_1(\partial V_L) \longrightarrow H_1(X_L).$$

Therefore the image of c_L in $\mathrm{Gal}(X_L^{\mathrm{ab}}/X_L) = H_1(X_L)$ is id for any c_L representing c. Hence the assertion follows.

$\mathrm{Ker}(\rho_{(M,\mathfrak{L})}) \subset P_{(M,\mathfrak{L})}$: Suppose $\mathbf{a} = (a_K)_{K \in S_{\mathfrak{L}}} \in \mathrm{Ker}(\rho_{(M,\mathfrak{L})})$. Since $\mathfrak{L}$ is a generic link and $\mathbf{a} \in J_{(M,\mathfrak{L})}$, there is a finite sublink $L \in F(\mathfrak{L})$ such that $a_K \in \mathbb{Z}[\alpha_K]$ for $K \in S_{\mathfrak{L} \setminus L}$ and $\langle [K] \mid K \in S_L \rangle = H_1(M)$. Let

$$p_L : J_{(M,\mathfrak{L})} \to H_1(\partial V_L) = \prod_{K \in S_L} H_1(\partial V_K) \quad \text{and} \quad \psi_L : \mathrm{Gal}(M, \mathfrak{L})^{\mathrm{ab}} \to H_1(X_L)$$

be the projections on the L-component and let $\varphi_L : H_2(M, L) \to H_2(M, \mathfrak{L})$ be the natural map. Consider the following commutative diagram:

$$
\begin{array}{ccccc}
H_2(M, \mathfrak{L}) & \xrightarrow{\partial_{(M,\mathfrak{L})}} & J_{(M,\mathfrak{L})} & \xrightarrow{\rho_{(M,\mathfrak{L})}} & \mathrm{Gal}(M, \mathfrak{L})^{\mathrm{ab}} \\
\varphi_L \uparrow & & \downarrow p_L & & \downarrow \varphi_L \\
H_2(M, L) & \xrightarrow{\partial_L} & H_1(\partial V_L) & \xrightarrow{\rho_{L,L}} & H_1(X_L),
\end{array}
$$

where the 2nd row is exact by the excision and the relative homology sequence, and $\rho_{L,L} := \sum_{K \in S_L} \rho_{K,L}$ so that $\rho_{L,L} \circ p_L = \rho_L$. Since $\rho_L \circ p_L(\mathbf{a}) = \varphi_L \circ \rho_{(M,\mathfrak{L})}(\mathbf{a}) = 0$, $(a_K)_{K \in S_L} = p_L(\mathbf{a}) \in \mathrm{Ker}(\rho_L)$ and so there is $A \in H_2(M, L)$ such that $\partial_L(A) = (a_K)_{K \in S_L}$. We set $B := \iota_L(A)$ and $\mathbf{b} = (b_K)_{K \in S_{\mathfrak{L}}} := \partial_{(M,\mathfrak{L})}(B)$. We claim $\mathbf{a} = \mathbf{b}$, from which the assertion follows. To show this, it suffices to show that $(a_K)_{K \in S_{L'}} = (b_K)_{K \in S_{L'}}$ for any $L' \in S_{\mathfrak{L}}$ with $L \subset L'$. Set $a' := (a_K)_{K \in S_{L'}}, b' := (b_K)_{K \in S_{L'}}$. Note that a' and b' are both in $H_1(\partial V_L) \times \prod_{K \in S_{L' \setminus L}} \mathbb{Z}[\alpha_K]$ and that $p_{L',L}(a') = (a_K)_{K \in S_L} = (b_K)_{K \in S_L} = p_{L',L}(b')$.

$$
\begin{array}{ccccc}
H_2(M, L') & \xrightarrow{\partial_{L'}} & H_1(\partial V_{L'}) & \xrightarrow{\rho_{L',L'}} & H_1(X_{L'}) \\
\iota_{L,L'} \uparrow & & \downarrow p_{L',L} & & \downarrow \varphi_{L',L} \\
H_2(M, L) & \xrightarrow{\partial_L} & H_1(\partial V_L) & \xrightarrow{\rho_{L,L}} & H_1(X_L).
\end{array}
$$

We set $d := b' - a'$. Then $d \in \prod_{K \in S_{L' \setminus L}} \mathbb{Z}[\alpha_K]$. Since $C_2(L) = C_2(L') = 0$,

$$Z_2(M, L) = \{c \in C_2(M) \mid \partial c \in C_1(L)\}$$
$$\subset Z_2(M, L') = \{c \in C_2(M) \mid \partial c \in C_1(L')\}.$$

Since $\rho_{L', L'}(d) = 0$, there is $D \in Z_2(M, L')$ such that $\partial_{L', L'}([D]) = d$. Then $\partial_{L'}([D]) \in \prod_{K \in S_{L' \setminus L}} \mathbb{Z}[\alpha_K]$ and so $D \in Z_2(M)$.

By the Poincaré duality $H_2(M) \simeq H^1(M)$ and the universal coefficient theorem,

$$H_2(M) \xrightarrow{\sim} \mathrm{Hom}(H_1(M), \mathbb{Z}); \quad c_2 \mapsto (c_1 \mapsto I(c_2, c_1)),$$

where $I : H_2(M) \times H_1(M) \to \mathbb{Z}$ is the intersection form. Since $\partial_{L'}([D]) \in \prod_{K \in S_{L' \setminus L}} \mathbb{Z}[\alpha_K]$, $\partial_L([D]) = 0$ and $I([D], [K]) = 0$ for any $K \in S_L$. Since $H_1(M)$ is generated by $[K]$ for $K \in S_L$, $[D] = 0$ and so $d = \partial_{L'}([D]) = 0$. Hence $a = b$.

(2) By the definitions above, we have

$$C_{(M, \mathfrak{L})} / f_*(C_{N, f^{-1}(\mathfrak{L})}) \simeq J_{(M, \mathfrak{L})} / (P_{(M, \mathfrak{L})} + f_*(J_{(N, f^{-1}(\mathfrak{L}))})).$$

Consider the composite map

$$\rho_{Y_L / X_L} : J_{(M, \mathfrak{L})} \xrightarrow{\rho_L} H_1(X_L) \longrightarrow H_1(X_L) / f_*(H_1(Y_L)),$$

which is surjective, since $\mathfrak{L}$ is a generic link in M. Since $\mathfrak{L}$ is a stably generic link, $\rho_{f^{-1}(L)} : J_{(N, f^{-1}(\mathfrak{L}))} \to H_1(Y_L)$ is surjective and hence $f_*(J_{(N, f^{-1}(\mathfrak{L}))}) \to f_*(H_1(Y_L))$ is surjective. Consider the following commutative diagram:

$$
\begin{array}{ccc}
f_*(J_{(N, f^{-1}(\mathfrak{L}))}) & \longrightarrow & f_*(H_1(Y_L)) \\
\downarrow & & \downarrow \\
J_{(M, \mathfrak{L})} & \xrightarrow{\rho_L} & H_1(X_L),
\end{array}
$$

from which $\rho_L^{-1}(f_*(H_1(Y_L))) = \mathrm{Ker}(\rho_L) + f_*(J_{(N, \mathfrak{L})})$. By Proposition 7.2.4 below, $\mathrm{Ker}(\rho_L) = P_{(M, \mathfrak{L})} + U^L$. Since $U^L \subset f_*(J_{(N, f^{-1}(\mathfrak{L}))})$, we have $\mathrm{Ker}(\rho_{Y_L / X_L}) = P_{(M, \mathfrak{L})} + f_*(J_{(N, f^{-1}(\mathfrak{L}))})$. Therefore,

$$J_{(M \mathfrak{L})} / (P_{(M, \mathfrak{L})} + f_*(J_{(N, f^{-1}(\mathfrak{L}))})) \simeq H_1(X_L) / H_1(Y_L) \simeq \mathrm{Gal}(N/M).$$

(3) By the definitions of the maps involved in the diagram, it is easy to see that the commutative diagram holds. The latter assertions follow from the

relation between the Hurewicz map $\sigma_{N/M}$ in Sect. 6.1 and the reciprocity homomorphism $\rho_{(N,f^{-1}(\mathfrak{L}))/(M,\mathfrak{L})}$, which is given in Theorem 7.2.5 below.

(4) For $L \in S_{\mathfrak{L}}$, let Cov_L be the set of finite Abelian covering of M ramified over a sublink of L, and let Sub^f_L be the set of subgroups of $C_{(M,\mathfrak{L})}$ of finite index which contains $\mathrm{Ker}(\rho_L : C_{(M,\mathfrak{L})} \twoheadrightarrow H_1(X_L)) = (U^L + P_{(M,\mathfrak{L})})/P_{(M,\mathfrak{L})}$. Note that Cov_L coincides with the set of finite Abelian coverings of X_L and that Sub^f_L corresponds bijectively to the set of subgroups of $H_1(X_L)$ of finite index. So, by Galois correspondence in covering theory, the correspondence $(f : N \to M) \mapsto f_*(C_{(N,f^{-1}(\mathfrak{L}))})$ gives the bijection

$$\mathrm{Cov}_L \xrightarrow{\sim} \mathrm{Sub}^f_L \qquad\qquad (\star)$$

such that $C_{(M,\mathfrak{L})}/f_*(C_{(N,f^{-1}(\mathfrak{L}))}) \simeq \mathrm{Gal}(N/M) \simeq H_1(X_L)/f_*(H_1(Y_{f^{-1}(L)}))$. Next we show that $\cup_{L \in F(\mathfrak{L})}\mathrm{Sub}^f_L$ coincides with the set Op^f of open subgroups of $C_{(M,\mathfrak{L})}$. Suppose $U \in \mathrm{Sub}^f_L$ for some $L \in S_{\mathfrak{L}}$. Then U contains $(U^L + P_{(M,\mathfrak{L})})/P_{(M,\mathfrak{L})}$ and is of finite index in $C_{(M,\mathfrak{L})}$. By the definition of the restricted topology of $C_{(M,\mathfrak{L})}$, U is open and hence $U \in \mathrm{Op}^f_L$. Conversely, suppose $U \in \mathrm{Op}^f_L$. Then the inverse image $q^{-1}(U)$ under the quotient map $q : J_{(M,\mathfrak{L})} \to C_{(M,\mathfrak{L})}$ is an open subgroup of $J_{(M,\mathfrak{L})}$ and so $q^{-1}(U)$ contains $\prod_{K \in S_L} n_K \mathbb{Z}[\alpha_K] \times \prod_{K \in S_{\mathfrak{L} \setminus L}} \mathbb{Z}[\alpha_K]$ for some $L \in S_{\mathfrak{L}}$ and $n_K \in \mathbb{N}$. Therefore U contains $(U^L + P_{(M,\mathfrak{L})})/P_{(M,\mathfrak{P})} = \mathrm{Ker}(\rho_L : C_{(M,\mathfrak{L})} \twoheadrightarrow H_1(X_L))$ and hence $U \in \mathrm{Sub}^f_L$. By $\star$, we obtain the bijection

$$\cup_{L \in S_{\mathfrak{L}}}\mathrm{Cov}_L \xrightarrow{\sim} \mathrm{Op}^f; \ (f : N \longrightarrow M) \longmapsto f_*(C_{(N,f^{-1}(\mathfrak{L}))}),$$

which is the desired assertion.

$\square$

We shall give the relation between idelic class field theory and classical formulation by Hurewicz theorem in Sect. 6.1. First, we give the relation between the idèle class group C_M and the homology group $H_1(M)$. For this, we set

$$U_{(M,\mathfrak{L})} := \prod_{K \in S_{\mathfrak{L}}} U_K,$$ called the *unit idèle group* of $(M, \mathfrak{L})$,

and, for $L \in F(\mathfrak{L})$,

$$U_L := \prod_{K \in S_L} U_K, \ U^L := \prod_{K \in S_{\mathfrak{L} \setminus S_L}} U_K \text{ so that } U_{(M,\mathfrak{L})} = U_L \times U^L.$$

Proposition 7.2.4 *Notations being as above, the homomorphism* $\rho_L : J_{(M,\mathfrak{L})} \to H_1(X_L)$ *induces the isomorphism*

$$J_{(M,\mathfrak{L})}/(P_{(M,\mathfrak{L})} + U^L) \simeq H_1(X_L).$$

In particular, we have the isomorphism

$$J_{(M,\mathfrak{L})}/(P_{(M,\mathfrak{L})} + U_{(M,\mathfrak{L})}) \simeq H_1(M).$$

Proof Since $\mathfrak{L}$ is a generic link in M, ρ_L is surjective. So it suffices to consider the kernel of ρ_L. Note that ρ_L coincides with the following composite map:

$$J_{(M,\mathfrak{L})} \longrightarrow J_{(M,\mathfrak{L})}/P_{(M,\mathfrak{L})} \xrightarrow{\rho_{(M,\mathfrak{L})}} \mathrm{Gal}(M, \mathfrak{L})^{\mathrm{ab}} \simeq \varprojlim_{L} H_1(X_L) \xrightarrow{\mathrm{pr}_L} H_1(X_L),$$

where pr_L is the projection. When $L \subset L'$ ($L' \in F(\mathfrak{L})$), $\mathrm{Ker}(\psi_{L',L} : H_1(X_{L'}) \to H_1(X_L))$ is generated by the meridian classes $[\alpha_K]$ for $K \in S_{L'\backslash L}$ by the Mayer–Vietoris exact sequence. Therefore

$$\mathrm{Ker}(J_{(M,\mathfrak{L})}/P_{(M,\mathfrak{L})} \longrightarrow H_1(X_L)) = (U^L + P_{(M,\mathfrak{L})})/P_{(M,\mathfrak{L})},$$

from which the assertion follows.   □

Now let M be an oriented, connected, closed 3-manifold with stably generic link $\mathfrak{L}$. Let $f : N \to M$ be a finite Abelian covering of oriented, connected, closed 3-manifolds ramified over a finite sublink $L \in F(\mathfrak{L})$. Set

$X := M \setminus L$, $Y := N \setminus f^{-1}(L)$ and $d :=$ the degree of the covering $f|Y$.

As in Sect. 6.1, where the case $M = S^3$ is treated, we have the Hurewicz map $\sigma_{M/N}$: $Z^1(X) \to \mathrm{Gal}(N/M)$ which factors through $H_1(X)$ and induces the isomorphism $\sigma_{N/M} : H_1(X)/f_*(H_1(Y)) \xrightarrow{\sim} \mathrm{Gal}(N/M)$.

By Proposition 7.2.4, we obtain the isomorphism

$$C^L_{(M,\mathfrak{L})} := J_{(M,\mathfrak{L})}/(U^L + P_{(M,\mathfrak{L})}) \simeq H_1(X).$$

Composing it with the natural homomorphism $C_{(M,\mathfrak{L})} \to C^L_{(M,\mathfrak{L})}$ gives the homomorphism

$$C_{(M,\mathfrak{L})} \longrightarrow H_1(X),$$

which is nothing but ρ_L. By the constructions, we have the following:

Theorem 7.2.5 *Notations being as above, the reciprocity homomorphism $\rho_{N/M/k}$: $C_{(M,\mathfrak{L})} \to \mathrm{Gal}(N/M)$ and the Hurewicz map $\sigma_{N/M} : H(X) \to \mathrm{Gal}(Y/X)$ are related by the following commutative diagram with exact rows:*

$$
\begin{array}{ccccccccc}
0 & \longrightarrow & f_*(C_{(N,f^{-1}(\mathfrak{L}))}) & \longrightarrow & C_{(M,\mathfrak{L})} & \xrightarrow{\rho_{N/M}} & \mathrm{Gal}(N/M) & \longrightarrow & 0 \\
 & & \downarrow & & \rho_L\downarrow & & \| & & \\
0 & \longrightarrow & f_*(H_1(Y)) & \longrightarrow & H_1(X) & \xrightarrow{\sigma_{N/M}} & \mathrm{Gal}(Y/X) & \longrightarrow & 0,
\end{array}
$$

where the left vertical arrow is the restriction of ρ_L.

   For a knot K in X, let f be the order of $[K] \in H_1(X)$ in $H_1(X)/f_*(H_1(Y))$. Then K is decomposed in F into $r := d/f$ different knots $f^{-1}(K) = \{\mathfrak{K}_1, \ldots, \mathfrak{K}_r\}$.

Finally, we give a topological analogue for knots of Chebotarev density theorem for primes (cf. Theorem 7.3.3 below) in the context of class field theory for 3-manifolds as follows.

Theorem 7.2.6 *Let* $\mathfrak{L} = \bigsqcup_{n \in \mathbb{N}} K_n$ *be a Chebotarev link in an oriented, connected, closed 3-manifold* M. *Let* $f : N \to M$ *be a finite Abelian covering of oriented, connected, closed 3-manifolds of degree* d, *ramified over a finite link* $L \in F(\mathfrak{L})$ *with Galois group* G. *Set* $X := M \setminus L, Y := N \setminus f^{-1}(L)$. *As in Sect. 6.1, by the Hurewicz theorem, we have the homomorphism* $\sigma_{N/M} : Z_1(X) \to G$. *Then, for any* $g \in G$,

$$\lim_{x \to \infty} \frac{1}{x} \#\{K_n \in S_{\mathfrak{L}} \mid \sigma_{N/M}(K_n) = g, n \le x\} = \frac{1}{d}.$$

Remark 7.2.7 As we mentioned in Remark 3.0.1 (3), a 3-dimensional foliated dynamical system $(M, \mathcal{F}, \phi)$ is regarded as a refined analogue of a number ring. To be precise, a triple $(M, \mathcal{F}, \phi)$ is called a 3-dimensional *foliated dynamical system* if M is an oriented, connected, closed smooth 3-manifold, $\mathcal{F}$ is a (complex) foliation by Riemann surfaces on M, and ϕ is a smooth dynamical system on M such that these data must satisfy the following conditions:

(i) there are a finite number of compact leaves $L_1^\infty, \ldots, L_r^\infty$, which may be empty, such that, for any i and t, $\phi^t(L_i^\infty) = L_i^\infty$ and that, any orbit of the flow ϕ is transverse to any leaf in $M \setminus \bigcup_{i=1}^r L_i^\infty$;
(ii) for each $t \in \mathbb{R}$, the diffeomorphism ϕ^t of M maps any leaf to a leaf.

Here closed orbits of ϕ and non-transverse compact leaves $L_1^\infty, \ldots, L_r^\infty$ may be seen as analogue of finite primes and infinite primes, respectively, of a number field k (cf. [43, 118]).

In fact, Example 7.1.6 gives rise to examples of such $(M, \mathcal{F}, \phi)$ as follows. We keep the same notations as in Example 7.1.6 and note that all continuous maps (homeomorphisms) there can be replaced by smooth maps (diffeomorphisms). We let $M = S^3$. We find the foliation on $S^3 \setminus \text{int}(V_K)$, which consists of ∂V_K and the leaves in $S^3 \setminus V_K$ obtained by turbulizing the fibers of π around ∂V_K. We fill in V_K with Reeb components to obtain the foliation $\mathcal{F}$ on all of S^3. We define the smooth flow on $S^3 \setminus \text{int}(V_K)$ by the suspension of the monodromy φ of π and the flow on $\text{int}(V_L)$ by the one transverse to any leaf of the Reeb foliation. Thus we have the foliated dynamical system $(S^3, \mathcal{F}, \phi)$. Here, moreover, the non-transverse compact leaf ∂V_K corresponds to the end of $S^3 \setminus K$. So it also fits with the analogy (3.7) in arithmetic topology. This construction of a 3-dimensional foliated dynamical system can be generalized to any orientable, connected, closed 3-manifold M, owing to Soma's theorem [211] asserting that any orientable, connected, closed 3-manifold contains a hyperbolic, fibered link.

It was shown in [105] that a 3-dimensional foliated dynamical system enjoys a property similar to Hilbert reciprocity law for a number field. So it is natural to expect that one can develop idèlic theory for 3-dimensional foliated dynamical systems and its connection with their dynamical L-functions, as in the case of number fields.

7.3 Idelic Class Field Theory for Number Fields

Let k be a number field of finite degree over $\mathbb{Q}$. We shall use the same notations as in Sect. 2.2 for the various notions concerning the number field k. Let $\mathcal{O}_k$ be the ring of integers in k and let $\mathrm{Max}(\mathcal{O}_k)$ be the set of maximal ideals of $\mathcal{O}_k$. Let S_k^∞ be the set of infinite primes of k and let $S_k := \mathrm{Max}(\mathcal{O}_k) \cup S_k^\infty$ be the set of all primes of k.

We start to recall local class field theory at each prime of k (cf. 2.3.2). Let $\mathfrak{p} \in \mathrm{Max}(\mathcal{O}_k)$ be a *finite prime* of k and let $k_\mathfrak{p}$ be the $\mathfrak{p}$-adic field. Let $\mathcal{O}_\mathfrak{p}$ be the ring of $\mathfrak{p}$-adic integers, ϖ a prime element of $\mathcal{O}_\mathfrak{p}$ and $\mathbb{F}_\mathfrak{p} = \mathcal{O}_\mathfrak{p}/(\pi)$ the finite residue field. Let $v_\mathfrak{p}$ be the $\mathfrak{p}$-adic additive valuation with $v_\mathfrak{p}(\varpi) = 1$. Let $U_\mathfrak{p}$ denote the unit group $\mathcal{O}_\mathfrak{p}^\times$. Then we have the exact sequence

$$0 \longrightarrow U_\mathfrak{p} \longrightarrow k_\mathfrak{p}^\times \xrightarrow{v_\mathfrak{p}} \mathbb{Z} \longrightarrow 0.$$

Let $k_\mathfrak{p}^{\mathrm{ab}}$ be the maximal Abelian extension of $k_\mathfrak{p}$ and let $k_\mathfrak{p}^{\mathrm{ur}}$ be the maximal unramified extension of $k_\mathfrak{p}$. Note that the Galois group $\mathrm{Gal}(k_\mathfrak{p}^{\mathrm{ur}}/k_\mathfrak{p})$ is identified with the absolute Galois group $\mathrm{Gal}(\overline{\mathbb{F}}_\mathfrak{p}/\mathbb{F}_\mathfrak{p})$, which is topologically generated by the Frobenius automorphism $\sigma_\mathfrak{p}$.

By Tate local duality, we obtain the reciprocity homomorphism for $k_\mathfrak{p}$

$$\rho_{k_\mathfrak{p}} : k_\mathfrak{p}^\times \longrightarrow \mathrm{Gal}(k_\mathfrak{p}^{\mathrm{ab}}/k_\mathfrak{p}).$$

For a finite Abelian extension $F_\mathfrak{P}/k_\mathfrak{p}$, we find the *reciprocity homomorphism* for $F_\mathfrak{P}/k_\mathfrak{p}$ by composing $\rho_{k_\mathfrak{p}}$ with the natural quotient map $\mathrm{Gal}(k_\mathfrak{p}^{\mathrm{ab}}/k_\mathfrak{p}) \to \mathrm{Gal}(F_\mathfrak{P}/k_\mathfrak{p})$

$$\rho_{F_\mathfrak{P}/k_\mathfrak{p}} : k_\mathfrak{p}^\times \longrightarrow \mathrm{Gal}(F_\mathfrak{P}/k_\mathfrak{p}).$$

The main results of local class field theory for the $\mathfrak{p}$-adic field are stated as follows.

Theorem 7.3.1 *Let the notations be as above.*

(1) *The reciprocity homomorphism $\rho_{k_\mathfrak{p}}$ is a continuous injective homomorphism with the dense image.*

(2) *We have the following commutative diagram:*

$$
\begin{array}{ccccccccc}
0 & \longrightarrow & U_\mathfrak{p} & \longrightarrow & k_\mathfrak{p}^\times & \xrightarrow{v_\mathfrak{p}} & \mathbb{Z} & \longrightarrow & 0 \\
 & & \wr\downarrow & & \downarrow\rho_{k_\mathfrak{p}} & & \cap\downarrow & & \\
0 & \longrightarrow & \mathrm{Gal}(k_\mathfrak{p}^{\mathrm{ab}}/k_\mathfrak{p}^{\mathrm{ur}}) & \longrightarrow & \mathrm{Gal}(k_\mathfrak{p}^{\mathrm{ab}}/k_\mathfrak{p}) & \longrightarrow & \mathrm{Gal}(k_\mathfrak{p}^{\mathrm{ur}}/k_\mathfrak{p}) = \hat{\mathbb{Z}} & \longrightarrow & 0,
\end{array}
$$

where the left vertical isomorphism is the restriction of $\rho_{k_{\mathfrak{p}}}$ to $U_{\mathfrak{p}}$ and the right vertical injection is the map sending 1 to the Frobenius automorphism $\sigma_{\mathfrak{p}}$ and hence $\rho_{k_{\mathfrak{p}}}(\varpi) = \sigma_{\mathfrak{p}}$.

(3) *For a finite Abelian extension $F_{\mathfrak{P}}/k_{\mathfrak{p}}$, the reciprocity homomorphism for $F_{\mathfrak{P}}/k_{\mathfrak{p}}$ induces the isomorphism*

$$\rho_{F_{\mathfrak{P}}/k_{\mathfrak{p}}} : k_{\mathfrak{p}}^{\times}/N_{F_{\mathfrak{P}}/k_{\mathfrak{p}}}(F_{\mathfrak{P}}^{\times}) \xrightarrow{\sim} \mathrm{Gal}(F_{\mathfrak{P}}/k_{\mathfrak{p}})$$

and $F_{\mathfrak{P}}/k_{\mathfrak{p}}$ is unramified if and only if $\rho_{F_{\mathfrak{P}}/k_{\mathfrak{p}}}(U_{\mathfrak{p}}) = \mathrm{id}_{F_{\mathfrak{P}}}$.

(4) *The map $F_{\mathfrak{P}} \mapsto N_{F_{\mathfrak{P}}/k_{\mathfrak{p}}}(F_{\mathfrak{P}}^{\times})$ gives the bijection between the set of finite Abelian extensions of $k_{\mathfrak{p}}$ and the set of open subgroups of $k_{\mathfrak{p}}^{\times}$ of finite index.*

Let us see the case of the local field k_v for an infinite prime v of k. When v is a complex prime, $k_v = \mathbb{C}$ is algebraically closed and so the reciprocity homomorphism $\rho_{\mathbb{C}}$ is trivial. When v is a real prime, $k_v = \mathbb{R}$ and the reciprocity homomorphism $\rho_{\mathbb{R}} : \mathbb{R}^{\times} \to \mathrm{Gal}(\mathbb{C}/\mathbb{R})$ induces the isomorphism $\rho_{\mathbb{C}/\mathbb{R}} : \mathbb{R}^{\times}/N_{\mathbb{C}/\mathbb{R}}(\mathbb{C}^{\times}) \xrightarrow{\sim} \mathrm{Gal}(\mathbb{C}/\mathbb{R})$, where $N_{\mathbb{C}/\mathbb{R}}(\mathbb{C}^{\times}) = \mathbb{R}_+$, the connected component of 1 consisting positive real numbers. We set $\mathcal{O}_v := k_v$ and $U_v := k_v^{\times}$ for $v \in S_k^{\infty}$.

Now we construct idelic class field theory for k by getting local theories together over all primes of k (cf. 2.3.3). We define the *idèle group* J_k by

$$J_k := \prod_{\mathfrak{p} \in \mathrm{Max}(\mathcal{O}_k)} k_{\mathfrak{p}}^{\times} \times \prod_{v \in S_k^{\infty}} k_v^{\times}$$

where $\prod_{\mathfrak{p} \in \mathrm{Max}(\mathcal{O}_k)}$ is the restricted direct product of $k_{\mathfrak{p}}^{\times}$'s with respect to $U_{\mathfrak{p}}$'s, namely,

$$J_k = \left\{ (a_{\mathfrak{p}})_{\mathfrak{p} \in S_k} \in \prod_{\mathfrak{p} \in S_k} k_{\mathfrak{p}}^{\times} \,\middle|\, a_{\mathfrak{p}} \in U_{\mathfrak{p}} \text{ except for finitely many } \mathfrak{p}\text{'s in } S_k \right\}.$$

An element of J_k is called an *idèle* of k. The idèle group J_k is a topological group equipped with the restricted topology. Namely, a basis of neighborhood of the identity $1 \in J_k$ is given by sets

$$\prod_{\mathfrak{p} \in S} \mathcal{U}_{\mathfrak{p}} \times \prod_{\mathfrak{p} \notin S} U_{\mathfrak{p}},$$

where S runs over finite sets of primes containing S_k^{∞} and $\mathcal{U}_{\mathfrak{p}}$ runs over a basis of neighborhood of $1 \in k_{\mathfrak{p}}^{\times}$.

Since for $a \in k^{\times}$, $v_{\mathfrak{p}}(a_{\mathfrak{p}}) = 0$ except for finitely many $\mathfrak{p}$'s, a is regarded as an idèle embedded diagonally, called an *principal idèle* of k. We define the *principal idèle group*, denoted by P_k, by the closed subgroup consisting of principal idèles, namely, the image of the diagonal embedding $\Delta_k : k^{\times} \hookrightarrow J_k$. We then define the *idèle class group* C_k of k by

$$C_k := J_k/P_k.$$

Let k^{ab} be the maximal Abelian extension of k. We note that for any prime $\mathfrak{p}$ of k (possibly an infinite prime), there is the natural homomorphism $\iota_{\mathfrak{p}} : \mathrm{Gal}(k_{\mathfrak{p}}^{\mathrm{ab}}/k_{\mathfrak{p}}) \to \mathrm{Gal}(k^{\mathrm{ab}}/k)$. We define the continuous homomorphism

$$\rho_k : J_k \longrightarrow \mathrm{Gal}(k^{\mathrm{ab}}/k)$$

by

$$\rho_k((a_{\mathfrak{p}})_{\mathfrak{p}\in S_k}) := \prod_{\mathfrak{p}\in S_k} \iota_{\mathfrak{p}}(\rho_{k_{\mathfrak{p}}}(a_{\mathfrak{p}})),$$

where the r.h.s is a finite product, since $\rho_{k_{\mathfrak{p}}}(a_{\mathfrak{p}}) = \mathrm{id}_{k_{\mathfrak{p}}^{\mathrm{ab}}}$ except for finitely many $\mathfrak{p}$'s. For a finite Abelian extension F/k, we have the homomorphism $\rho_{F/k}$ by composing ρ_k with the natural quotient map $\mathrm{Gal}(k^{\mathrm{ab}}/k) \to \mathrm{Gal}(F/k)$

$$\rho_{F/k} : J_k \longrightarrow \mathrm{Gal}(F/k).$$

The main results of idelic class field theory for the number field k are stated as follows.

Theorem 7.3.2 *Let the notations be as above.*

(1) $\rho_k(P_k) = \{\mathrm{id}_{k^{\mathrm{ab}}}\}$ *and hence* ρ_k *induces the continuous homomorphism, called the reciprocity homomorphism,*

$$\rho_k : C_k \longrightarrow \mathrm{Gal}(k^{\mathrm{ab}}/k),$$

which is surjective and whose kernel coincides with the connected component of 1 *of* C_k.

(2) *For a finite Abelian extension* F/k, $\rho_{K/k}$ *induces the isomorphism*

$$\rho_{F/k} : C_k/N_{F/k}(C_F) \xrightarrow{\sim} \mathrm{Gal}(F/k).$$

(3) *The relation with local class field theory is given as follows: Let* $\iota_{\mathfrak{p}} : k_{\mathfrak{p}}^{\times} \to C_k$ *be the map defined by* $\iota_{\mathfrak{p}}(a_{\mathfrak{p}}) = [(1, \ldots, 1, a_{\mathfrak{p}}, 1, \ldots)]$ *for* $\mathfrak{p} \in S_k$, *and let* $r_{\mathfrak{p}} : \mathrm{Gal}(k_{\mathfrak{p}}^{\mathrm{ab}}/k_{\mathfrak{p}}) \to \mathrm{Gal}(k^{\mathrm{ab}}/k)$ *be the restriction map. Then one has the following commutative diagram:*

$$\begin{array}{ccc} k_{\mathfrak{p}}^{\times} & \xrightarrow{\rho_{k_{\mathfrak{p}}}} & \mathrm{Gal}(k_{\mathfrak{p}}^{\mathrm{ab}}/k_{\mathfrak{p}}) \\ \iota_{\mathfrak{p}}\downarrow & & \downarrow \\ C_k & \xrightarrow{\rho_k} & \mathrm{Gal}(k^{\mathrm{ab}}/k). \end{array}$$

Further, we have, for $\mathfrak{p} \in \mathrm{Max}(\mathcal{O}_k)$,

$$\mathfrak{p} \text{ is completely decomposed in } F/k \iff \rho_{F/k} \circ \iota_{\mathfrak{p}}(k_{\mathfrak{p}}^{\times}) = \mathrm{id}_F,$$
$$\mathfrak{p} \text{ is unramified in } F/k \iff \rho_{F/k} \circ \iota_{\mathfrak{p}}(U_{\mathfrak{p}}) = \mathrm{id}_F.$$

(4) *The map* $C_F \mapsto N_{F/k}(C_F)$ *gives the bijection between the set of finite Abelian extensions of* k *and the set of open subgroups of* C_k *of finite index.*

Theorems 7.3.1 and 7.3.2 are the main results in local and global class field theory. The details of the proofs can be found in standard books on class field theory. We refer to [10], [28], [166], [174], [205] etc.

We give the relation between idelic class field theory and the (classical) ideal theoretic Artin reciprocity in Sect. 6.1. Let F/k be a finite Abelian extension of finite algebraic number fields ramified over a finite set S of finite primes of k. let $d := [F : k]$. Let $f : \mathrm{Spec}(\mathcal{O}_F) \to \mathrm{Spec}(\mathcal{O}_k)$ be the associated ramified covering of the prime spectra of the rings of integers. We set

$X := \mathrm{Spec}(\mathcal{O}_k) \setminus S, X_0 := \mathrm{Max}(\mathcal{O}_k) \setminus S$,
$S_F := f^{-1}(S)$,
$Y := \mathrm{Spec}(\mathcal{O}_F) \setminus S_K, Y_0 := \mathrm{Max}(\mathcal{O}_F) \setminus S_F$.

Further, we set

$I(X) := \bigoplus_{\mathfrak{p} \in X_0} \mathbb{Z}, \; P(X) := \{(a) \in P^+(k) \mid a \equiv 1 \bmod \mathfrak{q} \; (\forall \mathfrak{q} \in S)\}$,
$H(X) := I(X)/P(X)$,
$I(Y) := \bigoplus_{\mathfrak{P} \in Y_0} \mathbb{Z}, \; P(Y) := \{(\alpha) \in P^+(F) \mid \alpha \equiv 1 \bmod \mathfrak{Q} \; (\forall \mathfrak{Q} \in S_F)\}$,
$H(Y) := I(Y)/P(Y)$.

As in Sect. 6.1, where the case $k = \mathbb{Q}$ is treated, we have the Artin reciprocity map $\sigma_{K/k} : I(X) \to \mathrm{Gal}(K/k)$ which factors through $H(X)$ and induces the isomorphism $\sigma_{F/k} : H(X)/N_{F/k}(H(Y)) \xrightarrow{\sim} \mathrm{Gal}(F/k)$.

We let

$$J_{k,S} := \{(a_{\mathfrak{p}}) \in J_k \mid a_{\mathfrak{p}} \equiv 1 \bmod \mathfrak{p} \; (\forall \mathfrak{p} \in S), \; a_{\mathfrak{p}} > 0 \; (\forall \text{ real prime } \mathfrak{p})\}$$
$$J_k^S := \{(a_{\mathfrak{p}}) \in J_k \mid a_{\mathfrak{p}} = 1 \; (\forall \mathfrak{p} \in S \cup S_k^{\infty}) \; \mathfrak{p})\}$$

and set

$$C_k^S := J_k/(J_{k,S} \cdot P_k).$$

We define the homomorphism $\kappa_X : J_k^S \to I(X)$ by

$$\kappa_X((a_{\mathfrak{p}})_{\mathfrak{p} \in S_k}) := \prod_{\mathfrak{p} \in X_0} \mathfrak{p}^{v_{\mathfrak{p}}(a_{\mathfrak{p}})},$$

where the r.h.s is a finite product. By the approximation theorem,

$$C_k^S = J_{k,S} J_k^S P_l / J_{k,S} P_k \simeq J_k^S / (J_{k,S} P_k \cap J_k^S)$$

and κ_X induces

$$C_k^S \xrightarrow{\sim} H(X).$$

Composing it with the natural homomorphism $C_k \to C_k^S$, we have the homomorphism, denoted by the same κ_X,

$$\kappa_X : C_k \longrightarrow H(X).$$

By the above constructions, we obtain the following:

Theorem 7.3.3 *Notations being as above, the reciprocity homomorphism $\rho_{K/k} : C_k \to \mathrm{Gal}(F/k)$ and the Artin reciprocity homomorphism $\sigma_{F/k} : H(X) \to \mathrm{Gal}(K/k)$ are related by the following commutative diagram with exact rows:*

$$
\begin{array}{ccccccccc}
0 & \longrightarrow & \mathrm{N}_{F/k}(C_K) & \longrightarrow & C_k & \xrightarrow{\rho_{F/k}} & \mathrm{Gal}(F/k) & \longrightarrow & 0 \\
& & \downarrow & & {\scriptstyle \kappa_X}\downarrow & & \| & & \\
0 & \longrightarrow & \mathrm{N}_{F/k}(H(Y)) & \longrightarrow & H(X) & \xrightarrow{\sigma_{F/k}} & \mathrm{Gal}(F/k) & \longrightarrow & 0,
\end{array}
$$

where the left vertical arrow is the restriction of κ_X.

For a maximal ideal $\mathfrak{p} \in I(X)$, let f be the order of $[\mathfrak{p}] \in H(X)$ in $H(X)/\mathrm{N}_{F/k}(H(Y))$. Then $\mathfrak{p}$ is decomposed in F into a product $(p) = \mathfrak{P}_1 \cdots \mathfrak{P}_r$ of $r := d/f$ different maximal ideals.

Finally, we mention the *Chebotarev density theorem* for finite algebraic number fields in the context of class field theory.

Theorem 7.3.4 *Let F/k be a finite Abelian extension of finite algebraic number fields ramified over a finite set S of finite primes of k. let $d := [F : k]$. Notations being as above, let $\sigma_{K/k} : I(X) \to \mathrm{Gal}(F/k)$ be the Artin reciprocity map. Then, for any $g \in G$, we have*

$$\lim_{x \to \infty} \frac{\#\{\mathfrak{p} \in \mathrm{Max}(\mathcal{O}_k) \mid \sigma_{F/k}(\mathfrak{p}) = g, \mathrm{N}\mathfrak{p} \le x\}}{\#\{\mathfrak{p} \in \mathrm{Max}(\mathcal{O}_k) \mid \mathrm{N}\mathfrak{p} \le x\}} = \frac{1}{d}.$$

Theorem 7.3.4 is a generalization of Dirichlet's arithmetic progression theorem ($k = \mathbb{Q}$, F = cyclotomic field). It can also be formulated for any finite Galois extension of finite algebraic number fields (cf. [28, Chapter VIII], [127, Chapter XV], [175, Chapter VII, §13]). Its topological analogue for knots was discussed as Chebotarev link in Sects. 7.1 and 7.2.

Summary

Stably generic link, Chebotarev link $\mathfrak{L}$	The set of primes of a number field
Local class field theory	Local class field theory
for a torus ∂V_K	for a local field $k_\mathfrak{p}$
The local reciprocity isomorphism	The local reciprocity homomorphism
$\rho_{\partial V_K} : H_1(\partial V_K) \xrightarrow{\sim} \mathrm{Gal}(\partial V_K^{\mathrm{ab}}/\partial V_K)$	$\rho_{k_\mathfrak{p}} : k_\mathfrak{p}^\times \longrightarrow \mathrm{Gal}(k_\mathfrak{p}^{\mathrm{ab}}/k_\mathfrak{p})$
Idelic class field theory	Idelic class field theory
for $(M, \mathfrak{L})$	for a number field k
Global reciprocity isomorphism	Global reciprocity homomorphism
$\rho_{(M,\mathfrak{L})} : C_{(M,\mathfrak{L})} \xrightarrow{\sim} \mathrm{Gal}(M, \mathfrak{L})^{\mathrm{ab}}$	$\rho_k : C_k \longrightarrow \mathrm{Gal}(k^{\mathrm{ab}}/k)$
Chebotarev density theorem	Cheborarev density theorem
for knots in a 3-manifold	for primes in a number field

Chapter 8
Link Groups and Galois Groups with Restricted Ramification

As explained in Chap. 3, our basic idea is to regard a Galois group with restricted ramification $G_S = \pi_1(\mathrm{Spec}(\mathbb{Z}) \setminus S)$, $S = \{p_1, \cdots, p_r\}$, as an analogue of a link group $G_L = \pi_1(S^3 \setminus L)$, $L = K_1 \cup \cdots \cup K_r$ (cf. (3.11)). Since the profinite group G_S is too big in general, we consider a maximal pro-l quotient $G_S(l)$ for some prime number l to derive the information on how S is "linked". As for pro-l extensions of number fields, there are classical and extensive works due to I. Šafarevič and H. Koch, among others, and a theorem by Koch on the structure of $G_S(l)$ turns out to be an analogue of J. Milnor's theorem on the structure of the link group G_L.

8.1 Link Groups

Let $L = K_1 \cup \cdots \cup K_r$ be an r-component link in S^3 and let $G_L = \pi_1(S^3 \setminus L)$ be the link group of L. Let F be the free group on the words $x_1, \ldots, x_r$, where x_i corresponds to a meridian of K_i.

Theorem 8.1.1 (Milnor [148]) *For each $d \in \mathbb{N}$, there is $y_i^{(d)} \in F$ such that*

$$G_L/G_L^{(d)} = \langle x_1, \cdots, x_r \mid [x_1, y_1^{(d)}] = \cdots = [x_r, y_r^{(d)}] = 1, F^{(d)} = 1 \rangle,$$
$$y_i^{(d)} \equiv y_i^{(d+1)} \bmod F^{(d)},$$

where $y_i^{(d)}$ is a word representing a longitude β_i of K_i in $G_L/G_L^{(d)}$. We also have

$$\beta_j \equiv \prod_{i \neq j} \alpha_i^{\mathrm{lk}(K_i, K_j)} \bmod G_L^{(2)}.$$

Proof As in Example 2.1.6, we choose a regular projection of L into a hyperplane and divide K_i into arcs $\alpha_{i1}, \ldots, \alpha_{i\lambda_i}$. Fix a base point b above the hyperplane and

M. Morishita, *Knots and Primes*, Universitext,
https://doi.org/10.1007/978-981-99-9255-3_8

Fig. 8.1 The relations among x_{ij}'s

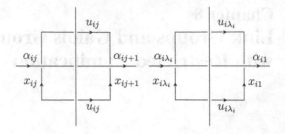

let x_{ij} be a loop coming down from b, passing below α_{ij} from right to left, and returning b. Then we have the following presentation for G_L:

$$G_L = \left\langle x_{ij}\,(1 \leq i \leq r, 1 \leq j \leq \lambda_i) \,\middle|\, \begin{array}{c} R_{ij} := x_{ij}u_{ij}x_{i\,j+1}^{-1}u_{ij}^{-1} = 1 \\ (1 \leq i \leq r, 1 \leq j < \lambda_i) \\ R_{i\,\lambda_i} := x_{i\,\lambda_i}u_{i\,\lambda_i}x_{i1}^{-1}u_{i\,\lambda_i}^{-1} = 1 \,(1 \leq i \leq r) \end{array} \right\rangle,$$

(8.1)

where u_{ij} is a word of x_{kl}'s (Fig. 8.1).

Let $v_{ij} := u_{i1} \cdots u_{ij}$. Then $v_{i\,\lambda_i}$ represents a parallel of K_i [148] and a longitude of K_i is represented by $v_{i\,\lambda_i}x_{i1}^{k_i} = u_{i1} \cdots u_{i\,\lambda_i}x_{i1}^{k_i}$ where the integer k_i is defined so that the sum of powers of x_{ij} in the word $v_{i\,\lambda_i}x_{i1}^{k_i}$ is 0. Set

$$\begin{cases} s_{ij} := x_{i1}v_{ij}x_{i\,j+1}^{-1}v_{ij}^{-1} \,(1 \leq j < \lambda_i, 1 \leq i \leq r), \\ s_{i\,\lambda_i} := x_{i1}v_{i\,\lambda_i}x_{i1}^{-1}v_{i\,\lambda_i}^{-1} \,(1 \leq i \leq r). \end{cases}$$

Then one has

$$\begin{cases} R_{i1} = s_{i1} \,(1 \leq i \leq r), \\ R_{ij} = v_{i\,j-1}^{-1}s_{i\,j-1}^{-1}s_{ij}v_{i\,j-1} \,(1 < j \leq \lambda_i, 1 \leq i \leq r). \end{cases}$$

(8.2)

By (8.1) and (8.2), we have

$$G_L = \langle x_{ij}(1 \leq i \leq r, 1 \leq j \leq \lambda_i) \,|\, s_{ij} = 1(1 \leq i \leq r, 1 \leq j \leq \lambda_i) \rangle.$$

(8.3)

Let $\overline{F}$ be the free group on the words x_{ij} $(1 \leq i \leq r, 1 \leq j \leq \lambda_i)$ and F the free group on the words x_{i1} $(1 \leq i \leq r)$. We regard F as a subgroup of $\overline{F}$ in the obvious way. For each $d \in \mathbb{N}$, we define the homomorphism $\eta_d : \overline{F} \to F$ inductively by

$$\begin{cases} \eta_1(x_{ij}) := x_{i1}, \\ \eta_{d+1}(x_{i1}) := x_{i1}, \\ \eta_{d+1}(x_{i\,j+1}) := \eta_d(v_{ij}^{-1}x_{i1}v_{ij}) \,(1 \leq j < \lambda_i). \end{cases}$$

Let $N := \langle\langle R_{ij}(1 \leq i \leq r, 1 \leq j \leq \lambda_i)(1 \leq i \leq r, 1 \leq j \leq \lambda_i)\rangle\rangle = \langle\langle s_{ij}(1 \leq i \leq r, 1 \leq j \leq \lambda_i)\rangle\rangle$. Then one has:

(1_d) $\eta_d(x_{ij}) \equiv x_{ij} \bmod \overline{F}^{(d)}N$ $(1 \leq i \leq r, 1 \leq j \leq \lambda_i)$,

(2_d) $\eta_d(x_{ij}) \equiv \eta_{d+1}(x_{ij}) \bmod F^{(d)}$ $(1 \leq i \leq r, 1 \leq j \leq \lambda_i)$.

Proof of (1_d) Since $\overline{F}^{(1)} = \overline{F}$, (1_d) is obviously true for $d = 1$. Assume that (1_d) holds for $d \geq 1$. We then have $\eta_d(v_{ij}) \equiv v_{ij} \bmod \overline{F}^{(d)}N$. Since $\eta_d(x_{i1}) = x_{i1}$, one has $\eta_d(v_{ij}^{-1}x_{i1}v_{ij}) \equiv v_{ij}^{-1}x_{i1}v_{ij} \bmod \overline{F}^{(d+1)}N$. Here we used the following (*):

(*) $x \in F, a \equiv b \bmod \overline{F}^{(d)}N \Rightarrow a^{-1}xa \equiv b^{-1}xb \bmod \overline{F}^{(d+1)}N$.

Therefore one has $\eta_{d+1}(x_{i\,j+1}) = \eta_d(v_{ij}^{-1}x_{i1}v_{ij}) \equiv v_{ij}^{-1}x_{i1}v_{ij} \bmod \overline{F}^{(d+1)}N$. Since $s_{ij} = x_{i1}v_{ij}x_{i\,j+1}^{-1}v_{ij}^{-1} \in N$, $v_{ij}^{-1}x_{i1}v_{ij} \equiv x_{i\,j+1} \bmod N$. Hence $\eta_{d+1}(x_{i\,j+1}) \equiv x_{i\,j+1} \bmod \overline{F}^{(d+1)}N$, and $\eta_{d+1}(x_{i1}) = x_{i1}$ by definition. So (1_{d+1}) holds.

Proof of (2_d) Since $F^{(1)} = F$, (2_d) is obvious for the cases of $d = 1$. Assume that (2_d) holds for $d \geq 1$. $\eta_d(v_{ij}) \equiv \eta_{d+1}(v_{ij}) \bmod F^{(d)}$. Since $\eta_d(x_{i1}) = \eta_{d+1}(x_{i1}) = x_{i1}$, we have $\eta_{d+1}(x_{i\,j+1}) = \eta_d(v_{ij}^{-1}x_{i1}v_{ij}) = \eta_{d+1}(v_{ij}^{-1}x_{i1}v_{ij}) = \eta_{d+2}(x_{i\,j+1}) \bmod F^{(d+1)}$. (Here we used the assertion obtained by replacing $\overline{F}^{(d)}N$ by $F^{(d)}$ in (*), and $\eta_d(x_{i1}) = \eta_{d+1}(x_{i1}) = x_{i1}$.) We also have $\eta_{d+1}(x_{i1}) = \eta_{d+2}(x_{i1}) = x_{i1}$ by definition. So (2_{d+1}) holds.

Therefore, by (8.3),

$$G_L/G_L^{(d)} = \langle x_{ij} \mid s_{ij} = 1, \overline{F}^{(d)} = 1\rangle$$
$$= \langle x_{ij} \mid s_{ij} = 1, x_{ij} = \eta_d(x_{ij}), \overline{F}^{(d)} = 1\rangle \text{ (by } (1_d))$$
$$\simeq \langle x_{i1} \mid \eta_d(s_{ij}) = 1, F^{(d)} = 1\rangle. \; (\eta_d(\overline{F}^{(d)}) = F^{(d)}).$$

Here we have, for $1 \leq j < \lambda_i$,

$$\eta_d(s_{ij}) = \eta_d(x_{i1})\eta_d(v_{ij})\eta_d(x_{ij+1})^{-1}\eta_d(v_{ij})^{-1}$$
$$\equiv x_{i1}\eta_d(v_{ij})\eta_{d+1}(x_{ij+1})^{-1}\eta_d(v_{ij})^{-1} \bmod F^{(d)} \text{ (by } (2_d))$$
$$= x_{i1}\eta_d(v_{ij})\eta_d(v_{ij}^{-1}x_{i1}v_{ij})^{-1}\eta_d(v_{ij})^{-1}$$
$$= 1,$$

and hence

$$G_L/G_L^{(d)} = \langle x_{i1} (1 \leq i \leq r) \mid \eta_d(s_{i\,\lambda_i}) = 1, F^{(d)} = 1\rangle.$$

Letting $x_i := x_{i1}$, $y_i^{(d)} := \eta_d(v_{i\,\lambda_i}x_{i1}^{k_i})$, we have $\eta_d(s_{i\,\lambda_i}) = \eta_d(x_{i1}v_{i\,\lambda_i}x_{i1}^{-1}v_{i\,\lambda_i}^{-1}) = \eta_d(x_{i1}v_{i\,\lambda_i}x_{i1}^{k_i}x_{i1}^{-1}x_{i1}^{-k_i}v_{i\,\lambda_i}^{-1}) = [x_i, y_i^{(d)}]$. Thus we obtain

$$G_L/G_L^{(d)} = \langle x_1, \cdots, x_r \mid [x_1, y_1^{(d)}] = \cdots = [x_r, y_r^{(d)}] = 1, F^{(d)} = 1\rangle,$$

and $y_i^{(d)} \equiv y_i^{(d+1)} \bmod F^{(d)}$ by (2_d). It follows from Proposition 4.1.1 that $\beta_j \equiv \prod_{i \neq j} \alpha_i^{\mathrm{lk}(K_i, K_j)} \bmod G_L^{(2)}$.

Remark 8.1.2

(1) Two r-component links $L = K_1 \cup \cdots \cup K_r$ and $L' = K_1' \cup \cdots \cup K_r'$ are called *isotopic* if there is a continuous family $h_t : rS^1 \to S^3$ such that $h_t : rS^1 \xrightarrow{\approx} h_t(rS^1)$ is a homeomorphism for all $t \in [0, 1]$, $L = h_0(rS^1)$ and $L' = h_1(rS^1)$, where rS^1 means the disjoint union of r copies of S^1. If L, L' are equivalent, then L, L' are isotopic. It is shown ([148]) that if L and L' are isotopic, then $G_L/G_L^{(d)}$ and $G_{L'}/G_{L'}^{(d)}$ are isomorphic so that for given pairs of a meridian and a longitude (α_i, β_i), (α_i', β_i') of K_i, K_i' respectively, (α_i, β_i) is sent to a simultaneous conjugate $(\gamma \alpha_i' \gamma^{-1}, \gamma \beta_i' \gamma^{-1})$ under this isomorphism.

(2) Theorem 8.1.1 can be extended for a link in any homology 3-sphere [226].

(3) When L is a *pure braid link* (a link obtained by closing a pure braid), it is known that $y_i^{(d)}$ is independent of d and G_L itself has the following presentation (E. Artin's theorem. [16, Theorem 2.2]):

$$G_L = \langle x_1, \cdots, x_r \mid [x_1, y_1] = \cdots = [x_r, y_r] = 1 \rangle.$$

Let l be a prime number. For a group G, $G^{(d,l)}$ denotes the d-th term of the l-lower central series of G defined by $G^{(1,l)} := G$, $G^{(d+1,l)} := (G^{(d,l)})^l [G^{(d,l)}, G]$. Now we note that for any normal subgroup N of G_L whose index $[G_L : N]$ is a power of l, there is a (sufficiently large) d such that $G_L^{(d,l)} \subset N$. Hence the pro-l completion $\hat{G}_L(l)$ of the link group G_L is given as the projective limit $\hat{G}_L(l) := \varprojlim_d G_L/G_L^{(d,l)}$. Similarly, we have $\hat{F}(l) = \varprojlim_d F/F^{(d,l)}$. Since $y_i^{(d)} \equiv y_i^{(d+1)} \bmod F^{(d,l)}$, $y_i := (y_i^{(d)} \bmod F^{(d,l)})$ defines an element of $\hat{F}(l)$ whose image under the natural map $\hat{F}(l) \to \hat{G}_L(l)$ represents a longitude of K_i. The following theorem asserts that any link looks like a pure braid link after the pro-l completion.

Theorem 8.1.3 ([84]) *The pro-l group $\hat{G}_L(l)$ has the following presentation:*

$$\hat{G}_L(l) = \langle x_1, \cdots, x_r \mid [x_1, y_1] = \cdots = [x_r, y_r] = 1 \rangle.$$

Proof Let $N_d := \langle\langle [x_i, y_i^{(d)}] \, (1 \leq i \leq r) \rangle\rangle$. Then we obtain, by Theorem 8.1.1,

$$G_L/G_L^{(d,l)} \simeq F/N_d F^{(d,l)}.$$

Taking the projective limit, we get our assertion.   □

8.2 Pro-*l* Galois Groups with Restricted Ramification

Let l be a fixed prime number and let $S = \{p_1, \cdots, p_r\}$ be a set of r distinct prime numbers such that $p_i \equiv 1 \bmod l$ ($1 \le i \le r$). We let $e_S := \max\{e \mid p_i \equiv 1 \bmod l^e$ ($1 \le i \le r$) $\}$ and fix $m = l^e$ ($1 \le e \le e_S$). Choose an algebraic closure of $\mathbb{Q}$ (a base point $\overline{x}$) and let $G_S(l)$ be the maximal pro-*l* quotient of $\pi_1(\mathrm{Spec}(\mathbb{Z}) \setminus S, \overline{x})$, namely, the Galois group $\mathrm{Gal}(\mathbb{Q}_{\overline{S}}(l)/\mathbb{Q})$ of the maximal pro-*l* extension $\mathbb{Q}_{\overline{S}}(l)(\subset \overline{\mathbb{Q}})$ of $\mathbb{Q}$, unramified outside $\overline{S} := S \cup \{\infty\}$ (Example 2.2.20). We fix an algebraic closure $\overline{\mathbb{Q}}_{p_i}$ of each $\mathbb{Q}_{p_i}$ and an embedding $\overline{\mathbb{Q}} \hookrightarrow \overline{\mathbb{Q}}_{p_i}$. Let $\mathbb{Q}_{p_i}(l)$ be the maximal pro-*l* extension of $\mathbb{Q}_{p_i}$ ($\subset \overline{\mathbb{Q}}_{p_i}$) (Example 0.2.24). Then

$$\mathbb{Q}_{p_i}(l) = \mathbb{Q}_{p_i}(\zeta_{l^n}, \sqrt[l^n]{p_i} \mid n \ge 1),$$

where $\zeta_{l^n} \in \overline{\mathbb{Q}}$ is a primitive l^n-th root of unity such that $\zeta_{l^t}^{l^s} = \zeta_{l^{t-s}}$ ($t \ge s$). Note that $\zeta_m \in \mathbb{Q}_{p_i}$ ($1 \le i \le r$) by our choice of m. Set $G_{\mathbb{Q}_{p_i}}(l) := \mathrm{Gal}(\mathbb{Q}_{p_i}(l)/\mathbb{Q}_{p_i})$. The local pro-*l* group $G_{\mathbb{Q}_{p_i}}(l)$ is then generated by the monodromy τ_i and the extension of the Frobenius automorphism σ_i defined by

$$\begin{aligned} \tau_i(\zeta_{l^n}) &= \zeta_{l^n}, \quad \tau_i(\sqrt[l^n]{p_i}) = \zeta_{l^n}\sqrt[l^n]{p_i}, \\ \sigma_i(\zeta_{l^n}) &= \zeta_{l^n}^{p_i}, \quad \sigma_i(\sqrt[l^n]{p_i}) = \sqrt[l^n]{p_i} \end{aligned} \tag{8.4}$$

and τ_i, σ_i are subject to the relations $\tau_i^{p_i-1}[\tau_i, \sigma_i] = 1$ (Example 2.2.23). Note that a choice of an extension of the Frobenius automorphism is determined modulo the inertia group and (8.4) gives a normalization such a choice. The embedding $\overline{\mathbb{Q}} \hookrightarrow \overline{\mathbb{Q}}_{p_i}$ induces the embedding $\mathbb{Q}_{\overline{S}}(l) \hookrightarrow \mathbb{Q}_{p_i}(l)$ (hence a prime of $\mathbb{Q}_{\overline{S}}(l)$ over p_i). From this we have the homomorphism $\eta_i : \mathrm{Gal}(\mathbb{Q}_{p_i}(l)/\mathbb{Q}_{p_i}) \to G_S(l)$. We denote by the same τ_i, σ_i the image of τ_i, σ_i under η_i. Let $\hat{F}(l)$ denote the free pro-*l* group on the words $x_1, \ldots, x_r$, where x_i represents τ_i.

Theorem 8.2.1 (Koch [112, 6]) *The pro-l group $G_S(l)$ has the following presentation:*

$$G_S(l) = \langle x_1, \cdots, x_r \mid x_1^{p_1-1}[x_1, y_1] = \cdots = x_r^{p_r-1}[x_r, y_r] = 1 \rangle,$$

where $y_i \in \hat{F}(l)$ is the pro-l word representing σ_i. Define $\mathrm{lk}(p_i, p_j) \in \mathbb{Z}_l$ *and* $\mathrm{lk}_m(p_i, p_j) \in \mathbb{Z}/m\mathbb{Z}$ *by*

$$\sigma_j \equiv \prod_{i \ne j} \tau_i^{\mathrm{lk}(p_i, p_j)} \bmod G_S(l)^{(2)}, \quad \mathrm{lk}_m(p_i, p_j) := \mathrm{lk}(p_i, p_j) \bmod m.$$

Then one has

$$\zeta_m^{\mathrm{lk}_m(p_i, p_j)} = \left(\frac{p_j}{p_i}\right)_m.$$

Here $\left(\frac{*}{p_i}\right)_m$ stands for the m-th power residue symbol in $\mathbb{Q}_{p_i}$ (2.6).

Proof Since an analogue of the Wirtinger presentation for the Galois group $G_S(l)$ is not known, we shall use the Tate–Poitou theorem in Sect. 2.3.3 instead in order to obtain an arithmetic analogue of Theorem 8.1.1 or 8.1.3. Let τ_i' denote the image of τ_i under the Abelianization $G_S(l) \to G_S(l)/G_S(l)^{(2)}$. Then one has, by Example 2.3.6,

$$G_S(l)/G_S(l)^{(2)} = \langle \tau_1' \rangle \times \cdots \times \langle \tau_r' \rangle \simeq \mathbb{Z}/l^{f_1}\mathbb{Z} \times \cdots \times \mathbb{Z}/l^{f_r}\mathbb{Z},$$

where f_i is defined by $p_i - 1 = l^{f_i} q_i$, $(l, q_i) = 1$. Therefore, by Proposition 2.2.5, the pro-l group $G_S(l)$ is generated topologically by τ_i $(1 \leq i \leq r)$. In the following, we denote simply by $H^i(\mathfrak{G})$ the cohomology group $H^i(\mathfrak{G}, \mathbb{F}_l)$ for a pro-finite group $\mathfrak{G}$. Recall that the minimal number of relations among the generators $\tau_1, \ldots, \tau_r$ is given by $\dim_{\mathbb{F}_l} H^2(G_S(l))$ (Proposition 2.2.5). These relations all come from the relations $\tau_i^{p_i-1}[\tau_i, \sigma_i] = 1$ of the local Galois groups $G_{\mathbb{Q}_{p_i}}(l)$ if the homomorphism on the Galois cohomology groups

$$\varphi : H^2(G_S(l)) \longrightarrow \prod_{i=1}^r H^2(G_{\mathbb{Q}_{p_i}}(l))$$

induced by $\eta_i : G_{\mathbb{Q}_{p_i}}(l) \to G_S(l)$ is injective ([112, Proposition 1.14]). We shall show this next. Let $G_{\mathbb{Q}} := \mathrm{Gal}(\overline{\mathbb{Q}}/\mathbb{Q})$ and let I_S be the kernel of the natural homomorphism $G_{\mathbb{Q}} \to G_S(l)$:

$$1 \longrightarrow I_S \longrightarrow G_{\mathbb{Q}} \longrightarrow G_S(l) \longrightarrow 1 \quad \text{(exact).}$$

We then have the following Hochschild–Serre exact sequence:

$$H^1(G_{\mathbb{Q}}) \xrightarrow{\mathrm{res}} H^1(I_S)^{G_S(l)} \xrightarrow{\mathrm{tra}} H^2(G_S(l)) \xrightarrow{\mathrm{inf}} H^2(G_{\mathbb{Q}}). \tag{8.5}$$

For each prime number p, let $G_{\mathbb{Q}_p} := \mathrm{Gal}(\overline{\mathbb{Q}}_p/\mathbb{Q}_p)$ and $I_{\mathbb{Q}_p}$ be the inertia group. Let $G_{\mathbb{Q}_p}(l)$ be the maximal pro-l quotient of $G_{\mathbb{Q}_p}$. Then we note that the natural homomorphism $G_{\mathbb{Q}_p} \to G_{\mathbb{Q}_p}(l)$ induces the isomorphism $H^i(G_{\mathbb{Q}_p}(l)) \xrightarrow{\sim} H^i(G_{\mathbb{Q}_p})$ $(i \geq 1)$, since $H^i(T) = 0$ $(i \geq 1)$ for $T := \mathrm{Ker}(G_{\mathbb{Q}_p} \to G_{\mathbb{Q}_p}(l))$. Then we consider the following diagram:

$$
\begin{array}{ccc}
H^2(G_S(l)) & \longrightarrow & H^2(G_\mathbb{Q}) \\
\varphi\downarrow & & \downarrow j \\
\displaystyle\bigoplus_{i=1}^{r} H^2(G_{\mathbb{Q}_{p_i}}(l)) & \xrightarrow{\iota} & \displaystyle\bigoplus_{p} H^2(G_{\mathbb{Q}_p}),
\end{array}
\tag{8.6}
$$

where ι is the natural injection and θ is the localization map. Since the composite $H^2(G_S(l)) \to H^2(G_\mathbb{Q}) \to H^2(G_{\mathbb{Q}_p})$ is 0-map if $p \notin S$, one sees that the above diagram is commutative. By the Tate–Poitou exact sequence 2.3.3, one has

$$
\mathrm{Ker}(j) \simeq \mathrm{Ker}(H^1(\mathbb{Q}, \mu_l) \longrightarrow \prod_p H^1(\mathbb{Q}_p, \mu_l))^*
$$

$$
\simeq \mathrm{Ker}(\mathbb{Q}^\times/(\mathbb{Q}^\times)^l \longrightarrow \prod_p \mathbb{Q}_p^\times/(\mathbb{Q}_p^\times)^l)^*
$$

$$
= 0,
$$

where μ_l denotes the group of *l*-th roots of unity. Hence j is injective. Therefore, by (8.5), (8.6), we have the exact sequence

$$
H^1(G_\mathbb{Q}) \longrightarrow H^1(I_S)^{G_S(l)} \longrightarrow \mathrm{Ker}(\varphi) \longrightarrow 0.
$$

Consider the following commutative exact diagram:

$$
\begin{array}{ccccc}
& & 0 & & 0 \\
& & \downarrow & & \downarrow \\
& & H^1(G_\mathbb{Q}) & \to & H^1(I_S)^{G_S(l)} \\
& & \downarrow & & \downarrow \\
0 \to \displaystyle\prod_{v\in\overline{S}} H^1(G_{\mathbb{Q}_v}) \times \prod_{p\notin\overline{S}} H^1_{\mathrm{ur}}(G_{\mathbb{Q}_p}) & \to & \displaystyle\prod_{v\in S_\mathbb{Q}} H^1(G_{\mathbb{Q}_v}) & \to & \displaystyle\bigoplus_{p\notin\overline{S}} H^1(I_{\mathbb{Q}_p})^{G_{\mathbb{Q}_p}} \to 0 \\
& & \downarrow & & \downarrow \\
H^1(\mathbb{Q}, \mu_l)^* & = & H^1(\mathbb{Q}, \mu_l)^* & & \\
& & \downarrow & & \\
& & 0, & &
\end{array}
$$

where we set $S_\mathbb{Q} := \mathrm{Max}(\mathbb{Z}) \cup \{\infty\}$, $\overline{S} := S \cup \{\infty\}$ and $G_{\mathbb{Q}_\infty} = \mathrm{Gal}(\mathbb{C}/\mathbb{R})$. The exact sequence in the middle row follows from the short exact sequence $1 \to I_{\mathbb{Q}_p} \to G_{\mathbb{Q}_p} \to \hat{\mathbb{Z}} \to 1$, and the exact sequences in the middle and right columns follow from the Tate–Poitou exact sequence and the definition of I_S respectively. Hence one has

$$
\mathrm{Ker}(\varphi) \hookrightarrow \mathrm{Coker}\left(\prod_{v\in\overline{S}} H^1(G_{\mathbb{Q}_v}) \times \prod_{p\notin\overline{S}} H^1_{\mathrm{ur}}(G_{\mathbb{Q}_p}) \to H^1(\mathbb{Q},\mu_l)^*\right)
$$
$$
\simeq \mathrm{Ker}\left(\mathbb{Q}^\times/(\mathbb{Q}^\times)^l \longrightarrow \prod_{v\in\overline{S}} \mathbb{Q}_v^\times/(\mathbb{Q}_v^\times)^l \times \prod_{p\notin\overline{S}} \mathbb{Q}_p^\times/\mathbb{Z}_p^\times(\mathbb{Q}_p^\times)^l\right)^*
$$
$$
= 0.
$$

Our assertion about the mod m linking number $\mathrm{lk}_m(p_i, p_j)$ of p_j, p_i is shown in the same manner as in Proposition 5.2.4.

The analogy between Theorem 8.1.1 (or 8.1.3) and Theorem 8.2.1 is clear, and we can see the group-theoretic analogy clearly between the linking number and the power residue symbol. We also note that the relations $[x_i, y_i] = 1$ for G_L and $x_i^{p_i-1}[x_i, y_i] = 1$ for $G_S(l)$ come from the relations (3.3) for the local fundamental groups $\pi_1(\partial V_{K_i})$ and $\pi_1^t(\mathrm{Spec}(\mathbb{Q}_{p_i}))$ respectively.

Remark 8.2.2

(1) Theorem 8.2.1 can be extended for a finite set of primes in a number field as follows (cf. [111, Theorem 11.8]). Let k be a number field of finite degree over $\mathbb{Q}$ which contains a primitive m-th root ζ of unity, where m is a power of l. Let $S = \{\mathfrak{p}_1, \ldots, \mathfrak{p}_r\}$ be a set of r distinct primes of k such that $N\mathfrak{p}_i \equiv 1$ mod l ($1 \le i \le r$). We assume that (1) the class number of k is prime to l and (2) $B_S := \{\alpha \in k^\times | (\alpha) = \mathfrak{a}^l$ ($\mathfrak{a}$ being an ideal), $\alpha \in (k_{\mathfrak{p}_i}^\times)^l$ ($1 \le i \le r$)$\}/(k^\times)^l = 1$. Then the maximal pro-l quotient $G_S(k)(l)$ of $\pi_1(\mathrm{Spec}(\mathcal{O}_k) \setminus S)$ has the following presentation:

$$
G_S(k)(l) = \langle x_{i_1}, \cdots, x_{i_s} \mid x_{i_1}^{N\mathfrak{p}_{i_1}-1}[x_{i_1}, y_{i_1}] = \cdots = x_{i_s}^{N\mathfrak{p}_{i_s}-1}[x_{i_s}, y_{i_s}] = 1\rangle,
$$
(8.7)

where x_{i_j} represents a monodromy over $\mathfrak{p}_{i_j}$ and y_{i_j} represents an extension of the Frobenius automorphism over $\mathfrak{p}_{i_j}$, and $s = r - r_2(k)$ ($r_2(k)$ is the number of complex primes. (cf. Sect. 2.2). The assumption $B_S = 1$ is a sufficient condition for the localization map $H^2(G_S(k)(l)) \to \prod_{i=1}^r H^2(D_{\mathfrak{p}_i}(l))$ to be injective [112, Theorem 4.2], which yields that relations of the pro-l group $G_S(k)(l)$ come from those of the local Galois groups $\mathrm{Gal}(k_{\mathfrak{p}_i}(l)/k_{\mathfrak{p}_i})$.

(2) Turaev [226] gave a homological proof of Theorem 8.1.1 using Stallings' result [212], which may be close to the proof of Theorem 8.2.1.

(3) Labute [123] gave a proof of Murasugi's conjecture on the structure of the graded quotients of the lower central series of a link group, using Milnor's Theorem 8.1.1 and mod prime linking diagram. As for the pro-l Galois group

$G_S(l)$, more generally of pro-*l* Galois groups of Koch type, some properties such as mildness, cohomological *l*-dimension were investigated using the arithmetic (triple) linking numbers, linking diagrams in [25], [61], [62], [124], [125] and [199]. Mizusawa [150] studied Koch type presentations for certain pro-*l* Galois groups allowing the ramification over *l* in connection with Iwasawa theory (see also [75]).

Summary

J. Milnor's theorem	H. Koch's theorem
$G_L/G_L^{(d)} = \langle x_1, \cdots, x_r \mid$	$G_S(l) = \langle x_1, \cdots, x_r \mid$
$[x_1, y_1^{(d)}] = \cdots = [x_r, y_r^{(d)}] = 1, F^{(d)} = 1\rangle$	$x_1^{p_1-1}[x_1, y_1] = \cdots = x_r^{p_r-1}[x_r, y_r] = 1\rangle$
x_i : meridian of K_i	x_i : monodromy over p_i
$y_i^{(d)}$: longitude of K_i	y_i : Frobenius auto. over p_i

Chapter 9
Milnor Invariants and Multiple Power Residue Symbols

The notion of higher linking numbers (Milnor $\overline{\mu}$-invariants) was introduced by J. Milnor [148]. By the analogy between link groups and Galois groups with restricted ramification in Chap. 8, we can introduce arithmetic analogues of the Milnor invariants for prime numbers. They may be regarded as multiple generalization of the power residue symbol and the Rédei triple symbol.

9.1 Fox Free Differential Calculus

For a group G and a commutative ring R, we set

$$R[G] := \left\{ \sum_{g \in G} a_g g \ \Big|\ a_g \in R,\ a_g = 0 \text{ except for a finite number of } g \right\}.$$

For $\alpha = \sum a_g g, \beta = \sum b_g g \in R[G]$ and $c \in R$, we define the sum, the action of R and the multiplication on $R[G]$ by

$$\begin{cases} \alpha + \beta := \sum (a_g + b_g)g \\ c\alpha := \sum (ca_g)g \\ \alpha \cdot \beta := \sum_g (\sum_h a_h b_{h^{-1}g})g. \end{cases}$$

Then $R[G]$ forms an R-algebra, called the *group algebra* of G over R. Note that if we identify an element $\alpha = \sum a_g g$ with an R-valued function $\alpha : g \mapsto a_g$ with finite support, the sum, R-action and multiplication in the above correspond to the usual sum, R-action and convolution for R-valued functions, respectively.

© The Author(s), under exclusive license to Springer Nature Singapore Pte Ltd. 2024   113
M. Morishita, *Knots and Primes*, Universitext,
https://doi.org/10.1007/978-981-99-9255-3_9

A homomorphism $\psi : G \to H$ of groups is naturally extended to an R-algebra homomorphism, which we also denote by the same ψ:

$$\psi : R[G] \longrightarrow R[H]; \quad \psi\left(\sum a_g g\right) := \sum a_g \psi(g).$$

In particular, when H is the unit group $\{e\}$, we have, by identifying $a \in R$ with ae, the R-algebra homomorphism

$$\epsilon_{R[G]} : R[G] \longrightarrow R; \quad \epsilon_{R[G]}\left(\sum a_g g\right) := \sum a_g$$

which is called the *augmentation map*. The kernel $\mathrm{Ker}(\epsilon_{R[G]})$ is called the *augmentation ideal* of $R[G]$ and is denoted by $I_{R[G]}$.

Let $\mathbb{Z}\langle\langle X_1, \ldots, X_r \rangle\rangle$ be the algebra of non-commutative formal power series of variables $X_1, \ldots, X_r$ over $\mathbb{Z}$:

$$\mathbb{Z}\langle\langle X_1, \ldots, X_r \rangle\rangle := \left\{ \sum_{1 \leq i_1, \ldots, i_n \leq r} a_{i_1 \cdots i_n} X_{i_1} \cdots X_{i_n} \,\middle|\, n \geq 0, a_{i_1 \cdots i_n} \in \mathbb{Z} \right\}.$$

The *degree* of $f = f(X_1, \ldots, X_r) = \sum a_{i_1 \cdots i_n} X_{i_1} \cdots X_{i_n}$ is the smallest integer n such that that $a_{i_1 \cdots i_n} \neq 0$ and is denoted by $\deg(f)$.

Let F be the free group on the letters $x_1, \ldots, x_r$. Define the homomorphism

$$M : F \longrightarrow \mathbb{Z}\langle\langle X_1, \ldots, X_r \rangle\rangle^{\times}$$

by

$$M(x_i) := 1 + X_i, \; M(x_i^{-1}) := 1 - X_i + X_i^2 - \cdots \quad (1 \leq i \leq r).$$

We extend, in the $\mathbb{Z}$-linear manner, M to a $\mathbb{Z}$-algebra homomorphism

$$M : \mathbb{Z}[F] \longrightarrow \mathbb{Z}\langle\langle X_1, \ldots, X_r \rangle\rangle,$$

which is also denoted by M.

Lemma 9.1.1 *The map M is injective. (M is called the Magnus embedding.)*

Proof Assume $f \in F$ is not the identity and let $f = x_{i_1}^{e_1} \cdots x_{i_n}^{e_n}$ $(1 \leq i_1, \ldots, i_n \leq r, i_j \neq i_{j+1}, e_j (\neq 0) \in \mathbb{Z})$ be a reduced word. Then we can write

$$M(x_{i_j}^{e_j}) = 1 + e_j X_{i_j} + X_{i_j}^2 g_j(X_{i_j}), \quad g_j(X_{i_j}) \in \mathbb{Z}\langle\langle X_{i_j} \rangle\rangle$$

and hence we have

$$M(f) = (1 + e_1 X_{i_1} + X_{i_1}^2 g_1(X_{i_1})) \cdots (1 + e_n X_{i_n} + X_{i_n}^2 g_n(X_{i_n})).$$

Here the coefficient $X_{i_1} \cdots X_{i_n}$ is $e_1 \cdots e_n \neq 0$ and so $M(f) \neq 1$. Therefore $M :$ $F \to \mathbb{Z}\langle\langle X_1, \ldots, X_r \rangle\rangle^{\times}$ is injective and hence the extended M is also injective. $\square$

For $\alpha \in \mathbb{Z}[F]$,

$$M(\alpha) = \epsilon_{\mathbb{Z}[F]}(\alpha) + \sum_{\substack{I=(i_1 \cdots i_n) \\ 1 \le i_1, \ldots, i_n \le r}} \mu(I; \alpha) X_I, \quad X_I := X_{i_1} \cdots X_{i_n}$$

is called the *Magnus expansion* of α and the coefficients $\mu(I; \alpha) (\in \mathbb{Z})$ are called the *Magnus coefficients*. As we shall show in the following, the Fox free differential calculus gives an interpretation of the Magnus expansion as the "Taylor expansion" with respect to the non-commutative variables $x_1, \ldots, x_r$.

Theorem 9.1.2 ([52]) *For any $\alpha \in \mathbb{Z}[F]$, there exists α_j uniquely for each j $(1 \le j \le r)$ such that one has*

$$\alpha = \epsilon_{\mathbb{Z}[F]}(\alpha) + \sum_{j=1}^{r} \alpha_j(x_j - 1).$$

We call α_j the Fox derivative *of α with respect to x_j and write $\alpha_j = \partial\alpha/\partial x_j$.*

Proof For $f = x_{i_1}^{e_1} \cdots x_{i_n}^{e_n}$ $(e_j = \pm 1)$, we set

$$\begin{cases} \dfrac{\partial f}{\partial x_j} := \dfrac{\partial x_{i_1}^{e_1}}{\partial x_j} + x_{i_1}^{e_1}\dfrac{\partial x_{i_2}^{e_2}}{\partial x_j} + \cdots + x_{i_1}^{e_1} \cdots x_{i_{n-1}}^{e_{n-1}}\dfrac{\partial x_{i_n}^{e_n}}{\partial x_j} \\ \dfrac{\partial x_i}{\partial x_j} := \delta_{ij}, \dfrac{\partial x_i^{-1}}{\partial x_j} := -x_i^{-1}\delta_{ij} \text{ where } \delta_{ij} = 1(i = j), = 0(i \neq j) \end{cases}$$

and for $\alpha = \sum a_f f \in \mathbb{Z}[F]$, we set

$$\frac{\partial\alpha}{\partial x_j} = \sum a_f \frac{\partial f}{\partial x_j}.$$

Noting $(\partial x_i^{e_i}/\partial x_j)(x_i - 1) = (x_i^{e_i} - 1)\delta_{ij}$, one has

$$\begin{aligned} 1 + \sum_{j=1}^{r}\frac{\partial f}{\partial x_j}(x_j - 1) &= 1 + \sum_{j=1}^{r}\frac{\partial x_{i_1}^{e_1}}{\partial x_j}(x_j - 1) + \sum_{j=1}^{r}x_{i_1}^{e_1}\frac{\partial x_{i_2}^{e_2}}{\partial x_j}(x_j - 1) + \cdots \\ &\quad + \sum_{j=1}^{r}x_{i_1}^{e_1} \cdots x_{i_{n-1}}^{e_{n-1}}\frac{\partial x_{i_n}^{e_n}}{\partial x_j}(x_j - 1) \\ &= 1 + (x_{i_1}^{e_1} - 1) + x_{i_1}^{e_1}(x_{i_2}^{e_2} - 1) + \cdots + x_{i_1}^{e_1} \cdots x_{i_{n-1}}^{e_{n-1}}(x_{i_n}^{e_n} - 1) \\ &= x_{i_1}^{e_1} \cdots x_{i_n}^{e_n} \\ &= f \end{aligned}$$

and hence, for $\alpha \in \mathbb{Z}[F]$,

$$\alpha = \epsilon_{\mathbb{Z}[F]}(\alpha) + \sum_{j=1}^{r} \frac{\partial \alpha}{\partial x_j}(x_j - 1).$$

Next, suppose

$$\alpha = \epsilon_{\mathbb{Z}[F]}(\alpha) + \sum_{j=1}^{r} \alpha_j(x_j - 1) = \epsilon_{\mathbb{Z}[F]}(\alpha) + \sum_{j=1}^{r} \alpha'_j(x_j - 1) \ (\alpha_j, \alpha'_j \in \mathbb{Z}[F]).$$

Applying the Magnus embedding, one has

$$\sum_{j=1}^{r} (M(\alpha_j) - M(\alpha'_j)) X_j = 0$$

and therefore

$$M(\alpha_j) = M(\alpha'_j) \ (1 \leq j \leq n).$$

By Lemma 9.1.1, we have $\alpha_j = \alpha'_j \ (1 \leq j \leq n)$.   □

Proposition 9.1.3 *The Fox derivative $\partial/\partial x_j : \mathbb{Z}[F] \to \mathbb{Z}[F]$ satisfies the following properties:*

(1) $\dfrac{\partial x_i}{\partial x_j} = \delta_{ij}.$

(2) $\dfrac{\partial(\alpha + \beta)}{\partial x_j} = \dfrac{\partial \alpha}{\partial x_j} + \dfrac{\partial \beta}{\partial x_j}, \ \dfrac{\partial(c\alpha)}{\partial x_j} = c\dfrac{\partial \alpha}{\partial x_j} \ (\alpha, \beta \in \mathbb{Z}[F], c \in \mathbb{Z}).$

(3) $\dfrac{\partial(\alpha\beta)}{\partial x_j} = \dfrac{\partial \alpha}{\partial x_j}\epsilon_{\mathbb{Z}[F]}(\beta) + \alpha\dfrac{\partial \beta}{\partial x_j} \ (\alpha, \beta \in \mathbb{Z}[F]).$

(4) $\dfrac{\partial f^{-1}}{\partial x_j} = -f^{-1}\dfrac{\partial f}{\partial x_j} \ (f \in F).$

Proof (1) follows from the definition. We leave the proof of (2) to readers. Let us prove (3) and (4).

(3) By Theorem 9.1.2, we have

$$
\alpha\beta = \left(\epsilon_{\mathbb{Z}[F]}(\alpha) + \sum_{j=1}^{r} \frac{\partial \alpha}{\partial x_j}(x_j - 1) \right) \cdot \left(\epsilon_{\mathbb{Z}[F]}(\beta) + \sum_{k=1}^{r} \frac{\partial \beta}{\partial x_k}(x_k - 1) \right)
$$

$$
= \epsilon_{\mathbb{Z}[F]}(\alpha\beta) + \sum_{j=1}^{r} \frac{\partial \alpha}{\partial x_j}\epsilon_{\mathbb{Z}[F]}(\beta)(x_j - 1) + \sum_{k=1}^{r} \epsilon_{\mathbb{Z}[F]}(\alpha)\frac{\partial \beta}{\partial x_k}(x_k - 1)
$$

$$
+ \sum_{j,k=1}^{r} \frac{\partial \alpha}{\partial x_j}(x_j - 1)\frac{\partial \beta}{\partial x_k}(x_k - 1)
$$

$$
= \epsilon_{\mathbb{Z}[F]}(\alpha\beta) + \sum_{j=1}^{r} \frac{\partial \alpha}{\partial x_j}\epsilon_{\mathbb{Z}[F]}(\beta)(x_j - 1)
$$

$$
+ \sum_{k=1}^{r} \left(\epsilon_{\mathbb{Z}[F]}(\alpha) + \sum_{j=1}^{r} \frac{\partial \alpha}{\partial x_j}(x_j - 1) \right) \frac{\partial \beta}{\partial x_k}(x_k - 1)
$$

$$
= \epsilon_{\mathbb{Z}[F]}(\alpha\beta) + \sum_{j=1}^{r} \left(\frac{\partial \alpha}{\partial x_j}\epsilon_{\mathbb{Z}[F]}(\beta) + \alpha\frac{\partial \beta}{\partial x_j} \right) (x_j - 1),
$$

which yields the assertion by the uniqueness of the Fox derivative.
(4) Taking the Fox derivative of both sides of $f \cdot f^{-1} = 1$ using (3),

$$
\frac{\partial f}{\partial x_j} + f\frac{\partial f^{-1}}{\partial x_j} = 0,
$$

which yields the assertion.   □

We define the higher derivatives inductively by

$$
\frac{\partial^n \alpha}{\partial x_{i_1} \cdots \partial x_{i_n}} := \frac{\partial}{\partial x_{i_1}} \left(\frac{\partial^{n-1} \alpha}{\partial x_{i_2} \cdots \partial x_{i_n}} \right) \quad (\alpha \in \mathbb{Z}[F]).
$$

For simplicity, we also denote this by $D_I(\alpha)$ ($I = (i_1 \cdots i_n)$). The relation with the Magnus coefficients is given as follows.

Proposition 9.1.4 *For $\alpha, \beta \in \mathbb{Z}[F]$ and $I = (i_1 \cdots i_n)$, we have the following:*

(1) $\mu(I; \alpha) = \epsilon_{\mathbb{Z}[F]}(D_I(\alpha))$.
(2) $\mu(I; \alpha\beta) = \sum_{I=JK} \mu(J; \alpha)\mu(K; \beta)$ where the sum rages over the pairs (J, K)
   such that $I = JK$.

Proof

(1) By Theorem 9.1.2, we have

$$\alpha = \epsilon_{\mathbb{Z}[F]}(\alpha) + \sum_{j=1}^{r} \frac{\partial \alpha}{\partial x_j}(x_j - 1)$$

$$= \epsilon_{\mathbb{Z}[F]}(\alpha) + \sum_{j=1}^{r} \left(\epsilon_{\mathbb{Z}[F]} \left(\frac{\partial \alpha}{\partial x_j} \right) + \sum_{i=1}^{r} \frac{\partial^2 \alpha}{\partial x_i \partial x_j}(x_i - 1) \right)(x_j - 1)$$

$$= \epsilon_{\mathbb{Z}[F]}(\alpha) + \sum_{j=1}^{r} \epsilon_{\mathbb{Z}[F]} \left(\frac{\partial \alpha}{\partial x_j} \right)(x_j - 1) + \sum_{1 \le i, j \le r} \frac{\partial^2 \alpha}{\partial x_i \partial x_j}(x_i - 1)(x_j - 1)$$

$$\cdots$$

$$= \epsilon_{\mathbb{Z}[F]}(\alpha) + \sum_{i_1=1}^{r} \epsilon_{\mathbb{Z}[F]} \left(\frac{\partial \alpha}{\partial x_{i_1}} \right)(x_{i_1} - 1) + \cdots$$

$$+ \sum_{1 \le i_1, \ldots, i_n \le r} \epsilon_{\mathbb{Z}[F]} \left(\frac{\partial^n \alpha}{\partial x_{i_1} \cdots \partial x_{i_n}} \right)(x_{i_1} - 1) \cdots (x_{i_n} - 1)$$

$$+ \sum_{1 \le i_1, \ldots, i_{n+1} \le r} \frac{\partial^{n+1} \alpha}{\partial x_{i_1} \cdots \partial x_{i_{n+1}}}(x_{i_1} - 1) \cdots (x_{i_{n+1}} - 1).$$

Hence

$$M(\alpha) = \epsilon_{\mathbb{Z}[F]}(\alpha) + \sum_{i_1=1}^{r} \epsilon_{\mathbb{Z}[F]} \left(\frac{\partial \alpha}{\partial x_{i_1}} \right) X_{i_1} + \cdots$$

$$+ \sum_{1 \le i_1, \ldots, i_n \le r} \epsilon_{\mathbb{Z}[F]} \left(\frac{\partial^n \alpha}{\partial x_{i_1} \cdots \partial x_{i_n}} \right) X_{i_1} \cdots X_{i_n}$$

$$+ \sum_{1 \le i_1, \ldots, i_{n+1} \le r} M \left(\frac{\partial^{n+1} \alpha}{\partial x_{i_1} \cdots \partial x_{i_{n+1}}} \right) X_{i_1} \cdots X_{i_{n+1}}.$$

Comparing the coefficients of $X_{i_1} \cdots X_{i_n}$, we obtain the assertion.
(2) By Proposition 9.1.3 (2), (3),

$$\frac{\partial^n (\alpha \beta)}{\partial x_{i_1} \cdots \partial x_{i_n}} = \frac{\partial^n \alpha}{\partial x_{i_1} \cdots \partial x_{i_n}} \epsilon_{\mathbb{Z}[F]}(\beta)$$

$$+ \sum_{m=1}^{n-1} \frac{\partial^m \alpha}{\partial x_{i_1} \cdots \partial x_{i_m}} \epsilon_{\mathbb{Z}[F]} \left(\frac{\partial^{n-m} \beta}{\partial x_{i_{m+1}} \cdots \partial x_{i_{n-m}}} \right)$$

$$+ \alpha \epsilon_{\mathbb{Z}[F]} \left(\frac{\partial^n \beta}{\partial x_{i_1} \cdots \partial x_{i_n}} \right).$$

Applying $\epsilon_{\mathbb{Z}[F]}$ to both sides, we get the assertion by (1).

$$\square$$

The relation between the Magnus coefficients and the lower central series of a free group is given as follows. For a multiple index $I = (i_1 \cdots i_n)$, let $|I| := n$.

Proposition 9.1.5 *For $d \geq 2$, the following conditions are equivalent:*

(1) $f \in F^{(d)}$.
(2) *For any I with $\leq |I| < d$, $\mu(I; f) = \epsilon_{\mathbb{Z}[F]}(D_I(f)) = 0$.*

Namely,

$$F^{(d)} = \{f \in F \mid \deg(M(f) - 1) \geq d\}.$$

Proof Set $F_{(d)} := \{f \in F \mid \deg(M(f) - 1) \geq d\}$. Then $\{F_{(d)}\}_{d \geq 1}$ forms a descending series of normal subgroups of F. First, we show $F^{(d)} \subset F_{(d)}$ by induction on d. By definition, $F^{(1)} = F_{(1)}$. Assume $F^{(d-1)} \subset F_{(d-1)}$. Let $f \in F, g \in F^{(d-1)}$. Writing $M(f) = 1 + P, M(g) = 1 + Q$ ($\deg(P) \geq 1, \deg(Q) \geq d - 1$),

$$
\begin{aligned}
M([f, g]) &= M(f)M(g)M(f)^{-1}M(g)^{-1} \\
&= (1 + P)(1 + Q)(1 - P + P^2 - \cdots)(1 - Q + Q^2 - \cdots) \\
&= 1 + (PQ - QP) + \text{(the terms of higher degree)}.
\end{aligned}
$$

Since $\deg(PQ - QP) \geq d$, we have $[f, g] \in F_{(d)}$. Hence $F^{(d)} \subset F_{(d)}$. From this, we obtain the natural homomorphism for $d \geq 1$,

$$\varphi_d : F^{(d)}/F^{(d+1)} \longrightarrow F_{(d)}/F_{(d+1)}.$$

It suffices to show that φ_d is injective for any $d \geq 1$, because $0 = \mathrm{Ker}(\varphi_d) = F_{(d+1)}/F^{(d+1)}$ implies $F^{(d+1)} = F_{(d+1)}$. The injectivity of φ_d is shown as follows. First, we define $\pi : F \to \bigoplus_{d \geq 1} F^{(d)}/F^{(d+1)}$ as follows: if $f \in F^{(m)}$ and $f \notin F^{(m+1)}$, set $\pi(f)_m := f \bmod F^{(m+1)}, \pi(f)_n := 0, n \neq m$. We then define $\pi(f)$ by $(\pi(f)_d)$. Next, define $\lambda : \bigoplus_{d \geq 1} F_{(d)}/F_{(d+1)} \to \mathbb{Z}\langle\langle X_1, \ldots, X_r \rangle\rangle$ as follows: For $f_d \in F_{(d)}$, letting $M(f_d) = 1 + P_d + P_{d+1} + \cdots$ (P_j = the sum of monomials of degree j), set $\lambda(f_d \bmod F_{(d+1)}) := P_d$. For $f = (f_d) \in \bigoplus_{d \geq 1} F_{(d)}/F_{(d+1)}$, we set $\lambda(f) := \sum_{d \geq 1} \lambda(f_d)$. Let $\varphi := \bigoplus_{d \geq 1} \varphi_d$. The composite

$$F \xrightarrow{\pi} \bigoplus_{d \geq 1} F^{(d)}/F^{(d+1)} \xrightarrow{\varphi} \bigoplus_{d \geq 1} F_{(d)}/F_{(d+1)} \xrightarrow{\lambda} \mathbb{Z}\langle\langle X_1, \ldots, X_r \rangle\rangle.$$

satisfies $(\lambda \circ \varphi \circ \pi)(f) = M(f) - 1$ ($f \in F$). Since M is injective, $\lambda \circ \varphi \circ \pi$ is so. Hence φ is injective. $\square$

Finally we shall show the shuffle relation among Magnus coefficients. For multiple indices $I = (i_1 \cdots i_m)$, $J = (j_1 \cdots j_n)$, a pair (a, b) of sequences of integers $a = (a(1), \ldots, a(m))$, $b = (b(1), \ldots, b(n))$ is called the *shuffle* of I and J, if one has

$$1 \leq a(1) < \cdots < a(|I|) \leq |I| + |J|, \quad 1 \leq b(1) < \cdots < b(|J|) \leq |I| + |J|$$

and there is a multiple index $H = (h_1 \cdots h_l)$ such that the following conditions hold:

$$\begin{cases} 1) & h_{a(s)} = i_s \ (s = 1, \ldots, m), \quad h_{b(t)} = j_t \ (t = 1, \ldots, n), \\ 2) & \text{for any } u = 1, \ldots, l, \text{ there is } s \text{ or } t \text{ such that } u = a(s) \text{ or } u = b(t) \\ & \text{(possibly } u = a(s) = b(t)) \end{cases}$$

A multiple index $H = (h_1 \cdots h_l)$ ($l \leq m + n$) determined from a shuffle of I and J as above is called a *result of a shuffle*. For example, when $I = (12)$ and $J = (123)$, both $(a = (12), b = (134))$ and $(a = (13), b = (124))$ are shuffles of I and J whose result is the same $H = (1223)$. Let $Sh(I, J)$ denote the set of results of shuffles of I and J allowing overlapping (it corresponds bijectively to the set of shuffles of I and J). A shuffle (a, b) is called a *proper shuffle* if $a(s) \neq b(t)$ ($1 \leq s \leq m, 1 \leq t \leq n$). The result H of a proper shuffle (a, b) has length $m + n$. We denote by $PSh(I, J)$ the set of results of proper shuffles of I and J. The Magnus coefficients satisfies the following *shuffle relation*.

Proposition 9.1.6 ([29]) *For multiple indices I, J ($|I|, |J| \geq 1$) and $f \in F$, we have the following formula:*

$$\mu(I; f)\mu(J; f) = \sum_{H \in Sh(I,J)} \mu(H; f).$$

Proof We shall prove the above equality by induction on the length of a word f.

When $f = x_i$: Both sides of the equality are 0 unless $I = J = (i)$. If $I = J = (i)$, $Sh((i), (i)) = \{(i), (ii), (ii)\}$, $\mu((i); x_i) = 1$, $\mu((ii); x_i) = 0$. Hence both sides are 1.

When $f = x_i^{-1}$: Both sides of the equality is 0 unless I and J are of the form $(i \cdots i)$. If $I = (i \cdots i)$ with $|I| = s$ and $J = (i \cdots i)$ with $|J| = t$, then the left-hand side $= \mu(I; x_i^{-1})\mu(J; x_i^{-1}) = (-1)^s(-1)^t = (-1)^{s+t}$. Letting $(s, t)_u$ denote the number of $H = (i \cdots i) \in Sh(I, J)$ with $|H| = u$, the right-hand side $= \sum_{u=\max\{s,t\}}^{s+t}(-1)^u(s, t)_u$. Therefore it suffices to show

$$(*)_{r,s} \qquad \sum_{u=\max\{s,t\}}^{s+t} (-1)^u(s, t)_u = (-1)^{s+t}. \tag{9.1}$$

Proof of (9.1): Induction on t. First, noting $(s, 1)_s = s$ and $(s, 1)_{s+1} = s + 1$, we have $(-1)^s s + (-1)^{s+1}(s + 1) = (-1)^{s+1}$ and so $(*)_{s,1}$ holds for any s. Let $t \geq 1$ and assume that $(*)_{s,t}$ holds for any s. Then

$$t(-1)^{s+t} + (t + 1) \sum_{u=\max\{s,t+1\}}^{s+t+1} (-1)^u(s, t + 1)_u$$

$$= \sum_{v=t}^{t+1} \sum_{u=\max\{s,v\}}^{s+v} (-1)^u(s, v)_u(1, t)_v \quad \text{(by the inductive hypothesis)}$$

$$= \sum_{w=s}^{s+1} \sum_{u=\max\{w,t\}}^{w+t} (-1)^u(s, 1)_w(w, t)_u \quad (Sh(I, Sh((i), J)) = Sh(Sh(I, (i)), J))$$

$$= \sum_{w=s}^{s+1} (-1)^{w+t}(s, 1)_w \quad \text{(by the inductive hypothesis)}$$

$$= (-1)^{s+t+1}.$$

Therefore

$$\sum_{u=\max\{s,t+1\}}^{s+t+1} (-1)^u(s, t + 1)_u = \frac{1}{t + 1}((-1)^{s+t+1} - t(-1)^{s+t}) = (-1)^{s+t+1}$$

and hence (9.1) holds.

When f has length > 1, we can write $f = f_1 f_2$ ($f_1, f_2 \in F$). Then we have

$$\mu(I; f)\mu(J; f)$$

$$= \left(\sum_{I=I_1 I_2} \mu(I_1; f_1)\mu(I_2; f_2) \right) \left(\sum_{J=J_1 J_2} \mu(J_1; f_1)\mu(J_2; f_2) \right)$$

$$\text{(by Proposition 9.1.4 (2))}$$

$$= \sum_{I=I_1 I_2, J=J_1 J_2} \mu(I_1; f_1)\mu(J_1; f_1)\mu(I_2; f_2)\mu(J_2; f_2)$$

$$= \sum_{I=I_1 I_2, J=J_1 J_2} \sum_{H_1 \in Sh(I_1, J_1), H_2 \in Sh(I_2, J_2)} \mu(H_1; f_1)\mu(H_2; f_2)$$

$$= \sum_{H \in Sh(I,J)} \sum_{H=H_1 H_2} \mu(H_1; f_1)\mu(H_2; f_2)$$

$$= \sum_{H \in Sh(I,J)} \mu(H; f) \quad \text{(Proposition 9.1.4 (2))}.$$

Hence the induction works and the assertion follows. $\square$

9.2 Milnor Invariants

Let $L = K_1 \cup \cdots \cup K_r$ be a link in S^3 and let $G_L = \pi_1(S^3 \setminus L)$ be the link group of L. In this section, we shall use the same notation as in Sect. 8.1. Let α_i be a meridian of K_i and let F be the free group on the words $x_1, \ldots, x_r$, where x_i represents α_i. By Theorem 8.1.1, for each $d \geq 1$, there is $y_i^{(d)} \in F$ such that

$$G_L/G_L^{(d)} = \langle x_1, \ldots, x_r \mid [x_1, y_1^{(d)}] = \cdots = [x_r, y_r^{(d)}] = 1, F^{(d)} = 1 \rangle,$$
$$y_i^{(d)} \equiv y_i^{(d+1)} \bmod F^{(d)} \quad (1 \leq i \leq r).$$

Here $y_i^{(d)}$ is the word which represents a longitude β_i of K_i in $G_L/G_L^{(d)}$. Let

$$M(y_i^{(d)}) = 1 + \sum_{\substack{I=(i_1\cdots i_n) \\ 1 \leq i_1,\ldots,i_n \leq r}} \mu^{(d)}(Ii)X_I$$

be the Magnus expansion of $y_i^{(d)}$. By Theorem 9.1.4 (1), we have

$$\mu^{(d)}(Ii) = \epsilon_{\mathbb{Z}[F]}(D_I(y_i^{(d)})).$$

By Proposition 9.1.5, $\mu^{(d)}(I)$ is independent of d if $d \geq |I|$ and so we define the *Milnor number* $\mu(I)$ to be $\mu^{(d)}(I)$ by taking a sufficiently large d. For a multi-index I with $|I| \geq 2$, we define $\Delta(I)$ to be the ideal of $\mathbb{Z}$ generated by $\mu(J)$ where J runs over cyclic permutations of proper subsequences of I. If $|I| = 1$, we set $\mu(I) := 0$. So $\Delta(I) = 0$ if $|I| = 1, 2$. We then define the *Milnor $\overline{\mu}$-invariant* by

$$\overline{\mu}(I) := \mu(I) \bmod \Delta(I).$$

Theorem 9.2.1 ([148])

(1) $\overline{\mu}(ij) = \mathrm{lk}(K_i, K_j)$ $(i \neq j)$.
(2) *If* $2 \leq |I| \leq d$, $\overline{\mu}(I)$ *is a link invariant* L *(in fact, it is an isotopy invariant).*
(3) (shuffle relation) *For any* I, J $(|I|, |J| \geq 1)$ *and* i $(1 \leq i \leq r)$, *we have*
$$\sum_{H \in PSh(I,J)} \overline{\mu}(Hi) \equiv 0 \bmod \mathrm{g.c.d}\{\Delta(Hi) \mid H \in PSh(I, J)\}.$$
(4) (cyclic symmetry) $\overline{\mu}(i_1 \cdots i_n) = \overline{\mu}(i_2 \cdots i_n i_1) = \cdots = \overline{\mu}(i_n i_1 \cdots i_{n-1}).$

Proof

(1) By Theorem 5.1.1, $\beta_j \equiv \prod_{i \neq j} \alpha_i^{\mathrm{lk}(K_i, K_j)} \bmod G_L^{(2)}$. Hence

$$M(y_j^{(d)}) = 1 + \sum_{i \neq j} \mathrm{lk}(K_i, K_j)X_i + (\text{higher terms}),$$

which yields $\overline{\mu}(ij) = \mu(ij) = \mathrm{lk}(K_i, K_j)$.

(2) We need to show that $\bar{\mu}(I)$ is determined by the link group G_L, namely, independent of the choice of a meridian and a longitude. Let $I = (i_1 \cdots i_n)$, $2 \le n \le d$. It then suffices to show the following:

(i) $\bar{\mu}(I)$ is not changed if $y_{i_n}^{(d)}$ is replaced by a conjugate.

(ii) $\bar{\mu}(I)$ is not changed if x_i is replaced by a conjugate.

(iii) $\bar{\mu}(I)$ is not changed if $y_{i_n}^{(d)}$ is multiplied by a conjugate of $[x_i, y_i^{(d)}]$.

(iv) $\bar{\mu}(I)$ is not changed if $y_{i_n}^{(d)}$ is multiplied by an element of $F^{(d)}$.

Let $I' := (i_1 \cdots i_{n-1})$.

Proof of (i): For $f \in F$, $M(x_i f x_i^{-1}) = (1 + X_i)M(f)(1 - X_i + X_i^2 + \cdots)$. Comparing the coefficients of $X_{I'}$ in both sides, $\mu(I'; x_i f x_i^{-1}) \equiv \mu(I'; f)$ mod $\mathfrak{a}(I')$, where $\mathfrak{a}(I')$ is the ideal of $\mathbb{Z}$ generated by $\mu(J; f)$, where J ranges over all proper subsequences of I'. Similarly, $\mu(I'; x_i^{-1} f x_i) \equiv \mu(I'; f)$ mod $\mathfrak{a}(I')$. Letting $f = y_{i_n}^{(d)}$ and noting $\mathfrak{a}(I') \subset \Delta(I)$, we obtain the assertion (i).

Proof of (ii): Suppose that x_i is replaced by $\bar{x}_i = x_j x_i x_j^{-1}$. As $x_i = x_j^{-1} \bar{x}_i x_j$, we have $1 + X_i = (1 - X_j + X_j^2 - \cdots)(1 + \bar{X}_i)(1 + X_j)$. Therefore, $X_i = \bar{X}_i + $ (terms containing $X_j \bar{X}_i$ or $\bar{X}_i X_j$). Each time X_i occurs in the Magnus expansion $M(y_j^{(d)})$, it is to be replaced by this last expansion. Then the coefficient of $\bar{X}_{i_1} \cdots \bar{X}_{i_{n-1}}$ in the new Magnus expansion of $y_{i_n}^{(d)}$ with respect to $\bar{X}_i$'s is given by

$$\mu(I) + \sum_J \mu(Ji_n) \quad (J \text{ is a proper subsequence of } I')$$

and so it equals $\mu(I)$ mod $\Delta(I)$. Similarly, $\bar{\mu}(I)$ is not changed when x_i is replaced by $\bar{x}_i = x_j^{-1} x_i x_j$.

Proof of (iii): Let $J = (i_1 \cdots i_s)$ $(1 \le s < n)$ be an initial segment of I' and let $J' = (j_1 \cdots j_t)$ be a subsequence of J. Comparing the coefficients of $X_{J'}$ in both sides of the equality

$$M([x_i, y_i^{(d)}]) = 1 + (M(x_i y_i^{(d)}) - M(y_i^{(d)} x_i))M(x_i^{-1})M(y_i^{(d)-1}),$$

we have

$$\mu(J'; [x_i, y_i^{(d)}]) \equiv \mu(J'; x_i y_i^{(d)}) - \mu(J'; y_i^{(d)} x_i) \text{ mod } \Delta(I), \qquad (9.2)$$

where

$$\mu(J'; x_i y_i^{(d)}) = \begin{cases} \mu(J'i) & (\text{if } i \ne j_1), \\ \mu(J' j_1) + \mu(j_2 \cdots j_t j_1) & (\text{if } i = j_1) \end{cases}$$

$$\mu(J'; y_i^{(d)} x_i) = \begin{cases} \mu(J'i) & (i \ne j_t), \\ \mu(J' j_t) + \mu(J') & (i = j_t) \end{cases}$$

and hence

$$\mu(J'; x_i y_i^{(d)}) - \mu(J'; y_i^{(d)} x_i) = \begin{cases} \mu(j_2 \cdots j_t j_1) - \delta_{j_1 j_t} \mu(J') & (i = j_1), \\ \mu(j_2 \cdots j_t j_1) \delta_{j_1 j_t} - \mu(J') & (i = j_t), \\ 0 & (\text{otherwise}) \end{cases}$$

$$\equiv 0 \mod \Delta(I).$$

Therefore, by (9.2),

$$\mu(J'; [x_i, y_i^{(d)}]) \equiv 0 \mod \Delta(I).$$

As in the proof of (i), one obtains

$$\mu(J; x_j^\varepsilon [x_i, y_i^{(d)}] x_j^{-\varepsilon}) \equiv \mu(J; [x_i, y_i^{(d)}]) \equiv 0 \mod \Delta(I) \ (\varepsilon = \pm 1). \qquad (9.3)$$

By Proposition 9.1.4 (2),

$$\mu(I'; x_j^\varepsilon [x_i, y_i^{(d)}] x_j^{-\varepsilon} y_{i_n}^{(d)}) = \sum_{I'=JK} \mu(J; x_j^\varepsilon [x_i, y_i^{(d)}] x_j^{-\varepsilon}) \mu(K i_n). \qquad (9.4)$$

By (9.3) and (9.4),

$$\mu(I'; x_j^\varepsilon [x_i, y_i^{(d)}] x_j^{-\varepsilon} y_{i_n}^{(d)}) \equiv \mu(I'; y_{i_n}^{(d)}) = \mu(I) \mod \Delta(I).$$

By the same argument as above applied for a tail segment J of I' and a subsequence J' of J, we have $\mu(I'; y_{i_n}^{(d)} x_j^\varepsilon [x_i, y_i^{(d)}] x_j^{-\varepsilon}) \equiv \mu(I'; y_{i_n}^{(d)}) = \mu(I) \mod \Delta(I)$.

Proof of (iv): If $n \le d$, $|I'| < d$ and hence $\mu(I'; y_{i_n}^{(d)} f) = \mu(I'; y_{i_n}^{(d)})$ for $f \in F^{(d)}$ by Proposition 9.1.5. Similarly, $\mu(I'; f y_{i_n}^{(d)}) = \mu(I'; y_{i_n}^{(d)})$ for $f \in F^{(d)}$. It follows from Remark 8.1.2 (1) that $\overline{\mu}(I)$ is an isotopy invariant.

(3) Shuffle relation: By Proposition 6.1.6, we have

$$\mu(Ii)\mu(Ji) = \sum_{H \in Sh(I,J)} \mu(Hi).$$

Here the left-hand side is congruent to 0 mod g.c.d $\{\Delta(Hi) \mid H \in PSh(I, J)\}$, while $\mu(Hi)$ in the right-hand side is congruent to 0 if $H \notin PSh(I, J)$. Hence the assertion follows.

(4) Cyclic symmetry: We use the same notations as in the proof of Theorem 8.1.1. Fix d such that $d > n$. By Example 2.1.6 (4), there are $z_{ij} \in \overline{F}$ such that

$$\prod_{i=1}^r \prod_{j=1}^{\lambda_i} z_{ij} r_{ij} z_{ij}^{-1} = 1. \qquad (9.5)$$

Set $z_i = \eta_d(z_{i\lambda_i})$. Since $\eta_d(r_{ij}) \equiv 1 \bmod F^{(d)}$ $(1 \le j < \lambda_i)$ and $\eta_d(r_{i\lambda_i}) \equiv [x_i, y_i^{(d)}] \bmod F^{(d)}$, (9.5) implies

$$\prod_{i=1}^{r} z_i[x_i, y_i^{(d)}]z_i^{-1} \in F^{(d)}. \tag{9.6}$$

Let $D := \{\sum_I c_I X_I \in \mathbb{Z}\langle\langle X_1, \ldots, X_r\rangle\rangle \mid c_I \equiv 0 \bmod \Delta(I), |I| \le d\}$. Note that D is a two-sided ideal of $\mathbb{Z}\langle\langle X_1, \ldots, X_r\rangle\rangle$. Let $M(y_i^{(d)}) = 1 + w_i$ $(1 \le i \le r)$. Since Ii is a cyclic permutation of a proper subsequence of any of jiI, jIi, iIj and Iij, $X_jX_iw_i, X_jw_iX_i, X_iw_iX_j, w_iX_iX_j \in D$. Therefore

$$\begin{aligned}
M(z_i[x_i, y_i^{(d)}]z_i^{-1}) &= 1 + M(z_i)(M(x_i)M(y_i^{(d)}) - M(y_i^{(d)})M(x_i)) \\
&\quad \times M(x_i^{-1})M(y_i^{(d)^{-1}})M(z_i^{-1}) \\
&= 1 + M(z_i)(X_iw_i - w_iX_i)M(x_i^{-1})M(y_i^{(d)^{-1}})M(z_i^{-1}) \\
&\equiv 1 + X_iw_i - w_iX_i \bmod D.
\end{aligned}$$

Hence, by (9.6), $\sum_{i=1}^{r}(X_iw_i - w_iX_i) \in D$. Since the coefficient of X_{iJ} in this sum is $\mu(Ji) - \mu(iJ)$, we have $\mu(Ji) \equiv \mu(iJ) \bmod \Delta(iJ)$, $|iJ| \le d$. This yields the cyclic symmetry. $\qquad\square$

As the linking number is an invariant associated to an abelian covering of X_L (i.e., abelian quotient of G_L), Milnor invariants are regarded as invariants associated to nilpotent coverings of X_L (i.e, nilpotent quotients of G_L). For a commutative ring R, let $N_n(R)$ be the group consisting of n by n unipotent uppertriangular matrices. For a multi-index $I = (i_1 \cdots i_n)$ $(n \ge 2)$, we define the map $\rho_I : F \to N_n(\mathbb{Z}/\Delta(I))$ by

$$\rho_I(f) := \begin{pmatrix} 1 & \epsilon\left(\dfrac{\partial f}{\partial x_{i_1}}\right) & \epsilon\left(\dfrac{\partial^2 f}{\partial x_{i_1}\partial x_{i_2}}\right) & \cdots & \epsilon\left(\dfrac{\partial^{n-1} f}{\partial x_{i_1}\cdots\partial x_{i_{n-1}}}\right) \\ & 1 & \epsilon\left(\dfrac{\partial f}{\partial x_{i_2}}\right) & \cdots & \epsilon\left(\dfrac{\partial^{n-2} f}{\partial x_{i_2}\cdots\partial x_{i_{n-1}}}\right) \\ & & \ddots & \ddots & \vdots \\ \text{\Large 0} & & & 1 & \epsilon\left(\dfrac{\partial f}{\partial x_{i_{n-1}}}\right) \\ & & & & 1 \end{pmatrix} \bmod \Delta(I),$$

where we set $\epsilon = \epsilon_{\mathbb{Z}[F]}$ for simplicity. By Proposition 6.1.4, we see that ρ_I is a homomorphism of groups

Theorem 9.2.2 (cf. [171])

(1) *The homomorphism ρ_I factors through the link group G_L. Further it is surjective if $i_1, \cdots, i_{n-1}$ are all distinct.*

(2) *Suppose that $i_1, \cdots, i_{n-1}$ are all distinct. Let $X_I \to X_L$ be the Galois covering corresponding to $\mathrm{Ker}(\rho_I)$ whose Galois group $\mathrm{Gal}(X_I/X_L) = N_n(\mathbb{Z}/\Delta(I))$. When $\Delta(I) \neq 0$, let $M_I \to S^3$ be the Fox completion of $X_I \to X_L$. Then $M_I \to S^3$ is a Galois covering ramified over the link $K_{i_1} \cup \cdots \cup K_{i_{n-1}}$. For a longitude β_{i_n} of K_{i_n}, one has*

$$
\rho_I(\beta_{i_n}) =
\begin{pmatrix}
1 & 0 & \cdots & 0 & \overline{\mu}(I) \\
  & 1 & \cdots &   & 0 \\
  &   & \ddots &   & \vdots \\
\text{\Large 0} &   &   & 1 & 0 \\
  &   &   &   & 1
\end{pmatrix}
$$

and hence the following holds:

$$
\overline{\mu}(I) = 0 \iff K_{i_n} \text{ is completely decomposed in } M_I \to S^3.
$$

Proof

(1) First, let us show that ρ_I factors through G_L. Take d such that $d > n$. By Theorem 8.1.1, it suffices to show that $\rho_I([x_i, y_i^{(d)}]) = I$ $(1 \le i \le r)$ and $\rho_I(f) = I$ $(f \in F^{(d)})$. The former can be shown in the manner similar to the proof of (iii) in the proof of Theorem 9.2.1 (1). The latter follows from Proposition 6.1.5. Next, suppose that $i_1, \ldots, i_{n-1}$ are all distinct. Since

$$
\rho_I(x_{i_t}) =
\begin{pmatrix}
1 &   &   &   &   \\
  & \ddots &   &   &   \\
  &   & 1\ 1 & \cdots \cdots &   \\
  &   &   & 1 &   \\
  &   &   &   & \ddots \\
  &   &   &   &   & 1
\end{pmatrix}
$$

and so $\rho_I(x_{i_1}), \ldots, \rho_I(x_{i_{n-1}})$ generate $N_n(\mathbb{Z}/\Delta(I))$, ρ_I is surjective.

(2) By (1), we obtain $\mathrm{Gal}(M_I/S^3) = N_n(\mathbb{Z}/\Delta(I))$. Since $\rho_I(x_j) = I$ $(j \neq i_1, \ldots, i_{n-1})$, $M_I \to S^3$ is a covering ramified over $K_{i_1} \cup \cdots \cup K_{i_{n-1}}$ when $\Delta(I) \neq 0$. For $J = (i_p \cdots i_q)$ $(|J| < n - 1)$, $\mu(J; y_{i_n}) \equiv 0 \bmod \Delta(I)$ by definition of $\Delta(I)$. Hence the latter assertion follows.

□

Fig. 9.1 Borromean rings

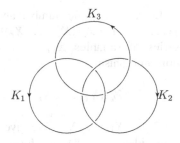

Example 9.2.3   (1) Let $L = K_1 \cup K_2 \cup K_3$ be the *Borromean rings* (Fig. 9.1).

We easily see that $\mu(ij) = \text{lk}(K_i, K_j) = 0$ for any i, j. We also see that $\beta_1 = \alpha_2\alpha_3\alpha_2^{-1}\alpha_3^{-1}$, $\beta_2 = \alpha_3\alpha_1\alpha_3^{-1}\alpha_1^{-1}$ and $\beta_3 = \alpha_1\alpha_2\alpha_1^{-1}\alpha_2^{-1}$. Hence, for instance, we have

$$M(y_3) = (1 + X_1)(1 + X_2)(1 - X_1 + X_1^2 - \cdots)(1 - X_2 + X_2^2 - \cdots)$$
$$= 1 + X_1X_2 - X_2X_1 + \text{(terms of degree} \geq 3)$$

and so $\mu(123) = 1, \mu(213) = -1$ and other $\mu(ij3)$'s are all 0. Likewise, we see that $\mu(ijk) = \pm 1$ if ijk is a permutation of 123 and $\mu(ijk) = 0$ otherwise.

Remark 9.2.4 Milnor invariants can be described in terms of the Massey products in the cohomology of X_L [189], [226]. This may be seen as a higher-order generalization of the cup product interpretation of the linking number (Remark 4.2.2). It follows immediately from this interpretation that Milnor invariants depends only on the space X_L.

9.3 Pro-*l* Fox Free Differential Calculus

Let $\mathfrak{R}$ be a compact complete local ring and let $\mathfrak{m}$ be its maximal ideal: $\mathfrak{R} = \varprojlim_i \mathfrak{R}/\mathfrak{m}^i$. Let $\mathfrak{G}$ be a pro-finite group and let $\{\mathfrak{N}_j | j \in J\}$ be the set of open normal subgroups of $\mathfrak{G}$. For $i' \geq i$ and $\mathfrak{N}_{j'} \subset \mathfrak{N}_j$, let $\varphi_{(i,j)}^{(i',j')}$ denote the natural ring homomorphism $\mathfrak{R}/\mathfrak{m}^{i'}[\mathfrak{G}/\mathfrak{N}_{j'}] \to \mathfrak{R}/\mathfrak{m}^i[\mathfrak{G}/\mathfrak{N}_j]$. Then $\{\mathfrak{R}/\mathfrak{m}^i[\mathfrak{G}/\mathfrak{N}_j], \varphi_{(i,j)}^{(i',j')}\}$ forms a projective system of finite rings. The projective limit $\varprojlim_{i,j} \mathfrak{R}/\mathfrak{m}^i[\mathfrak{G}/\mathfrak{N}_j]$ is called the *complete group algebra* of $\mathfrak{G}$ over $\mathfrak{R}$ and is denoted by $\mathfrak{R}[[\mathfrak{G}]]$. So $\mathfrak{R}[[\mathfrak{G}]]$ is a pro-finite algebra, in particular, a compact topological algebra. A continuous homomorphism $f : \mathfrak{G} \to \mathfrak{H}$ of pro-finite groups induces a continuous homomorphism $f : \mathfrak{R}[[\mathfrak{G}]] \to \mathfrak{R}[[\mathfrak{H}]]$ of completed group algebras. When $\mathfrak{H}$ is the unit group $\{e\}$, the induced map, denoted by $\epsilon_{\mathfrak{R}[[\mathfrak{G}]]} : \mathfrak{R}[[\mathfrak{G}]] \to \mathfrak{R}$, is called the *augmentation map*, and its kernel $I_{\mathfrak{R}[[\mathfrak{G}]]} := \text{Ker}(\epsilon_{\mathfrak{R}[[\mathfrak{G}]]})$ is called the *augmentation ideal* of $\mathfrak{R}[[\mathfrak{G}]]$.

Let l be a prime number and let $\hat{F}(l)$ be the free pro-l group on the letters $x_1, \ldots, x_r$. Let $\mathbb{Z}_l\langle\langle X_1, \ldots, X_r\rangle\rangle$ be the algebra of non-commutative formal power series of variables $X_1, \ldots, X_r$ over $\mathbb{Z}_l$. Let $\mathfrak{I}$ denote the kernel of the ring homomorphism

$$\mathbb{Z}_l\langle\langle X_1, \ldots, X_r\rangle\rangle \longrightarrow \mathbb{Z}_l; \quad f(X_1, \ldots, X_r) \longmapsto f(0, \ldots, 0).$$

So $\mathfrak{I} = (X_1, \ldots, X_r)$. We give a topology on $\mathbb{Z}_l\langle\langle X_1, \ldots, X_r\rangle\rangle$ so that the two-sided ideals $(l^j, \mathfrak{I}^d)_{j,d \geq 1}$ form a fundamental system of neighborhood of 0, and regard $\mathbb{Z}_l\langle\langle X_1, \ldots, X_r\rangle\rangle$ as a compact $\mathbb{Z}_l$-algebra. Let $M' : F \to \mathbb{Z}_l\langle\langle X_1, \ldots, X_r\rangle\rangle$ be the composite of the Magnus embedding $M : F \hookrightarrow \mathbb{Z}\langle\langle X_1, \ldots, X_r\rangle\rangle$ with the inclusion $\mathbb{Z}\langle\langle X_1, \ldots, X_r\rangle\rangle \subset \mathbb{Z}_l\langle\langle X_1, \ldots, X_r\rangle\rangle$. By Proposition 6.1.5, M' induces a $\mathbb{Z}_l$-algebra homomorphism

$$\mathbb{Z}/l^j\mathbb{Z}[F/F^{(d,l)}] \longrightarrow \mathbb{Z}_l\langle\langle X_1, \ldots, X_r\rangle\rangle/(l^j, \mathfrak{I}^d).$$

Taking the projective limit $\varprojlim_{j,d}$ and noting $\hat{F}(l) = \varprojlim_d F/F^{(l,d)}$, we obtain a continuous $\mathbb{Z}_l$-algebra homomorphism

$$\hat{M} : \mathbb{Z}_l[[\hat{F}(l)]] \longrightarrow \mathbb{Z}_l\langle\langle X_1, \ldots, X_r\rangle\rangle.$$

Note that the restriction of $\hat{M}$ to $\mathbb{Z}[F]$ is the Magnus embedding M.

Lemma 9.3.1 *The map $\hat{M}$ gives an isomorphism of topological $\mathbb{Z}_l$-algebras*

$$\mathbb{Z}_l[[\hat{F}(l)]] \simeq \mathbb{Z}_l\langle\langle X_1, \ldots, X_r\rangle\rangle.$$

$\hat{M}$ is called the *pro-l Magnus isomorphism*.

Proof Since $(x_i - 1)^d$ converges to 0 in $\mathbb{Z}_l[[\hat{F}(l)]]$ as $d \to \infty$, the map $X_i \to x_i - 1$ gives a continuous homomorphism $\hat{N} : \mathbb{Z}_l\langle\langle X_1, \ldots, X_r\rangle\rangle \to \mathbb{Z}_l[[\hat{F}(l)]]$. Since $\hat{M}$ and $\hat{N}$ are inverse maps of each other, $\hat{M}$ is an isomorphism of topological $\mathbb{Z}_l$-algebras. $\square$

For $\alpha \in \mathbb{Z}_l[[\hat{F}(l)]]$,

$$\hat{M}(\alpha) = \epsilon_{\mathbb{Z}_l[[\hat{F}(l)]]}(\alpha) + \sum_{\substack{I = (i_1 \cdots i_n) \\ 1 \leq i_1, \ldots, i_n \leq r}} \hat{\mu}(I; \alpha) X_I, \quad X_I := X_{i_1} \cdots X_{i_n}$$

is called the *pro-l Magnus expansion* of α and the coefficients $\hat{\mu}(I; \alpha) (\in \mathbb{Z}_l)$ the *pro-l Magnus coefficients*. For the case of pro-l groups, the analogue of Theorem 9.1.2 follows easily from Lemma 9.3.1.

Theorem 9.3.2 ([90], [184]) *For any* $\alpha \in \mathbb{Z}_l[[\hat{F}(l)]]$, *there exists uniquely* $\alpha_j \in \mathbb{Z}_l[[\hat{F}(l)]]$ *for each* j $(1 \leq j \leq r)$ *such that*

$$\alpha = \epsilon_{\mathbb{Z}_l[[\mathfrak{F}]]}(\alpha) + \sum_{j=1}^{r} \alpha_j(x_j - 1).$$

We call α_j the *pro-l Fox free derivative* of α with respect to x_j and write $\alpha_j = \partial\alpha/\partial x_j$.

Proof Since $\hat{M}(\alpha) = f(X_1, \ldots, X_r) \in \mathbb{Z}_l\langle\langle X_1, \ldots, X_r\rangle\rangle$ is written in a unique manner as

$$f(X_1, \ldots, X_r) = f(0, \ldots, 0) + \sum_{j=1}^{n} f_j X_j, \quad f_j \in \mathbb{Z}_l\langle\langle X_1, \ldots, X_r\rangle\rangle,$$

the assertion follows from Lemma 9.3.1.   □

Note that the pro-*l* Fox derivative $\partial/\partial x_j : \mathbb{Z}_l[[\hat{F}(l)]] \to \mathbb{Z}_l[[\hat{F}(l)]]$ is a continuous map whose restriction to $\mathbb{Z}[F]$ is the Fox derivative. The basic properties of pro-*l* Fox free derivatives are similar to those of Fox free derivatives given in Proposition 9.1.3 (we omit the proof):

Proposition 9.3.3 *The pro-l Fox derivative* $\partial/\partial x_j : \mathbb{Z}_l[[\hat{F}(l)]] \to \mathbb{Z}_l[[\hat{F}(l)]]$ *satisfies the following properties:*

(1) $\dfrac{\partial x_i}{\partial x_j} = \delta_{ij}.$

(2) $\dfrac{\partial(\alpha + \beta)}{\partial x_j} = \dfrac{\partial\alpha}{\partial x_j} + \dfrac{\partial\beta}{\partial x_j}, \quad \dfrac{\partial(c\alpha)}{\partial x_j} = c\dfrac{\partial\alpha}{\partial x_j} \quad (\alpha, \beta \in \mathbb{Z}_l[[\hat{F}(l)]], c \in \mathbb{Z}_l).$

(3) $\dfrac{\partial(\alpha\beta)}{\partial x_j} = \dfrac{\partial\alpha}{\partial x_j}\epsilon_{\mathbb{Z}_l[[\hat{F}(l)]]}(\beta) + \alpha\dfrac{\partial\beta}{\partial x_j} \quad (\alpha, \beta \in \mathbb{Z}_l[[\hat{F}(l)]]).$

(4) $\dfrac{\partial f^{-1}}{\partial x_j} = -f^{-1}\dfrac{\partial f}{\partial x_j} \quad (f \in \hat{F}(l)).$

We define the higher pro-*l* Fox derivatives inductively by

$$\frac{\partial^n \alpha}{\partial x_{i_1} \cdots \partial x_{i_n}} := \frac{\partial}{\partial x_{i_1}}\left(\frac{\partial^{n-1}\alpha}{\partial x_{i_2} \cdots \partial x_{i_n}}\right) \quad (\alpha \in \mathbb{Z}_l[[\hat{F}(l)]]),$$

which is also denoted by $D_I(\alpha)$ $(I = (i_1 \cdots i_n))$. The relations of the pro-*l* Fox derivatives with the pro-*l* Magnus coefficients and the lower central series are similar to those given in Propositions 9.1.4 and 9.1.5 (the proofs are omitted).

Proposition 9.3.4 *For $\alpha, \beta \in \mathbb{Z}_l[[\hat{F}(l)]]$ and a multi-index I, we have the following:*

(1) $\hat{\mu}(I; \alpha) = \epsilon_{\mathbb{Z}_l[[\hat{F}(l)]]}(D_I(\alpha))$.

(2) $\hat{\mu}(I; \alpha\beta) = \displaystyle\sum_{I=JK} \hat{\mu}(J; \alpha)\hat{\mu}(K; \beta)$, *where the sum ranges over all pairs of multi-indices (J, K) such that $I = JK$.*

Proposition 9.3.5 *For $d \geq 2$, the following conditions are equivalent:*

(1) $f \in \hat{F}(l)^{(d)}$.
(2) *For any I such that $1 \leq |I| < d$, $\hat{\mu}(I; f) = \epsilon_{\mathbb{Z}_l[[\hat{F}(l)]]}(D_I(f)) = 0$.*

Namely,

$$\hat{F}(l)^{(d)} = \{f \in \hat{F}(l) \mid \deg(\hat{M}(f) - 1) \geq d\}.$$

Proposition 9.3.6 *For multi-indices I, J ($|I|, |J| \geq 1$) and $f \in \hat{F}(l)$, we have*

$$\hat{\mu}(I; f)\hat{\mu}(J; f) = \sum_{H \in Sh(I,J)} \hat{\mu}(H; f).$$

Proof Since F is a dense subgroup of $\hat{F}(l)$ and $\hat{\mu}(I; *)$ coincides with $\mu(I; *)$ on F, the assertion follows from Proposition 9.1.6. $\square$

Remark 9.3.7 In [91, Appendix], the Fox free derivative is defined for a free profinite group and similar properties are shown.

We fix $m = l^e$ ($e \geq 1$). Taking mod m in the pro-l Magnus isomorphism, we obtain the *mod m Magnus isomorphism*

$$M_m : \mathbb{Z}/m\mathbb{Z}[[\hat{F}(l)]] \simeq \mathbb{Z}/m\mathbb{Z}\langle\langle X_1, \ldots, X_r \rangle\rangle.$$

For $\alpha \in \mathbb{Z}/m\mathbb{Z}[[\hat{F}(l)]]$, we have the *mod m Magnus expansion*

$$M_m(\alpha) = \epsilon_{\mathbb{Z}/m\mathbb{Z}[[\hat{F}(l)]]}(\alpha) + \sum_I \mu_m(I; \alpha)X_I.$$

The coefficients $\mu_m(I; \alpha)$ are called the *mod m Magnus coefficients*. For a pro-l group $\mathfrak{G}$ and $d \geq 1$, we define a normal subgroup of $\mathfrak{G}$ by

$$\mathfrak{G}_{(m,d)} := \{g \in \mathfrak{G} \mid g - 1 \in (I_{\mathbb{Z}/m\mathbb{Z}[[\mathfrak{G}]]})^d\}.$$

Then $\{\mathfrak{G}_{(m,d)}\}_{d \geq 1}$ forms a lower central series of $\mathfrak{G}$, called the *Zassenhaus filtration* of $\mathfrak{G}$. By definition, one sees, for $f \in \hat{F}(l)$ and $d \geq 2$,

$$f \in \hat{F}(l)_{(m,d)} \iff \mu_m(I; f) = 0 \text{ for any } I \text{ with } 1 \leq |I| < d. \tag{9.7}$$

9.4 Multiple Power Residue Symbols

Let l be a given prime number. Let $S = \{p_1, \cdots, p_r\}$ be a set of r distinct prime numbers such that $p_i \equiv 1 \bmod l$ ($1 \le i \le r$). Let $G_S(l) = \pi_1(\mathrm{Spec}(\mathbb{Z}) \setminus S)(l) = \mathrm{Gal}(\mathbb{Q}_S(l)/\mathbb{Q})$, where $\mathbb{Q}_S(l)$ is the maximal pro-l extension of $\mathbb{Q}$ unramified outside $S \cup \{\infty\}$. Set $e_S := \max\{e \mid p_i \equiv 1 \bmod l^e \ (1 \le i \le r)\}$ and fix $m = l^e$ ($1 \le e \le e_S$). In the following, we keep the same notation as in Sect. 8.2. Let x_i be the word representing a monodromy τ_i over p_i, $1 \le i \le r$, and let $\hat{F}(l)$ be the free pro-l group on $x_1, \ldots, x_r$. By Theorem 8.2.1, there is a pro-l word $y_i \in \hat{F}(l)$ representing an extension of the Frobenius automorphism over p_i for each i such that

$$G_S(l) = \langle x_1, \ldots, x_r \mid x_1^{p_1-1}[x_1, y_1] = \cdots = x_r^{p_r-1}[x_r, y_r] = 1 \rangle.$$

Let

$$\hat{M}(y_i) = 1 + \sum \hat{\mu}(Ii)X_I$$

be the pro-l Magnus expansion of y_i. By Theorem 6.3.4 (1), we have

$$\hat{\mu}(Ii) = \epsilon_{\mathbb{Z}_l[[\hat{F}(l)]]}(D_I(y_i)).$$

We call the coefficient $\hat{\mu}(I)$ the *l-adic Milnor number*. Similarly, let

$$M_m(y_i) = 1 + \sum \mu_m(Ii)X_I$$

be the mod m Magnus expansion of y_i and we call the coefficient

$$\mu_m(I) = \hat{\mu}(I) \bmod m$$

the *mod m Milnor number*. For a multi-index I with $1 \le |I| \le l^{e_S}$, let $\Delta_m(I)$ be the ideal of $\mathbb{Z}/m\mathbb{Z}$ generated by $\binom{l^{e_S}}{t}$ ($1 \le t < |I|$) and $\mu_m(J)$ (J running over cyclic permutations of proper subsequences of I). Then we define the *Milnor $\bar{\mu}_m$-invariant* by

$$\bar{\mu}_m(I) := \mu_m(I) \bmod \Delta_m(I).$$

Theorem 9.4.1 ([154], [157], [160])

(1) $\zeta_m^{\mu_m(ij)} = \left(\dfrac{p_j}{p_i}\right)_m$ *where ζ_m is the primitive m-th root of unity given in (8.4).*

(2) *If $2 \le |I| \le l^{e_S}$, $\bar{\mu}_m(I)$ is an invariant depending only on S and l.*

(3) *Let r be an integer such that $2 \leq r \leq l^{es}$. For multi-indices I, J ($|I| + |J| = r - 1$) and i ($1 \leq i \leq r$),*

$$\sum_{H \in PSh(I,J)} \overline{\mu}_m(Hi) \equiv 0 \mod \text{g.c.d}\{\Delta_m(Hi) | H \in PSh(I, J)\}.$$

Proof

(1) By Theorem 8.2.1, $\sigma_j \equiv \prod_{i \neq j} \tau_i^{\text{lk}(p_i, p_j)} \mod G_S(l)^{(2)}$. Therefore

$$\hat{M}(y_j) = 1 + \sum_{i \neq j} \text{lk}(p_i, p_j) X_i + (\text{terms of degree} \geq 2).$$

Hence $\mu_m(ij) = \text{lk}_m(p_i, p_j)$ and the assertion follows from Theorem 8.2.1.

(2) We must show that $\overline{\mu}_m(I)$ is independent of the choices of a monodromy over p_i and an extension of the Frobenius automorphism over p_i, namely, independent of the choice of a prime of $\mathbb{Q}_S(l)$ over p_i. Let $I = (i_1 \cdots i_n)$, $2 \leq n \leq l^{es}$. It suffices to show the following:

(i) $\overline{\mu}_m(I)$ is not changed if y_{i_n} is replaced by a conjugate.

(ii) $\overline{\mu}_m(I)$ is not changed if x_i is replaced by a conjugate.

(iii) $\overline{\mu}_m(I)$ is not changed if y_{i_n} is multiplied by a conjugate of $x_i^{p_i-1}[x_i, y_i]$.

Let $I' := (i_1 \cdots i_{n-1})$.

The proofs of (i) and (ii) are similar to those of (i) and (ii) in the proof of Theorem 9.2.1 (2) respectively.

Proof of (iii): Let J be an initial segment of I' and let J' be a subsequence of J. Then as in the proof of (iii) in the proof of Theorem 9.2.1 (2), we have

$$\mu_m(J'; [x_i, y_i]) \equiv 0 \mod \Delta_m(I).$$

By definition of $\Delta_m(I)$,

$$\hat{M}(x_i^{p_i-1}) = (1 + X_i)^{p_i-1}$$
$$\equiv 1 + (\text{terms of deg} \geq |I|) \mod \Delta_m(I).$$

Therefore $\mu_m(J'; x_i^{p_i-1}[x_i, y_i]) \equiv 0 \mod \Delta_m(I)$. It follows from this that

$$\mu_m(J; x_j^\varepsilon x_i^{p_i-1}[x_i, y_i] x_j^{-\varepsilon}) \equiv \mu_m(J; x_i^{p_i-1}[x_i, y_i])$$
$$\equiv 0 \mod \Delta_m(I) \ (\varepsilon = \pm 1).$$

Hence, by Proposition 9.3.4 (2),

$$\mu_m(I'; x_j^\varepsilon x_i^{p_i-1}[x_i, y_i] x_j^{-\varepsilon} y_{i_n}) \equiv \mu_m(I) \mod \Delta_m(I).$$

By the same argument as above applied for a tail segment J of I' and a subsequence J' of J, we can show $\mu_m(I'; y_{i_n} x_j^\epsilon x_i^{p_i-1}[x_i, y_i]x_j^{-\epsilon}) \equiv \mu_m(I'; y_{i_n}) = \mu_m(I) \mod \Delta_m(I)$.

(3) By Proposition 9.3.6, this is shown in the same manner as in the proof of Proposition 9.2.1 (4).

$\square$

Remark 9.4.2 Our arithmetic Milnor invariants $\overline{\mu}_m(I)$ do not satisfy the cyclic symmetry in general, since $\mathbb{Q}$ does not contain a primitive l-th root of unity if $l > 2$. When $l = 2$, the cyclic symmetry holds if $|I| = 2$ (quadratic reciprocity law), or if $I = (ijk)$ and ijk are all distinct. (This is Rédei's reciprocity law. See Theorem 9.4.11.)

As in the case of links, Milnor $\overline{\mu}_m$-invariants describe the decomposition law of a prime number in certain nilpotent extensions of $\mathbb{Q}$. Let $I = (i_1 \cdots i_n), 2 \leq n \leq l^{es}$ and assume $\Delta_m(I) \neq \mathbb{Z}/m\mathbb{Z}$. Define a group homomorphism $\rho_{(m,I)} : \hat{F}(l) \longrightarrow N_n((\mathbb{Z}/m\mathbb{Z})/\Delta_m(I))$ by

$\rho_{(m,I)}(f)$

$$:= \begin{pmatrix} 1 & \epsilon\left(\dfrac{\partial f}{\partial x_{i_1}}\right)_m & \epsilon\left(\dfrac{\partial^2 f}{\partial x_{i_1}\partial x_{i_2}}\right)_m & \cdots & \epsilon\left(\dfrac{\partial^{n-1} f}{\partial x_{i_1}\cdots \partial x_{i_{n-1}}}\right)_m \\ & 1 & \epsilon\left(\dfrac{\partial f}{\partial x_{i_2}}\right)_m & \cdots & \epsilon\left(\dfrac{\partial^{n-2} f}{\partial x_{i_2}\cdots \partial x_{i_{n-1}}}\right)_m \\ & & \ddots & \ddots & \vdots \\ & \text{\huge 0} & & 1 & \epsilon\left(\dfrac{\partial f}{\partial x_{i_{n-1}}}\right)_m \\ & & & & 1 \end{pmatrix} \mod \Delta_m(I),$$

where we set for simplicity $\epsilon(\alpha)_m = \epsilon_{\mathbb{Z}_l[[\hat{F}(l)]]}(\alpha) \mod m$ for $\alpha \in \mathbb{Z}_p[[\hat{F}(l)]]$.

Theorem 9.4.3 ([160])

(1) *The homomorphism $\rho_{(m,I)}$ factors through the Galois group $G_S(l)$. Further it is surjective if $i_1, \cdots, i_{n-1}$ are all distinct.*

(2) *Suppose that $i_1, \cdots, i_{n-1}$ are all distinct. Let $k_{(m,I)}$ be the extension over $\mathbb{Q}$ corresponding to $\mathrm{Ker}(\rho_{(m,I)})$. Then $k_{(m,I)}$ is a Galois extension of $\mathbb{Q}$ ramified over $p_{i_1}, \cdots, p_{i_{n-1}}$ with Galois group $\mathrm{Gal}(k_{(m,I)}/\mathbb{Q}) = N_n((\mathbb{Z}/m\mathbb{Z})/\Delta_m(I))$, and each ramification index over p_{i_t} is $\#((\mathbb{Z}/m\mathbb{Z})/\Delta_m(I))$. For a Frobenius*

automorphism σ_{i_n} over p_{i_n}, one has

$$\rho_{(m,I)}(\sigma_{i_n}) = \begin{pmatrix} 1 & 0 & \cdots & 0 & \overline{\mu}_m(I) \\ & 1 & \cdots & & 0 \\ & & \ddots & & \vdots \\ \mathbf{0} & & & 1 & 0 \\ & & & & 1 \end{pmatrix}$$

and hence the following holds:

$$\overline{\mu}_m(I) = 0 \iff p_n \text{ is completely decomposed in } k_{(m,I)}/\mathbb{Q}.$$

Proof (1) It suffices to show by Theorem 8.2.1 that $\rho_{(m,I)}(x_i^{p_i-1}[x_i, y_i]) = I$ ($1 \leq i \leq r$). This is proved as in the proof of (iii) in the proof of Theorem 6.4.1 (1). The latter is shown in the same manner as in the proof of Theorem 9.2.2 (1). The proof of (2) is also similar to that of Theorem 9.2.2 (2). The ramification index over p_{i_t} is the order of $\rho_{(m,I)}(x_{i_t})$, which is the same as the order the cyclic group $(\mathbb{Z}/m\mathbb{Z})/\Delta_m(I)$. $\qquad\square$

Since roots of unity contained in the national number field $\mathbb{Q}$ are only ± 1, it is natural to consider the case that $m = 2$ for arithmetic Milnor invariants, in order to obtain globally defined arithmetic invariants. Thus, by Theorem 9.4.1, we define the n-tuple *multiple Legendre symbol* for prime numbers $p_1, \ldots, p_n$ with each $p_i \equiv 1$ mod 4 by

$$[p_1, \ldots, p_n] := (-1)^{\mu_2(1\cdots n)}, \tag{9.8}$$

under the assumption that all $\mu_2(I) = 0$ for $|I| < n$. By Theorem 9.4.3, we have the following:

Corollary 9.4.4 *Notations and assumptions being as above, there is a Galois extension R_n of $\mathbb{Q}$ ramified over $p_1, \ldots, p_{n-1}$ with each ramification index being 2 and the Galois group $\mathrm{Gal}(R_n/\mathbb{Q}) = N_n(\mathbb{F}_2)$ such that*

$$[p_1, \ldots, p_n] = 1 \iff p_n \text{ is completely decomposed in } R_n/\mathbb{Q}.$$

In fact, for $n = 2$,

$$R_2 = \mathbb{Q}(\sqrt{p_1}), \quad [p_1, p_2] = \left(\frac{p_1}{p_2}\right).$$

The next example deals with the case that $n = 3$.

Example 9.4.5 (The Rédei Symbol) Let $l = 2$ and let $S := \{p_1, p_2, p_3\}$ be a triple of distinct prime numbers such that

$$p_i \equiv 1 \bmod 4, \quad \left(\frac{p_j}{p_i}\right) = 1 \ (1 \le i \ne j \le 3). \tag{9.9}$$

Set $k_i = \mathbb{Q}(\sqrt{p_i})$ $(i = 1, 2)$.

Lemma 9.4.6 ([192])

(1) *There is $\alpha_2 \in \mathcal{O}_{k_1}$ such that the following conditions hold:*
  (i) $N_{k_1/\mathbb{Q}}(\alpha_2) = p_2 z^2$ *(z is a non-zero integer),*
  (ii) $N(d_{k_1(\sqrt{\alpha_2})/k_1}) = p_2$ *($d_{k_1(\sqrt{\alpha_2})/k_1}$ is the relative discriminant).*
(2) *Let $\mathfrak{p}_3$ be a prime ideal of $\mathcal{O}_{k_1}$ over p_3. For such an α_2 as above, one has the Frobenius automorphism $\sigma_{\mathfrak{p}_3} = \left(\frac{k_1(\sqrt{\alpha_2})/k_1}{\mathfrak{p}_3}\right) \in \mathrm{Gal}(k_1(\sqrt{\alpha_2})/k_1)$, since $\mathfrak{p}_3$ is unramified in $k_1(\sqrt{\alpha_2})/k_1$.*

Then $\sigma_{\mathfrak{p}_3}$ is independent of the choices of α_2 and $\mathfrak{p}_3$.

Remark 9.4.7 One can find α_2 in Lemma 9.4.6 (1) as follows: By the assumption (9.9) and the computation of the Hilbert symbols, we can find a non-trivial integral solution (x, y, z) of $x^2 - p_1 y^2 - p_2 z^2 = 0$ [204, Chapter 3]. Set $\alpha_2 = x + y\sqrt{p_1}$. Then (1) is satisfied. Furthermore, by an elementary argument, we may assume g.c.d.$(x, y, z) = 1$, $y \equiv 0 \bmod 2$, $x - y \equiv 1 \bmod 4$. We then see that (2) is satisfied.

Definition 9.4.8 ([192]) Notation being as in Lemma 9.4.6, we define the *Rédei symbol* by

$$[p_1, p_2, p_3]_R = \begin{cases} 1 & \text{if } \sigma_{\mathfrak{p}_3} = \mathrm{id}_{k_1(\sqrt{\alpha_2})} \\ -1 & \text{otherwise.} \end{cases}$$

We set $\alpha_1 := \alpha_2 + \bar{\alpha}_2 + 2\sqrt{p_2}z = (\sqrt{\alpha_2} + \sqrt{\bar{\alpha}_2})^2 \in k_2$ and $k := k_1 k_2(\sqrt{\alpha_2}) = \mathbb{Q}(\sqrt{p_1}, \sqrt{p_2}, \sqrt{\alpha_2})$. Then the extension $k/\mathbb{Q}$ is a Galois extension with Galois group being the dihedral group of order 8 and it is unramified outside p_1, p_2, ∞ with ramification index over p_i being 2 by Lemma 9.4.6 (1). The intermediate fields of $k/\mathbb{Q}$ are given as follows:

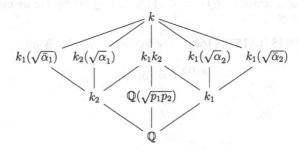

Define $s, t \in \mathrm{Gal}(k/\mathbb{Q})$ by

$$s(\sqrt{p_1}) = \sqrt{p_1}, s(\sqrt{p_2}) = -\sqrt{p_2}, s(\sqrt{\alpha_2}) = \sqrt{\alpha_2}$$
$$t(\sqrt{p_1}) = -\sqrt{p_1}, t(\sqrt{p_2}) = -\sqrt{p_2}, t(\sqrt{\alpha_2}) = -\sqrt{\bar{\alpha}_2}.$$

The Galois group $\mathrm{Gal}(k/\mathbb{Q})$ is then generated by s, t and the relations are given by

$$s^2 = t^4 = 1, \quad sts^{-1} = t^{-1}.$$

The subfields $k_1(\sqrt{\alpha_2})$ and $\mathbb{Q}(\sqrt{p_1 p_2})$ correspond to the subgroups generated by s and t respectively, and the subfields $k_1 k_2 = \mathbb{Q}(\sqrt{p_1}, \sqrt{p_2})$ and $k_2(\sqrt{\alpha_1})$ correspond to the subgroups generated by t^2 and st respectively. By the assumption (9.9), p_3 is completely decomposed in the extension $k_1 k_2/\mathbb{Q}$. Let $\mathfrak{P}_3$ be a prime ideal in $k_1 k_2$ over p_3. Since $\mathfrak{P}_3$ is decomposed in $k/k_1 k_2$ if and only if $\mathfrak{p}_3$ is decomposed in $k_1(\sqrt{\alpha_2})/k_1$, we have, by Definition 9.4.8,

$$[p_1, p_2, p_3]_R = \begin{cases} 1 & \sigma_{\mathfrak{P}_3} = \mathrm{id}_k \\ -1 & \text{otherwise.} \end{cases} \qquad (9.10)$$

By Theorem 5.2.2,

$$G_S(2) = \mathrm{Gal}(\mathbb{Q}_S(2)/\mathbb{Q})$$
$$= \langle x_1, x_2, x_3 \mid x_1^{p_1 - 1}[x_1, y_1] = x_2^{p_2 - 1}[x_2, y_2] = x_3^{p_3 - 1}[x_3, y_3] = 1 \rangle.$$

Let $\hat{F}(2)$ be the free pro-2 group on x_1, x_2, x_3 and let $\pi : \hat{F}(2) \to G_S(2)$ be the natural homomorphism. Since $k \subset \mathbb{Q}_S(2)$, we have the natural homomorphism $\psi : G_S(2) \to \mathrm{Gal}(k/\mathbb{Q})$. Let $\varphi := \psi \circ \pi : \hat{F}(2) \to \mathrm{Gal}(k/\mathbb{Q})$. We then see that

$$\varphi(x_1) = st, \quad \varphi(x_2) = s, \quad \varphi(x_3) = 1.$$

Therefore the relations among s, t are equivalent to the following relations:

$$\varphi(x_1)^2 = \varphi(x_2)^2 = 1, \quad \varphi(x_1 x_2)^4 = 1, \quad \varphi(x_3) = 1. \qquad (9.11)$$

On the other hand, since $\mu_2(ij) = 0$ $(1 \leq i, j \leq 3)$ by the assumption (9.9), $\bar{\mu}_2(123) = \mu_2(123) \in \mathbb{F}_2$.

Theorem 9.4.9 ([154], [157], [160]) *The following equality holds:*

$$(-1)^{\mu_2(123)} = [p_1, p_2, p_3]_R.$$

Proof By (9.10), we have

$$\varphi(y_3) = \begin{cases} 1 & (\{[p_1, p_2, p_3] = 1), \\ t^2 = \varphi((x_1 x_2)^2) & (\{p_1, p_2, p_3\} = -1). \end{cases}$$

By (9.11), $\mathrm{Ker}(\varphi)$ is generated as a normal subgroup of $\hat{F}(2)$ by $x_1^2, x_2^2, (x_1 x_2)^4, x_3$ and one has

$$
\begin{aligned}
M_2(x_1^2) &= (1 + X_1)^2 = 1 + X_1^2, \\
M_2(x_2^2) &= (1 + X_2)^2 = 1 + X_2^2, \\
M_2((x_1 x_2)^4) &= ((1 + X_1)(1 + X_2))^4 \equiv 1 \mod \deg \geq 4, \\
M_2(x_3) &= 1 + X_3.
\end{aligned}
$$

Therefore $\mu_2((1); *)$, $\mu_2((2); *)$ and $\mu_2((12); *)$ take their values 0 on $\mathrm{Ker}(\varphi)$.

If $\varphi(y_3) = 1$, $\mu_2(123) = \mu_2((12); y_3) = 0$ by $y_3 \in \mathrm{Ker}(\varphi)$.

If $\varphi(y_3) = t^2 = \varphi((x_1 x_2)^2)$, we can write $y_3 = (x_1 x_2)^2 f$, $f \in \mathrm{Ker}(\varphi)$. Then comparing the coefficients of $X_1 X_2$ in $M_2(y_3) = M_2((x_1 x_2)^2) M_2(f)$, we have

$$
\begin{aligned}
\mu_2(123) &= \mu_2((12); y_3) \\
&= \mu_2((12); (x_1 x_2)^2) + \mu_2((12); f) + \mu_2((1); (x_1 x_2)^2) \mu_2((2); f) \\
&= 1.
\end{aligned}
$$

This yields our assertion. □

We may note that Theorem 9.4.9 implies Lemma 9.4.6 (2). Namely, the mod 2 Milnor invariants are regarded as "universal" invariants determining the Redei symbol as a special case. We also note that the correspondence

$$s \longmapsto \begin{pmatrix} 1 & 0 & 0 \\ 0 & 1 & 1 \\ 0 & 0 & 1 \end{pmatrix}, \quad t \longmapsto \begin{pmatrix} 1 & 1 & 0 \\ 0 & 1 & 1 \\ 0 & 0 & 1 \end{pmatrix}$$

gives the isomorphism $\mathrm{Gal}(k/\mathbb{Q}) \simeq N_3(\mathbb{F}_2)$ and that $\rho_{(2,(123))}$ in Theorem 9.4.3 is nothing but the composite map $\hat{F}(2) \xrightarrow{\varphi} \mathrm{Gal}(k/\mathbb{Q}) \simeq N_3(\mathbb{F}_2)$:

$$
\begin{array}{ccccc}
\rho_{(2,(123))} : \hat{F}(2) & \longrightarrow & \mathrm{Gal}(k_{(2,(123))}/\mathbb{Q}) & \simeq & N_3(\mathbb{F}_2) \\
y_3 & \longmapsto & \sigma_{\mathfrak{P}_3} & \longmapsto & \begin{pmatrix} 1 & 0 & \bar{\mu}_2(123) \\ 0 & 1 & 0 \\ 0 & 0 & 1 \end{pmatrix}.
\end{array}
$$

By (9.8) and Corollary 9.4.4, we obtain the following:

Corollary 9.4.10 R_2 *coincides with Rédei's extension k and the triple Legendre symbol $[p_1, p_2, p_3]$ is equal to the Rédei symbol $[p_1, p_2, p_3]_R$.*

Fig. 9.2 Borromean primes

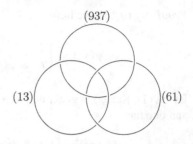

Rédei proved the invariance of the Rédei symbol under permutations of p_1, p_2 and p_3 [4, 192]. By Corollary 9.4.10, we have the following:

Theorem 9.4.11 *For any permutation ijk of 123, one has*

$$[p_i, p_j, p_k] = [p_1, p_2, p_3].$$

By Theorem 9.4.9 and Theorem 9.4.11, $\mu_2(ijk)$ is invariant under permutations of ijk.

Example 9.4.12 D. Vogel [237], [238] showed that for $S = \{(13), (61), (937)\}$,

$$\mu_2(ij) = 0 \ (1 \le i, j \le 3),$$
$$\mu_2(ijk) = 1 \ (ijk \text{ is a permutation of } 123), \quad \mu_2(ijk) = 0 \text{ (otherwise)}.$$

In view of Example 9.2.3, this triple of prime numbers may be called the *Borromean primes* (Fig. 9.2).

He also computes $\mu_2(ijk)$ for the case that ijk may not be pairwise distinct ([ibid.]).

Remark 9.4.13

(1) As in the case of links, arithmetic Milnor invariants are described in terms of the Massey products in the étale cohomology of $X_S = \operatorname{Spec}(\mathbb{Z}) \setminus S$ [160], [237], [238]. This may be seen as a higher-order generalization of the relation between the power residue symbol and the cup product [2], [111, 8.11], [165, 2], [240]. In particular, we have an interpretation of the Rédei symbol as a triple Massey product. It follows from this interpretation that arithmetic Milnor invariants depend only on X_S. For an application of Massey products in Galois cohomology to arithmetic, we may also refer to [207].

(2) It is a basic open problem to construct concretely in a systematic manner $N_n(\mathbb{F}_2)$-extensions R_n of $\mathbb{Q}$ in Corollary 9.4.4, which reveals the arithmetic meaning of the multiple Legendre symbol $[p_1, \ldots, p_n]$.

As for the case that $n = 4$, Amano [5] constructed a Galois extension $R_4/\mathbb{Q}$ whose Galois group is $N_4(\mathbb{F}_2)$ of order 64 and which is unramified outside p_1, p_2, p_3, ∞ with ramification index over p_i being 2 (under a certain condition) and introduced the 4-th multiple symbol $[p_1, p_2, p_3, p_4] \in \{\pm 1\}$, where p_1, p_2, p_3, p_4 are distinct prime numbers such that $p_i \equiv 1 \bmod 4$ $(1 \le i \le 4)$, $\left(\dfrac{p_i}{p_j}\right) = 1 (1 \le i \ne j \le 4)$ and $[p_i, p_j, p_k] = 1 (i, j, k$ are distinct from each other). The symbol $[p_1, p_2, p_3, p_4]$ describes the decomposition law of p_4 in $k/\mathbb{Q}$, just like the Rédei symbol. He then showed the equality

$$(-1)^{\mu_2(1234)} = [p_1, p_2, p_3, p_4],$$

which extends Theorem 9.4.9. For example, when $(p_1, p_2, p_3, p_4) = (5, 101, 8081, 449)$, the conditions $\left(\dfrac{p_i}{p_j}\right) = 1 (1 \le i \ne j \le 4)$ and $[p_i, p_j, p_k] = 1 (i, j, k$ are distinct from each other) are satisfied and $[p_1, p_2, p_3, p_4] = -1$ and

$$R_4 = \mathbb{Q}(\sqrt{5}, \sqrt{101}, \sqrt{8081}, \sqrt{241 + 100\sqrt{5}},$$
$$\sqrt{1009 + 100\sqrt{101}}, \sqrt{25 + 2\sqrt{5} + 2\sqrt{101}}).$$

So these 4 primes may be called *Milnor primes*, for they look like the following *Milnor link* of 4 components:

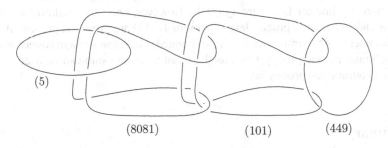

(5)

(8081)   (101)   (449)

(3) Let k be a finite algebraic number field containing a primitive m-th root of unity ζ_m, where m is a power of a prime number l, and let $\mathfrak{p}_1, \ldots, \mathfrak{p}_r$ be a finite set of finite primes of k with $N\mathfrak{p}_i \equiv 1 \bmod m$. It is also a natural problem to extend the mod 2 arithmetic Milnor invariants and multiple Legendre symbols in $\mathbb{Q}$ to the arithmetic Milnor invariant $\overline{\mu_m}(i_1 \cdots i_n) \in \mathbb{Z}/m\mathbb{Z}$ and the *multiple m-th power residue symbol*

$$[\mathfrak{p}_1, \cdots, \mathfrak{p}_n]_m := \zeta_m^{\mu_m(1\cdots n)}$$

in a number field k and construct concretely the extensions $R_n(k)/k$ generalizing R_n. It seems that there are some difficulties in deriving well-defined arithmetic Milnor invariants for primes from the Galois group $G_S(k)(l)$, as Remark 8.2.2 (1) indicates.

As for the case that $m = 3$ and $k = \mathbb{Q}(\zeta_3) = \mathbb{Q}(\sqrt{-3})$, Amano, Mizusawa and the author [8] introduced well-defined mod 3 triple Milnor invariants and triple cubic power residue symbols together with a concrete construction of $R_3(k)$, for 3 primes $\mathfrak{p}_1, \mathfrak{p}_2, \mathfrak{p}_3$ satisfying $\mathfrak{p}_i = (\varpi_i)$, $\varpi_i \equiv 1 \bmod (3\sqrt{-3})$ and $\left(\dfrac{\varpi_i}{\varpi_j}\right)_3 = 1$ $(i \neq j)$. We have

$$[\mathfrak{p}_1, \mathfrak{p}_2, \mathfrak{p}_3]_3 = \zeta_3^{\mu_3(123)},$$
$$R_3(k) = \mathbb{Q}(\zeta_3)(\sqrt[3]{\varpi_1}, \sqrt[3]{\varpi_2}, \sqrt[3]{(x + y\zeta_3\sqrt[3]{\varpi_1})^2(x + y\zeta_3^2\sqrt[3]{\varpi_1})}),$$

where $x, y \in \mathbb{Z}[\zeta_3]$ are certain solutions of $x^3 + \varpi_1 y^3 = \varpi_2 z^3$. For example, $(\mathfrak{p}_1, \mathfrak{p}_2, \mathfrak{p}_3) = ((-17), (-53), (-71))$, the conditions $[\mathfrak{p}_i, \mathfrak{p}_j]_3 = 1$ $(i \neq j)$ are satisfied and $[\mathfrak{p}_1, \mathfrak{p}_2, \mathfrak{p}_3]_3 = \zeta_3^2$. So these prime ideals may be called a *Borromean primes* in $\mathbb{Q}(\zeta_3)$.

In [75] and [150], Koch type presentations were studied for certain pro-l Galois groups allowing ramification over l, together with applications to triple symbols. For the approach using ramified coverings over the projective line $\mathbb{P}^1$ over k, we refer to [87], [114], [208].

(4) It is an interesting and important problem to show reciprocity type laws among $[\mathfrak{p}_{\pi(1)}, \ldots \mathfrak{p}_{\pi(n)}]$'s for permutations π of $1, \ldots, n$, which generalize the quadratic reciprocity law for Legendre symbols. However, only the quadratic reciprocity law and Rédei's reciprocity law (Theorem 9.4.11) are proved at present. It seems also interesting to introduce multiple (iterated) Gaussian sums in connection with such multiple reciprocity laws, as the usual Gaussian sum can be used to prove the quadratic reciprocity law.

Summary

Fox free derivative	pro-l (pro-finite) Fox free derivative
Milnor numbers of a link	l-adic Milnor numbers of primes
$\mu(i_1 \cdots i_n) = \epsilon_{\mathbb{Z}[F]}\left(\dfrac{\partial y_{i_n}^{(d)}}{\partial x_1 \cdots \partial x_{i_{n-1}}}\right)$	$\hat{\mu}(i_1 \cdots i_n) = \epsilon_{\mathbb{Z}_l[[\hat{F}]]}\left(\dfrac{\partial y_{i_n}}{\partial x_1 \cdots \partial x_{i_{n-1}}}\right)$
Milnor invariants of a link	mod m Milnor invariants of primes
$\overline{\mu}(i_1 \cdots i_n)$	$\overline{\mu}_m(i_1 \cdots i_n)$

Chapter 10
Alexander Modules and Iwasawa Modules

In this chapter, we shall introduce the differential module for a group homomorphism and show the Crowell exact sequence associated to a short exact sequence of groups. Applying these constructions to the Abelianization map of a link group, we obtain the Alexander module of a link and the exact sequence relating the Alexander module with the link module. The argument is purely group-theoretical and can be applied to pro-finite (pro-l) groups in a parallel manner to obtain the complete differential module and the complete Crowell exact sequence. Applying these constructions to a homomorphism from a Galois group with restricted ramification, we obtain the complete Alexander module for a set of primes and the exact sequence relating the complete Alexander module with a Galois (Iwasawa) module.

10.1 Differential Modules

Let G and H be groups and let $\psi : G \to H$ be a homomorphism. We also denote by the same ψ for the algebra homomorphism $\mathbb{Z}[G] \to \mathbb{Z}[H]$ of group algebras induced by ψ.

Definition 10.1.1 The ψ-*differential module* A_ψ is defined to be the quotient module of the left free $\mathbb{Z}[H]$-module $\bigoplus_{g \in G} \mathbb{Z}[H]dg$ on the symbols dg ($g \in G$) by the left $\mathbb{Z}[H]$-submodule generated by elements of the form $d(g_1 g_2) - dg_1 - \psi(g_1)dg_2$ ($g_1, g_2 \in G$):

$$A_\psi := \left(\bigoplus_{g \in G} \mathbb{Z}[H] \, dg \right) / \langle d(g_1 g_2) - dg_1 - \psi(g_1) \, dg_2 \, (g_1, g_2 \in G) \rangle_{\mathbb{Z}[H]}.$$

© The Author(s), under exclusive license to Springer Nature Singapore Pte Ltd. 2024   141
M. Morishita, *Knots and Primes*, Universitext,
https://doi.org/10.1007/978-981-99-9255-3_10

By definition, the map $d : G \to A_\psi$ defined by the correspondence $g \mapsto dg$ is a ψ-differential, namely, for $g_1, g_2 \in G$, one has

$$d(g_1 g_2) = d(g_1) + \psi(g_1) d(g_2)$$

and the following universal property holds:

(10.1.2) For any left $\mathbb{Z}[H]$-module A and any ψ-differential $\partial : G \to A$, there exists a unique $\mathbb{Z}[H]$-homomorphism $\varphi : A_\psi \to A$ such that $\varphi \circ d = \partial$.

Example 10.1.2 Let $H = G$ and $\psi = \mathrm{id}_G$. The map $\delta : G \to I_{\mathbb{Z}[G]}$ defined by $\delta(g) := g - 1$ is an id_G-differential as $g_1 g_2 - 1 = g_1 - 1 + g_1(g_2 - 1)$ $(g_1, g_2 \in G)$. Further, δ satisfies the universal property (10.1.2). (Take φ to be $\varphi(dg) := g - 1$.) Hence $A_{\mathrm{id}_G} = I_{\mathbb{Z}[G]}$.

We set $N := \mathrm{Ker}(\psi : G \to H)$.

Lemma 10.1.3 *One has*

$$\mathrm{Ker}(\psi \; : \; \mathbb{Z}[G] \longrightarrow \mathbb{Z}[H]) = I_{\mathbb{Z}[N]} \mathbb{Z}[G].$$

If ψ is surjective, we have an isomorphism of right $\mathbb{Z}[G]$-modules

$$\mathbb{Z}[G] / I_{\mathbb{Z}[N]} \mathbb{Z}[G] \simeq \mathbb{Z}[H].$$

Here $\mathbb{Z}[G]$ acts on $\mathbb{Z}[H]$ by the right multiplication via ψ.

Proof Since $\psi(I_{\mathbb{Z}[N]}) = 0$, we have $I_{\mathbb{Z}[N]} \mathbb{Z}[G] \subset \mathrm{Ker}(\psi)$. Let $\alpha = \sum_{g \in G} a_g g \in \mathrm{Ker}(\psi)$. Then

$$0 = \psi(\alpha) = \sum_{g \in G} a_g \psi(g) = \sum_{h \in \psi(G)} \left(\sum_{\psi(g) = h} a_g \right) h$$

and so $\sum_{\psi(g) = h} a_g = 0$ for any $h \in \psi(G)$. Let Ng_h denote the element of $N \backslash G$ corresponding to $h \in \psi(G)$ under the isomorphism $N \backslash G \simeq \psi(G)$. Then

$$
\begin{aligned}
\sum_{\psi(g) = h} a_g g &= \sum_{g \in Ng_h} a_g (g - 1) \\
&= \sum_{n \in N} a_{ng_h} (ng_h - 1) \\
&= \sum_{n \in N} a_{ng_h} \{ (n - 1) g_h + (g_h - 1) \} \\
&= \sum_{n \in N} a_{ng_h} (n - 1) g_h \in I_{\mathbb{Z}[N]} \mathbb{Z}[G].
\end{aligned}
$$

Hence $\alpha = \sum_{g \in G} a_g g = \sum_{h \in \psi(G)} (\sum_{\psi(g) = h} a_g g) \in I_{\mathbb{Z}[N]} \mathbb{Z}[G]$. The assertion of the latter half is easily verified. $\square$

In the rest of this section, we assume that ψ is surjective:

$$1 \longrightarrow N \longrightarrow G \overset{\psi}{\longrightarrow} H \longrightarrow 1 \quad \text{(exact)}.$$

Proposition 10.1.4 *The correspondence $dg \mapsto g - 1$ gives rise to the following isomorphism of left $\mathbb{Z}[H]$-modules:*

$$A_\psi \simeq I_{\mathbb{Z}[G]}/I_{\mathbb{Z}[N]}I_{\mathbb{Z}[G]}.$$

Here $\beta \in \mathbb{Z}[H]$ acts on the right-hand side by multiplication by any $\alpha \in \psi^{-1}(\beta)$.

Proof Via ψ, we regard $\mathbb{Z}[H]$ as a right $\mathbb{Z}[G]$-module. By Definition 10.1.1, we then have

$$A_\psi = \mathbb{Z}[H] \otimes_{\mathbb{Z}[G]} A_{\mathrm{id}_G}.$$

Hence, by Example 10.1.2 and Lemma 10.1.3, we have the following isomorphism of left $\mathbb{Z}[H]$-modules:

$$A_\psi \simeq (\mathbb{Z}[G]/I_{\mathbb{Z}[N]}\mathbb{Z}[G]) \otimes_{\mathbb{Z}[G]} I_{\mathbb{Z}[G]} \simeq I_{\mathbb{Z}[G]}/I_{\mathbb{Z}[N]}I_{\mathbb{Z}[G]}.$$

$\square$

Next, we suppose that G is a finitely presented group and choose a presentation

$$G = \langle x_1, \ldots, x_r \mid R_1 = \cdots = R_s = 1 \rangle.$$

Then we shall describe the ψ-differential module A_ψ using the Fox free differential calculus. Let F be the free group on $x_1, \ldots, x_r$ and let $\pi : F \to G$ be the natural homomorphism. Consider the $\mathbb{Z}[H]$-homomorphism

$$d_2 : \mathbb{Z}[H]^s \longrightarrow \mathbb{Z}[H]^r ; \quad (\beta_i) \longmapsto \left(\sum_{j=1}^{r} \beta_i (\psi \circ \pi) \left(\frac{\partial R_i}{\partial x_j} \right) \right).$$

Theorem 10.1.5 *The correspondence $dg \mapsto ((\psi \circ \pi)(\partial f/\partial x_j))$ gives rise to an isomorphism of left $\mathbb{Z}[H]$-modules:*

$$A_\psi \simeq \mathrm{Coker}(d_2),$$

where $f \in F$ is any element such that $\pi(f) = g$.

Proof Define the $\mathbb{Z}[H]$-homomorphism $\xi : \bigoplus_{g \in G} \mathbb{Z}[H] \, dg \to \mathrm{Coker}(d_2)$ by

$$\xi(dg) := \left((\psi \circ \pi) \left(\frac{\partial f}{\partial x_j} \right) \right) \bmod \mathrm{Im}(d_2) \quad (\pi(f) = g).$$

Since we have, for $k \in \ker(\pi)$,

$$\psi\left(\pi\left(\frac{\partial(fk)}{\partial x_j}\right)\right) = \psi\left(\pi\left(\frac{\partial f}{\partial x_j} + f\frac{\partial k}{\partial x_j}\right)\right) \equiv \psi\left(\pi\left(\frac{\partial f}{\partial x_j}\right)\right) \quad \text{mod} \ \ \text{Im}(d_2),$$

ξ is independent of the choice of f such that $\pi(f) = g$. For $g_1 = \pi(f_1)$, $g_2 = \pi(f_2) \in G$, by Proposition 9.1.3 (3),

$$\xi(d(g_1 g_2) - dg_1 - \psi(g_1)\,dg_2) = \psi\left(\pi\left(\frac{\partial(f_1 f_2)}{\partial x_j}\right)\right) - \psi\left(\pi\left(\frac{\partial f_1}{\partial x_j}\right)\right)$$
$$-\psi(g_1)\psi\left(\pi\left(\frac{\partial f_2}{\partial x_j}\right)\right)$$
$$= 0$$

and so ξ induces the $\mathbb{Z}[H]$-homomorphism

$$\xi : A_\psi \longrightarrow \text{Coker}(d_2).$$

On the other hand, we define $\eta : \mathbb{Z}[H]^r \to A_\psi$ by

$$\eta((\alpha_j)) := \left[\sum_{j=1}^r \alpha_j d\pi(x_j)\right].$$

Then

$$\eta\left(\psi\left(\pi\left(\frac{\partial R_i}{\partial x_j}\right)\right)\right) = \left[\sum_{j=1}^r \psi\left(\pi\left(\frac{\partial R_i}{\partial x_j}\right)\right) d\pi(x_j)\right].$$

Let μ be the $\mathbb{Z}[H]$-homomorphism $I_{\mathbb{Z}[G]} \to A_\psi$ induced by the isomorphism in Proposition 10.1.4. Noting $d\pi(x_j) = \mu(\pi(x_j) - 1)$, we have

$$\sum_{j=1}^r \psi\left(\pi\left(\frac{\partial R_i}{\partial x_j}\right)\right) d\pi(x_j) = \sum_{j=1}^r \psi\left(\pi\left(\frac{\partial R_i}{\partial x_j}\right)\right)(\mu(\pi(x_j) - 1))$$
$$= \mu\left(\sum_{j=1}^r \pi\left(\frac{\partial R_i}{\partial x_j}\right)(\pi(x_j) - 1)\right)$$
$$= \mu\left(\pi\left(\sum_{j=1}^r \frac{\partial R_i}{\partial x_j}(x_j - 1)\right)\right)$$
$$= \mu(\pi(R_i - 1))$$
$$= 0.$$

Hence η induces the $\mathbb{Z}[H]$-homomorphism

$$\eta : \text{Coker}(d_2) \longrightarrow A_\psi,$$

which gives

$$(\eta \circ \xi)(dg) = \eta\left(\left(\psi\left(\pi\left(\frac{\partial f}{\partial x_j}\right)\right)\right)\right) \quad (\pi(f) = g)$$

$$= \psi\left(\sum_{j=1}^{r} \pi\left(\frac{\partial f}{\partial x_j}\right) d\pi(x_j)\right)$$

$$= \sum_{j=1}^{r} \psi\left(\pi\left(\frac{\partial f}{\partial x_j}\right)\right) \mu(\pi(x_j) - 1)$$

$$= \mu(\pi(f - 1))$$

$$= \mu(g - 1)$$

$$= dg.$$

Hence $\eta \circ \xi = \text{id}_{A_\psi}$. $\xi \circ \eta = \text{id}_{\text{Coker}(d_2)}$ is also proved easily. $\qquad\square$

Corollary 10.1.6 *The ψ-differential module A_ψ has a free resolution over $\mathbb{Z}[H]$:*

$$\mathbb{Z}[H]^s \xrightarrow{Q_\psi} \mathbb{Z}[H]^r \longrightarrow A_\psi \longrightarrow 0$$

whose presentation matrix Q_ψ is given by

$$Q_\psi := \left((\psi \circ \pi)\left(\frac{\partial R_i}{\partial x_j}\right)\right).$$

When G is a free group, we have $A_\psi \simeq \mathbb{Z}[H]^r$.

For a commutative ring Z and a finitely generated Z-module M, let

$$Z^s \xrightarrow{Q} Z^r \longrightarrow M \longrightarrow 0$$

be a free resolution of M over Z with presentation matrix Q. For $d \geq 0$, we define $E_d(M)$ by the ideal of Z generated by $(r - d)$-minors of Q if $0 < r - d \leq s$, and set $E_d(M) := Z$ if $r - d \leq 0$ and $E_d(M) := 0$ if $r - d > s$. It is known [36] that $E_d(M)$ is independent of the choice of a free resolution of M and is called the d-th *elementary ideal (Fitting ideal)*. For the above case that $Z = \mathbb{Z}[H]$ and $M = A_\psi$, $E_d(A_\psi)$ can be defined if H is an Abelian group. Furthermore, if Z is a Noetherian unique factorization domain, a generator of the minimal principal ideal containing $E_d(M)$ (the intersection of all principal ideal containing $E_d(M)$) is defined up to the multiplication by an element of $Z^\times$. We denote such a generator by $\Delta_d(M)$.

Example 10.1.7 Let $L = K_1 \cup \cdots \cup K_r \subset S^3$ be an r-component link. The link group $G_L = \pi(S^3 \setminus L)$ has a Wirtinger presentation

$$G_L = \langle x_1, \ldots, x_n \mid R_1 = \cdots = R_{n-1} = 1 \rangle$$

with deficiency 1 (Example 2.1.6). Take a quotient group H of G_L and let ψ : $G_L \to H$ be the natural homomorphism. We call the ψ-differential module A_ψ the ψ-*Alexander module* of L. In particular, consider the Abelianization map $\psi : G_L \to H = G_L^{ab} = G_L / G_L^{(2)}$. Since H is the free Abelian group generated by the homology classes of meridians α_i of K_i ($1 \le i \le r$), $\mathbb{Z}[H]$ is identified with the Laurent polynomial ring $\Lambda_r := \mathbb{Z}[t_1^{\pm 1}, \ldots, t_r^{\pm 1}]$, where t_i is a variable corresponding to α_i. The Λ_r-module A_ψ is called the *Alexander module* of L and is denoted by A_L. The presentation matrix Q_L of A_L defined in Corollary 10.1.6 is then an $(n-1) \times n$ matrix over Λ_r and is called the *Alexander matrix* of L. (It depends on the choice of a Wirtinger presentation.) Since Λ_r is a Noetherian unique factorization domain, $E_d(A_L)$ and $\Delta_d(A_L)$ are defined for $d \ge 1$ and are called the *d-th Alexander ideal* and the *d-th Alexander polynomial* of L respectively. Next, consider the homomorphism $\psi : G_L \to H = \mathbb{Z}$ defined by $\psi(\alpha_i) = 1$ ($1 \le i \le r$). Then $\mathbb{Z}[H]$ is identified with the Laurent polynomial ring $\Lambda := \Lambda_1 = \mathbb{Z}[t^{\pm 1}]$ ($t \leftrightarrow 1 \in \mathbb{Z}$) and hence A_ψ becomes a Λ-module. This Λ-module A_ψ is called the *reduced Alexander module* of L and is denoted by A_L^{red}. The presentation matrix Q_L^{red} of A_L^{red} is then an $(n-1) \times n$ matrix over Λ and is called the *reduced Alexander matrix*. When L is a knot K, we have $A_K = A_K^{red}$. We can regard Λ as a Λ_r-module via the ring homomorphism $\eta : \Lambda_r \to \Lambda$ defined by $\eta(t_i) := t$. We then have $A_L^{red} = A_L \otimes_{\Lambda_r} \Lambda$. $\eta(E_d(A_L)) = E_d(A_L^{red})$ and $\eta(\Delta_d(A_L))$ ($d \ge 0$) are called the *d-th reduced Alexander ideal* and the *d-th reduced Alexander polynomial* of L respectively.

10.2   The Crowell Exact Sequence

As in Sect. 10.1, suppose that we are given a short exact sequence of groups:

$$1 \longrightarrow N \longrightarrow G \overset{\psi}{\longrightarrow} H \longrightarrow 1. \tag{10.1}$$

Theorem 10.2.1 *We have the exact sequence of left $\mathbb{Z}[H]$-modules*

$$0 \longrightarrow N^{ab} \overset{\theta_1}{\longrightarrow} A_\psi \overset{\theta_2}{\longrightarrow} \mathbb{Z}[H] \overset{\epsilon_{\mathbb{Z}[H]}}{\longrightarrow} \mathbb{Z} \longrightarrow 0.$$

Here N^{ab} is the Abelianization $N/N^{(2)}$ of N, θ_1 is the homomorphism induced by $n \mapsto dn$ ($n \in N$) and θ_2 is the homomorphism induced by $dg \mapsto \psi(g) - 1$ ($g \in G$). This exact sequence is called the Crowell exact sequence attached to (10.2.1) [35].

Proof Taking the N-homology sequence of the short exact sequence of left $\mathbb{Z}[N]$-modules

$$0 \longrightarrow I_{\mathbb{Z}[G]} \longrightarrow \mathbb{Z}[G] \xrightarrow{\epsilon_{\mathbb{Z}[G]}} \mathbb{Z} \longrightarrow 0,$$

we obtain the exact sequence

$$H_1(N, \mathbb{Z}[G]) \to H_1(N, \mathbb{Z}) \to H_0(N, I_{\mathbb{Z}[G]}) \to H_0(N, \mathbb{Z}[G]) \to H_0(N, \mathbb{Z}).$$

Here,

$H_0(N, \mathbb{Z}) = \mathbb{Z}$.
$H_0(N, \mathbb{Z}[G]) = \mathbb{Z}[G]/I_{\mathbb{Z}[N]}\mathbb{Z}[G] \simeq \mathbb{Z}[H]$ by Lemma 10.1.3.
$H_0(N, I_{\mathbb{Z}[G]}) = I_{\mathbb{Z}[G]}/I_{\mathbb{Z}[N]}\mathbb{Z}[G] \simeq A_\psi$ by Proposition 10.1.4.
$H_1(N, \mathbb{Z}) = N^{\mathrm{ab}}$.
$H_1(N, \mathbb{Z}[G]) = 0$ since $H_1(N, \mathbb{Z}[G]) = H_1(G, \mathbb{Z}[G/N] \otimes_{\mathbb{Z}} \mathbb{Z}[G])$ by Shapiro's lemma and $\mathbb{Z}[G/N] \otimes_{\mathbb{Z}} \mathbb{Z}[G]$ is a free $\mathbb{Z}[G]$-module.

Therefore we have the exact sequence

$$0 \longrightarrow N^{\mathrm{ab}} \to A_\psi \longrightarrow \mathbb{Z}[H] \xrightarrow{\epsilon_{\mathbb{Z}[H]}} \mathbb{Z} \longrightarrow 0.$$

It is easy to see that each map θ_i is the $\mathbb{Z}[H]$-homomorphism given in the statement.
□

Next, suppose that G is a finitely presented group with presentation

$$G = \langle x_1, \ldots, x_r \mid R_1 = \cdots = R_s = 1 \rangle.$$

Let us describe the Crowell exact sequence in terms of the Fox derivatives. Consider the $\mathbb{Z}[H]$-homomorphism

$$d_1 : \mathbb{Z}[H]^r \longrightarrow \mathbb{Z}[H]; \quad (\alpha_j) \mapsto \sum_{j=1}^{r} \alpha_j(\psi \circ \pi(x_j) - 1).$$

First, we see easily $\mathrm{Im}(d_1) = I_{\mathbb{Z}[H]}$. Since

$$
\begin{aligned}
(d_1 \circ d_2)((\beta_i)) &= d_1 \left(\sum_{i=1}^{s} \beta_i \psi \left(\pi \left(\frac{\partial R_i}{\partial x_j} \right) \right) \right) \\
&= \sum_{j=1}^{r} \left(\sum_{i=1}^{s} \beta_i \psi \left(\pi \left(\frac{\partial R_i}{\partial x_j} \right) \right) \right) (\psi \pi (x_j) - 1) \\
&= \sum_{i=1}^{s} \beta_i \psi \left(\pi \left(\sum_{j=1}^{r} \frac{\partial R_i}{\partial x_j} (x_j - 1) \right) \right) \\
&= \sum_{i=1}^{r} \beta_i \psi (\pi (R_i - 1)) \\
&= 0,
\end{aligned}
$$

we have a complex

$$
\mathbb{Z}[H]^s \xrightarrow{d_2} \mathbb{Z}[H]^r \xrightarrow{d_1} \mathbb{Z}[H]
$$

from which we obtain the exact sequence of left $\mathbb{Z}[H]$-modules

$$
0 \longrightarrow \mathrm{Ker}(d_1)/\mathrm{Im}(d_2) \longrightarrow \mathrm{Coker}(d_2) \xrightarrow{\overline{d}_1} \mathbb{Z}[H] \xrightarrow{\epsilon_{\mathbb{Z}[H]}} \mathbb{Z} \longrightarrow 0.
$$

We identify $\mathrm{Coker}(d_2)$ with A_ψ by the isomorphism in Theorem 10.1.5. Since

$$
d_1 \left(\psi \left(\pi \left(\frac{\partial f}{\partial x_i} \right) \right) \right) = \sum_{j=1}^{r} \psi \left(\pi \left(\frac{\partial f}{\partial x_j} \right) \right) d\pi(x_j) = \psi (\pi(f) - 1),
$$

$\overline{d}_1$ coincides with θ_2. Hence

$$
N^{\mathrm{ab}} \simeq \mathrm{Ker}(\theta_2) \simeq \mathrm{Ker}(\overline{d}_1) \simeq \mathrm{Ker}(d_1)/\mathrm{Im}(d_2),
$$

where $n \bmod N^{(2)}$ is mapped to $(\psi(\pi(\partial f/\partial x_j))) \bmod \mathrm{Im}(d_2)$ $(\pi(f) = g)$.

When G is a free group, the Crowell exact sequence boils down to the following *Blanchfield–Lyndon exact sequence*:

$$
0 \longrightarrow N^{\mathrm{ab}} \longrightarrow \mathbb{Z}[H]^r \xrightarrow{d_1} \mathbb{Z}[H] \xrightarrow{\epsilon_{\mathbb{Z}[H]}} \mathbb{Z} \longrightarrow 0.
$$

Let $L = K_1 \cup \cdots \cup K_r \subset S^3$ be an r-component link, X_L the link exterior and $G_L = \pi_1(X_L)$ the link group. For the case that $G = G_L$, the Crowell exact sequence has the following topological interpretation as follows. Let $h : X_H \to X_L$ be the covering corresponding to N: $\mathrm{Gal}(X_H/X_L) = H$. We fix a base point $x_0 \in X_L$

such that $G_L = \pi_1(X_L, x_0)$. Fix $y_0 \in h^{-1}(x_0)$ so that $N = \pi_1(X_H, y_0)$. Then we have the exact sequence

$$1 \longrightarrow N \xrightarrow{h_*} G_L \xrightarrow{\psi} H \longrightarrow 1$$

and the attached Crowell exact sequence is given by

$$0 \longrightarrow N^{\mathrm{ab}} \xrightarrow{\theta_1} A_\psi \xrightarrow{\theta_2} \mathbb{Z}[H] \xrightarrow{\epsilon_{\mathbb{Z}[H]}} \mathbb{Z} \longrightarrow 0. \tag{10.2}$$

On the other hand, one has the relative homology sequence for the pair $(X_H, h^{-1}(x_0))$:

$$0 \longrightarrow H_1(X_H) \xrightarrow{j} H_1(X_H, h^{-1}(x_0)) \xrightarrow{\delta} H_0(h^{-1}(x_0)) \xrightarrow{i} H_0(X_H) \longrightarrow 0. \tag{10.3}$$

The sequences (10.2) and (10.3) are identified as follows.

- The correspondence $1 \mapsto [y_0]$ gives a $\mathbb{Z}$-isomorphism

$$\varphi_0 : \mathbb{Z} \simeq H_0(X_H).$$

Since X_H is arcwise-connected, $[\sigma(y_0)] = [y_0]$ for $\sigma \in H$. Hence φ_0 is a $\mathbb{Z}[H]$-isomorphism.

- Since $H_0(h^{-1}(x_0)) = \bigoplus_{y \in h^{-1}(x_0)} H_0(\{y\}) = \bigoplus_{y \in h^{-1}(x_0)} \mathbb{Z}$ and the correspondence $\sigma \mapsto \sigma(y_0)$ induces the bijection $H \to h^{-1}(x_0)$, we have a $\mathbb{Z}$-isomorphism

$$\varphi_1 : \mathbb{Z}[H] \simeq H_0(h^{-1}(x_0)); \quad \sigma \longmapsto [\sigma(y_0)].$$

Since, for $\sigma_1, \sigma_2 \in H$,

$$\varphi_1(\sigma_1 \sigma_2) = [\sigma_1 \sigma_2(y_0)] = \sigma_1([\sigma_2(y_0)]) = \sigma_1 \varphi_1(\sigma_2),$$

φ_1 is a $\mathbb{Z}[H]$-isomorphism.

- For $g = [l] \in G_L$, let $\tilde{l}$ denote a lift of l with starting point y_0. Then $\tilde{l} \in C_1(X_H, h^{-1}(x_0))$ and we obtain the map

$$\partial : G_L \longrightarrow H_1(X_H, h^{-1}(x_0)); \quad \partial(g) := [\tilde{l}].$$

We claim that this ∂ is a ψ-differential. In fact, for $g_1 = [l_1], g_2 = [l_2] \in G_L$, let $\tilde{l}_1, \tilde{l}_2$ and $\widetilde{l_1 \vee l_2}$ be the lifts of l_1, l_2 and $l_1 \vee l_2$ with starting point y_0 respectively, and let $\tilde{l}_2'$ be the lift of l_2 with starting point $\tilde{l}_1(1)$. Then $\partial(g_1 g_2) = [\widetilde{l_1 \vee l_2}] = [\tilde{l}_1] + [\tilde{l}_2']$ and $\partial(g_1) + \psi(g_1)\partial(g_2) = [\tilde{l}_1] + \psi(g_1)[\tilde{l}_2] = [\tilde{l}_1] + [\tilde{l}_2']$. Hence

$\partial(g_1 g_2) = \partial(g_1) + \psi(g_1)\partial(g_2)$. By the universal property of the ψ-differential module, we obtain a $\mathbb{Z}[H]$-homomorphism

$$\varphi_2 : A_\psi \longrightarrow H_1(X_H, h^{-1}(x_0)); \quad dg \longmapsto [\tilde{l}].$$

- By Hurewicz's theorem, one has the $\mathbb{Z}[H]$-isomorphism:

$$\varphi_3 : N^{ab} \simeq H_1(X_H).$$

Putting all these together, we have the following diagram:

$$
\begin{array}{ccccccccc}
0 & \longrightarrow & N^{ab} & \stackrel{\theta_1}{\longrightarrow} & A_\psi & \stackrel{\theta_2}{\longrightarrow} & \mathbb{Z}[H] & \stackrel{\epsilon_{\mathbb{Z}[H]}}{\longrightarrow} & \mathbb{Z} & \longrightarrow & 0 \\
& & \downarrow{\varphi_3} & & \downarrow{\varphi_2} & & \downarrow{\varphi_1} & & \downarrow{\varphi_0} & & \\
0 & \longrightarrow & H_1(X_H) & \stackrel{j}{\longrightarrow} & H_1(X_H, h^{-1}(x_0)) & \stackrel{\delta}{\longrightarrow} & H_0(h^{-1}(x_0)) & \stackrel{i}{\longrightarrow} & H_0(X_H) & \longrightarrow & 0.
\end{array}
$$

This diagram is commutative.

- The right square: For $\sigma \in H$, $(i \circ \varphi_1)(\sigma) = i([\sigma(y_0)]) = [y_0]$ and $(\varphi_0 \circ \epsilon_{\mathbb{Z}[H]})(\sigma) = \varphi_0(1) = [y_0]$. Hence $i \circ \varphi_1 = \varphi_0 \circ \epsilon_{\mathbb{Z}[H]}$.
- The middle square: Since all maps are $\mathbb{Z}[H]$-homomorphism, it suffices to look at the images of $[dg] \in A_\psi$. One has $(\delta \circ \varphi_2)([dg]) = \delta([\tilde{l}]) = [\tilde{l}(1)] - [\tilde{l}(0)] = [\tilde{l}(1)] - [y_0]$ and $(\varphi_1 \circ \theta_2)([dg]) = \varphi_1(\psi(g) - 1) = [(\psi(g) - 1)(y_0)] = [\tilde{l}(1)] - [y_0]$. Hence $\delta \circ \varphi_2 = \varphi_1 \circ \theta_2$.
- The left square: Let $n = [\tilde{l}] \in N$. Then $(\varphi_2 \circ \theta_1)(n) = \varphi_2[dn] = [\tilde{l}]$ and $(j \circ \varphi_3)(n) = j([\tilde{l}]) = [\tilde{l}]$. Hence $\varphi_2 \circ \theta_1 = j \circ \varphi_3$.

Since φ_0, φ_1 and φ_3 are isomorphisms, φ_2 is isomorphic. Thus we see that the Crowell exact sequence (10.2) is nothing but the relative homology sequence (10.3).

Example 10.2.2 Let $\psi : G_L \to H = G_L^{ab}$ be the Abelianization map. Then X_H is the maximal Abelian covering of X_L (Example 2.1.13) and the Λ_r-module $N^{ab} = H_1(X_L^{ab})$ is called the *link module* of L. If $\psi : G_L \to H = \langle t \rangle = \mathbb{Z}$ is defined by sending each meridian of K_i to t, then X_H is the total linking number covering X_∞ of X_L (Example 2.1.15) and the Λ-module $N^{ab} = H_1(X_\infty)$ is called the *reduced link module* of L. When L is a knot K, the link module coincides with the reduced link module and is called the *knot module*. By Theorem 10.2.1 and $I_\Lambda \simeq \Lambda$, we obtain a Λ-isomorphism $A_L^{red} \simeq H_1(X_\infty) \oplus \Lambda$. Hence we have $E_d(H_1(X_\infty)) = E_{d+1}(A_L^{red})$ and $\Delta_d(H_1(X_\infty)) = \Delta_{d+1}(A_L^{red})$ $(d \geq 0)$.

10.3 Complete Differential Modules

Let $\mathfrak{G}$ and $\mathfrak{H}$ be pro-finite groups and let $\psi : \mathfrak{G} \to \mathfrak{H}$ be a continuous homomorphism. Let l a prime number fixed throughout this section. We also denote by the same ψ the algebra homomorphism $\mathbb{Z}_l[[\mathfrak{G}]] \to \mathbb{Z}_l[[\mathfrak{H}]]$ of complete group algebras over $\mathbb{Z}_l$ induced by ψ.

Definition 10.3.1 The *complete ψ-differential module* $\mathfrak{A}_\psi$ is defined to be the quotient module of the left free $\mathbb{Z}_l[[\mathfrak{H}]]$-module $\bigoplus_{g \in \mathfrak{G}} \mathbb{Z}_l[[\mathfrak{H}]]dg$ on the symbols dg $(g \in \mathfrak{G})$ by the left $\mathbb{Z}_l[[\mathfrak{H}]]$-submodule generated by elements of the form $d(g_1 g_2) - dg_1 - \psi(g_1)dg_2$ $(g_1, g_2 \in \mathfrak{G})$:

$$\mathfrak{A}_\psi := \left(\bigoplus_{g \in \mathfrak{G}} \mathbb{Z}_l[[\mathfrak{H}]]\, dg \right) / \langle d(g_1 g_2) - dg_1 - \psi(g_1)\, dg_2 \ (g_1, g_2 \in \mathfrak{G}) \rangle_{\mathbb{Z}_l[[\mathfrak{H}]]}.$$

By definition, the map $d : \mathfrak{G} \to \mathfrak{A}_\psi$ defined by the correspondence $g \mapsto dg$ is a ψ-differential, namely, one has

$$d(g_1 g_2) = d(g_1) + \psi(g_1)\, d(g_2) \ \ (g_1, g_2 \in \mathfrak{G})$$

and the following universal property holds:

(10.3.2) For any left $\mathbb{Z}_l[[\mathfrak{H}]]$-module $\mathfrak{A}$ and any ψ-differential $\partial : \mathfrak{G} \to \mathfrak{A}$, there exists a unique $\mathbb{Z}_l[[\mathfrak{H}]]$-homomorphism $\varphi : \mathfrak{A}_\psi \to \mathfrak{A}$ such that $\varphi \circ d = \partial$.

Example 10.3.2 Let $\mathfrak{H} = \mathfrak{G}$ and $\psi = \mathrm{id}_\mathfrak{G}$. Let $\mathfrak{G} = \varprojlim_i G_i$ (each G_i being a finite group). The map $\delta_i : G_i \to I_{\mathbb{Z}_l[G_i]}$ defined by $\delta_i(g) = g - 1$ is an id_{G_i}-differential (Example 10.1.2). Taking the projective limit $\varprojlim_i$, we obtain an $\mathrm{id}_\mathfrak{G}$-differential $\delta : \mathfrak{G} \to I_{\mathbb{Z}_l[[\mathfrak{G}]]}$ and we easily see that δ satisfies the universal property (10.3.2) ($\varphi(dg) = g - 1$).

We set $\mathfrak{N} := \mathrm{Ker}(\psi : \mathfrak{G} \to \mathfrak{H})$.

Lemma 10.3.3 *One has*

$$\mathrm{Ker}(\psi \ : \ \mathbb{Z}_l[[\mathfrak{G}]] \longrightarrow \mathbb{Z}_l[[\mathfrak{H}]]) = I_{\mathbb{Z}_l[[\mathfrak{N}]]}\mathbb{Z}_l[[\mathfrak{G}]].$$

If ψ is surjective, we have an isomorphism of right $\mathbb{Z}_l[[\mathfrak{G}]]$-modules:

$$\mathbb{Z}_l[[\mathfrak{G}]]/I_{\mathbb{Z}_l[[\mathfrak{N}]]}\mathbb{Z}_l[[\mathfrak{G}]] \simeq \mathbb{Z}_l[[\mathfrak{H}]].$$

Here $\mathbb{Z}_l[[\mathfrak{G}]]$ acts on $\mathbb{Z}_l[[\mathfrak{H}]]$ by the right multiplication via ψ.

Proof Let $\mathfrak{G} = \varprojlim_i G_i$ and $\mathfrak{H} = \varprojlim_j H_j$ (G_i, H_j being finite groups). Let ψ_{ij} be the composite $G_i \to \mathfrak{G} \xrightarrow{\psi} \mathfrak{H} \to H_j$ and set $N_{ij} := \mathrm{Ker}(\psi_{ij})$. By Lemma 7.1.4, we have $\mathrm{Ker}(\psi_{ij} : \mathbb{Z}_l[G_i] \to \mathbb{Z}_l[H_j]) = I_{\mathbb{Z}_l[N_{ij}]}\mathbb{Z}_l[G_i]$. Taking the projective limit $\varprojlim_{i,j}$, we get our assertion. The assertion of the latter half is easily verified. $\square$

In the rest of this section, we assume that ψ is surjective:

$$1 \longrightarrow \mathfrak{N} \longrightarrow \mathfrak{G} \xrightarrow{\psi} \mathfrak{H} \longrightarrow 1 \ \ \ (\text{exact}).$$

The following proposition can be proved in the same manner as in Proposition 10.1.4.

Proposition 10.3.4 *The correspondence $dg \mapsto g - 1$ induces an isomorphism of left $\mathbb{Z}_l[[\mathfrak{H}]]$-modules:*

$$\mathfrak{A}_\psi \simeq I_{\mathbb{Z}_l[[\mathfrak{G}]]}/I_{\mathbb{Z}_l[[\mathfrak{N}]]}I_{\mathbb{Z}_l[[\mathfrak{G}]]}.$$

Here $\beta \in \mathbb{Z}_l[[\mathfrak{H}]]$ acts on the right-hand side by the multiplication by any $\alpha \in \psi^{-1}(\beta)$.

Next, suppose that $\mathfrak{G}$ is a finitely presented pro-l group and choose a presentation

$$\mathfrak{G} = \langle x_1, \ldots, x_r \mid R_1 = \cdots = R_s = 1 \rangle.$$

Then we shall describe the complete ψ-differential module $\mathfrak{A}_\psi$ using the pro-l Fox free differential calculus. (We would also have a similar description for a pro-finite group with finite presentation using the pro-finite Fox free differential calculus. Cf. Remark 9.3.7.) Let $\hat{F}(l)$ be the free pro-l group on $x_1, \ldots, x_r$ and let $\pi : \hat{F}(l) \to \mathfrak{G}$ be the natural homomorphism. Consider the $\mathbb{Z}_l[[\mathfrak{H}]]$-homomorphism

$$d_2 \; : \; \mathbb{Z}_l[[\mathfrak{H}]]^s \longrightarrow \mathbb{Z}_l[[\mathfrak{H}]]^r; \;\; (\beta_i) \longmapsto \left(\sum_{i=1}^s \beta_i \left(\psi \left(\pi \left(\frac{\partial R_i}{\partial x_j} \right) \right) \right) \right).$$

Similarly to Theorem 10.1.5, we have the following theorem.

Theorem 10.3.5 ([157]) *The correspondence $dg \mapsto ((\psi \circ \pi)(\partial f / \partial x_j))$ gives rise to an isomorphism of left $\mathbb{Z}_l[[\mathfrak{H}]]$-modules:*

$$\mathfrak{A}_\psi \simeq \operatorname{Coker}(d_2),$$

where $f \in \hat{F}(l)$ is any element such that $\pi(f) = g$.

Corollary 10.3.6 *The complete ψ-differential module $\mathfrak{A}_\psi$ has a free resolution over $\mathbb{Z}_l[[\mathfrak{H}]]$:*

$$\mathbb{Z}_l[[\mathfrak{H}]]^s \xrightarrow{\mathfrak{Q}_\psi} \mathbb{Z}_l[[\mathfrak{H}]]^r \longrightarrow \mathfrak{A}_\psi \longrightarrow 0,$$

whose presentation matrix $\mathfrak{Q}_\psi$ is given by

$$\mathfrak{Q}_\psi := \left((\psi \circ \pi) \left(\frac{\partial R_i}{\partial x_j} \right) \right).$$

When $\mathfrak{G}$ a free pro-l group, we have $\mathfrak{A}_\psi \simeq \mathbb{Z}_l[[\mathfrak{H}]]^r$.

When $\mathfrak{H}$ is an Abelian group, the *d-th elementary ideal (Fitting ideal)* $E_d(\mathfrak{A}_\psi)$ of $\mathfrak{A}_\psi$ is defined for an integer $d \geq 0$ as in the discrete case.

Example 10.3.7 Let k be a number field of finite degree over $\mathbb{Q}$ and let S be the finite set of maximal ideals of $\mathcal{O}_k$. Let us take $\mathfrak{G}$ to be a (standard) quotient group of $G_S(k) = \pi_1(\mathrm{Spec}(\mathcal{O}_k) \setminus S)$. Choose a quotient $\mathfrak{H}$ of $\mathfrak{G}$ and consider the natural homomorphism $\psi : \mathfrak{G} \to \mathfrak{H}$. The we call the completed ψ-differential module $\mathfrak{A}_\psi$ the *complete ψ-Alexander module* of S. When $\mathfrak{H}$ is an Abelian group, we call $E_d(\mathfrak{A}_\psi)$ the *d-th complete ψ-Alexander ideal* of S. What $\mathfrak{H}$ we take depends on the situation we are considering. For example, if $\mathfrak{G}$ is the Galois group $G_S(l)$ ($k = \mathbb{Q}$, $S = \{p_1, \ldots, p_r\}$, $p_i \equiv 1 \bmod l$), dealt with in Chaps. 8 and 9, we can take $\mathfrak{H}$ to be $\mathbb{Z}/m\mathbb{Z}$ ($m = l^e$, $p_i \equiv 1 \bmod m$) and ψ defined by $\psi(\tau_i) = 1 \bmod m$. For this case, $\mathfrak{A}_\psi$ is a module over $\mathbb{Z}_l[[\mathbb{Z}/m\mathbb{Z}]] = \mathbb{Z}_l[[X]]/((1 + X)^m - 1)$. (This will be dealt with in Chap. 11.) When S contains the set of primes of k over l, we can take $\mathfrak{H}$ to be any $\mathbb{Z}_p$-extension $\mathrm{Gal}(k_\infty/k) = \mathbb{Z}_l$. For this case, $\mathfrak{A}_\psi$ becomes a module over $\hat{\Lambda} := \mathbb{Z}_l[[T]] \simeq \mathbb{Z}_l[[\mathrm{Gal}(k_\infty/k)]]$ by the pro-l Magnus isomorphism. (This case will be dealt with in Chaps. 12–14). The algebra $\hat{\Lambda}$ is called the *Iwasawa algebra*. Since $\hat{\Lambda}$ is a Noetherian unique factorization domain, $\Delta_d(\mathfrak{A}_\psi)$ ($d \geq 0$) is defined and in fact we can take $\Delta_d(\mathfrak{A}_\psi)$ to be a polynomial over $\mathbb{Z}_l$ as we will see in Lemma 12.2.3.

10.4 The Complete Crowell Exact Sequence

In this section, we present a complete version of the Crowell exact sequence in Sect. 10.2 ([176, §6], [178, 1]). Suppose that we are given a short exact sequence of profinite groups

$$1 \longrightarrow \mathfrak{N} \longrightarrow \mathfrak{G} \overset{\psi}{\longrightarrow} \mathfrak{H} \longrightarrow 1. \tag{10.4}$$

Theorem 10.4.1 *We have the exact sequence of left $\mathbb{Z}_l[[\mathfrak{H}]]$-modules*

$$0 \longrightarrow \mathfrak{N}^{\mathrm{ab}}(l) \overset{\theta_1}{\longrightarrow} \mathfrak{A}_\psi \overset{\theta_2}{\longrightarrow} \mathbb{Z}_l[[\mathfrak{H}]] \overset{\epsilon_{\mathbb{Z}_l[[\mathfrak{H}]]}}{\longrightarrow} \mathbb{Z}_l \longrightarrow 0.$$

Here $\mathfrak{N}^{\mathrm{ab}}(l)$ is the maximal pro-l quotient of the Abelianization $\mathfrak{N}/\mathfrak{N}^{(2)}$ of $\mathfrak{N}$, θ_1 is the homomorphism induced by $n \mapsto dn$ ($n \in \mathfrak{N}$) and θ_2 is the homomorphism induced by $dg \mapsto \psi(g) - 1$ ($g \in \mathfrak{G}$). This exact sequence is called the complete Crowell exact sequence attached to (10.4).

Proof Taking the $\mathfrak{N}$-homology sequence of the short exact sequence of left $\mathbb{Z}_l[[\mathfrak{N}]]$-modules

$$0 \longrightarrow I_{\mathbb{Z}_l[[\mathfrak{G}]]} \longrightarrow \mathbb{Z}_l[[\mathfrak{G}]] \overset{\epsilon_{\mathbb{Z}_l[[\mathfrak{G}]]}}{\longrightarrow} \mathbb{Z}_l \longrightarrow 0,$$

we have the exact sequence

$$H_1(\mathfrak{N}, \mathbb{Z}_l[[\mathfrak{G}]]) \longrightarrow H_1(\mathfrak{N}, \mathbb{Z}_l) \longrightarrow H_0(\mathfrak{N}, I_{\mathbb{Z}_l[[\mathfrak{G}]]})$$
$$\longrightarrow H_0(\mathfrak{N}, \mathbb{Z}_l[[\mathfrak{G}]]) \longrightarrow H_0(\mathfrak{N}, \mathbb{Z}_l).$$

Here $H_1(\mathfrak{N}, \mathbb{Z}_l) = \mathfrak{N}^{ab}(l)$. Since the other terms are described as in the proof of Proposition 10.2.2 by using Lemma 10.3.3 and Proposition 10.3.4, we obtain the desired exact sequence.   □

When $\mathfrak{G}$ is a finitely presented pro-l-group, as in the discrete case, we can give a description of the complete Crowell exact sequence in terms of the pro-l Fox derivatives by using Theorem 10.3.5.

Example 10.4.2 Let k be a number field of finite degree over $\mathbb{Q}$ and let S be a finite set of maximal ideals of $\mathcal{O}_k$. Let $\mathfrak{G}$ be a quotient of $G_S(k) = \pi_1(\mathrm{Spec}(\mathcal{O}_k) \setminus S)$, and we take $\mathfrak{H}$ to be a quotient of $\mathfrak{G}$ and consider the natural homomorphism $\psi : \mathfrak{G} \to \mathfrak{H}$. Then $\mathfrak{N}^{ab}(l)$ is called the ψ-*Galois module*. When S contains the set of primes over l and $\mathfrak{H} = \mathbb{Z}_l$, we call the ψ-Galois module $\mathfrak{N}^{ab}(l)$ the *Iwasawa module*. By Theorem 10.4.1 and $I_{\hat{\Lambda}} \simeq \hat{\Lambda}$, we have a $\hat{\Lambda}$-isomorphism $\mathfrak{A}_\psi \simeq \mathfrak{N}^{ab}(l) \oplus \hat{\Lambda}$. Therefore $E_d(\mathfrak{N}^{ab}(l)) = E_{d+1}(\mathfrak{A}_\psi)$, $\Delta_d(\mathfrak{N}^{ab}(l)) = \Delta_{d+1}(\mathfrak{A}_\psi)$ $(d \geq 0)$. $\Delta_0(\mathfrak{N}^{ab}(l))$ may be regarded as an analogue of the Alexander polynomial in the context of Iwasawa theory. Note, however, that we use the term *Iwasawa polynomial* of the $\hat{\Lambda}$-module $\mathfrak{N}^{ab}(l)$ in a different sense, according to the convention in Iwasawa theory (see Sect. 12.2). If $\mathfrak{N}^{ab}(l)$ is a finitely generated and torsion $\hat{\Lambda}$-module and has no non-trivial finite $\hat{\Lambda}$-submodule, it is known that $\Delta_0(\mathfrak{N}^{ab}(l))$ coincides with the Iwasawa polynomial of $\mathfrak{N}^{ab}(l)$ [241, p. 299, Ex.(3)] and [141, Appendix]. For example, this condition on $\mathfrak{N}^{ab}(l)$ is known to be satisfied if k is totally real (i.e., any infinite prime of k is a real prime) and $\mathfrak{H}$ is the Galois group of the cyclotomic $\mathbb{Z}_p$-extension of k [93, Theorem 18], [176, 11.3.2] and [241, Theorem 13.31].

Summary

Alexander module A_ψ	Complete Alexander module $\mathfrak{A}_\psi$
$\psi : G \longrightarrow H$	$\psi : \mathfrak{G} \longrightarrow \mathfrak{H}$
Crowell exact sequence	Complete Crowell exact sequence
$0 \longrightarrow N^{ab} \longrightarrow A_\psi \longrightarrow I_{\mathbb{Z}[H]} \longrightarrow 0$	$0 \longrightarrow \mathfrak{N}^{ab} \longrightarrow \mathfrak{A}_\psi \longrightarrow I_{\mathbb{Z}_l[[\mathfrak{H}]]} \longrightarrow 0$
$(N = \mathrm{Ker}(\psi : G \twoheadrightarrow H))$	$(\mathfrak{N} = \mathrm{Ker}(\mathfrak{G} \twoheadrightarrow \mathfrak{H}))$
Link module N^{ab}	Iwasawa module $\mathfrak{N}^{ab}$
$(G = G_L, H = \mathbb{Z})$	$(\mathfrak{G} = G_S(k), \mathfrak{H} = \mathbb{Z}_l)$

Chapter 11
Homology Groups and Ideal Class Groups II: Higher-Order Genus Theory

Let M be a rational homology 3-sphere which is a double covering of S^3 ramified over a r-component link and let k be a quadratic extension of $\mathbb{Q}$ ramified over r odd prime numbers. By the genus theory in Chap. 4, the 2-part of the homology group $H_1(M)$ or the 2-part of the narrow ideal class group $H^+(k)$ has the form

$$\bigoplus_{i=1}^{r-1} \mathbb{Z}/2^{a_i}\mathbb{Z} \quad (a_i \geq 1).$$

Since Gauss' time, it has been a problem to determine the 2^d-rank of $H^+(k)$ for $d > 1$ in terms of some quantities related to ramified prime numbers [246]. Among many works on this problem, L. Rédei [191] and [192, §4] expressed the 4-rank in terms of a matrix whose entries are given by the Legendre symbols involving p_i's. According to the analogy between the linking number and the Legendre symbol in Chap. 2, we find that Rédei's matrix is nothing but an arithmetic analogue of the mod 2 linking matrix. Therefore it would be a natural generalization to express the 2^d-rank in terms of a "higher linking matrix" whose entries are defined by using the Milnor numbers in Chap. 7 (M. Kapranov's question [98]). In this chapter, we shall show such a formula for a link, and then, imitating the method for a link, we shall show a higher-order generalization of Gauss' genus theory.

Throughout this chapter, let l be a fixed prime number.

11.1 The Universal Linking Matrix for a Link

Let $L = K_1 \cup \cdots \cup K_r$ be an r-component link in S^3, $X_L := S^3 \setminus \text{int}(V_L)$ the link exterior and $G_L = \pi_1(X_L)$ the link group. By Theorem 8.1.3, the pro-l completion of G_L has the following presentation:

$$\hat{G}_L(l) = \langle x_1, \ldots, x_r \mid [x_1, y_1] = \cdots = [x_r, y_r] = 1 \rangle,$$

where x_i is the word representing a meridian of K_i and y_i is the pro-l word representing a longitude of K_i. Let $\hat{\psi} : \hat{G}_L(l) \to \hat{G}_L(l)^{ab} = \mathbb{Z}_l^r$ be the Abelianization map and let $\hat{A}_L$ be the complete $\hat{\psi}$-differential module. Letting $\tau_i := \hat{\psi}(x_i)$, we have $\mathbb{Z}_l^r = \overline{\langle \tau_1 \rangle} \times \cdots \overline{\langle \tau_r \rangle}$, $\overline{\langle \tau_i \rangle} = \mathbb{Z}_l$. By the correspondence $\tau_i \mapsto 1 + T_i$, we identify the complete group algebra $\mathbb{Z}_l[[\hat{G}_L(l)^{ab}]] = \mathbb{Z}_l[[\mathbb{Z}_l^r]]$ with the commutative formal power series ring $\hat{\Lambda}_r := \mathbb{Z}_l[[T_1, \ldots, T_r]]$, and so $\hat{A}_L$ becomes a $\hat{\Lambda}_r$-module. Let A_L be the Alexander module of L. Then $\hat{A}_L = A_L \otimes_{\Lambda_r} \hat{\Lambda}_r$, where we regard Λ_r as a subring of $\hat{\Lambda}_r$ by the correspondence $t_i \mapsto 1 + T_i$. We call $\hat{A}_L$ the *complete Alexander module* of L. Similarly, let $\hat{\psi}^{red} : \hat{G}_L(l) \to \mathbb{Z}_l$ be the homomorphism defined by $\hat{\psi}^{red}(x_i) = 1$ $(1 \leq i \leq r)$ and let $\hat{A}_L^{red}$ be the complete $\hat{\psi}^{red}$-differential module. Then $\hat{A}_L^{red}$ is a module over the Iwasawa algebra $\hat{\Lambda} = \mathbb{Z}_l[[T]]$ and one has $\hat{A}_L^{red} = A_L^{red} \otimes_{\Lambda} \hat{\Lambda}$, where A_L^{red} is the reduced Alexander module of L. We call $\hat{A}_L^{red}$ the *reduced complete Alexander module*. Regarding $\hat{\Lambda}$ as a $\hat{\Lambda}_r$-module by the ring homomorphism $\eta : \hat{\Lambda}_r \to \hat{\Lambda}$ defined by $\eta(T_i) = T$ $(1 \leq i \leq r)$, one has $\hat{A}_L^{red} = \hat{A}_L \otimes_{\hat{\Lambda}_r} \hat{\Lambda}$.

Now let

$$\hat{M}(y_i) = 1 + \sum_{\substack{I=(i_1 \cdots i_n) \\ 1 \leq i_1, \ldots, i_n \leq r}} \hat{\mu}(Ii) X_I, \quad X_I := X_{i_1} \cdots X_{i_n}$$

be the pro-l Magnus expansion of y_i. Here $\hat{\mu}(I)$ is the image of the Milnor number $\mu(I) = \mu^{(d)}(I)$ $(d \geq |I|)$ under the natural inclusion $\mathbb{Z} \hookrightarrow \mathbb{Z}_l$.

Definition 11.1.1 We define the *universal linking matrix* $\hat{Q}_L = (\hat{Q}_L(ij))_{1 \leq i, j \leq r}$ of L over $\hat{\Lambda}_r$ by

$$\hat{Q}_L(ij) := \begin{cases} -\displaystyle\sum_{n \geq 1} \sum_{\substack{1 \leq i_1, \cdots, i_n \leq r \\ i_n \neq i}} \hat{\mu}(i_1 \cdots i_n i) T_{i_1} \cdots T_{i_n} & (i = j), \\ \hat{\mu}(ji) T_i + \displaystyle\sum_{n \geq 1} \sum_{1 \leq i_1, \cdots, i_n \leq r} \hat{\mu}(i_1 \cdots i_n ji) T_i T_{i_1} \cdots T_{i_n} & (i \neq j). \end{cases}$$

We also define the *reduced universal linking matrix* $\hat{Q}_L^{red}$ of L over $\hat{\Lambda}$ by $\eta(\hat{Q}_L)$.

Theorem 11.1.2 ([84]) *The universal linking matrix $\hat{Q}_L$ gives a presentation matrix for $\hat{A}_L$ over $\hat{\Lambda}_r$:*

$$(\hat{\Lambda}_r)^r \xrightarrow{\hat{Q}_L} (\hat{\Lambda}_r)^r \longrightarrow \hat{A}_L \longrightarrow 0 \quad \text{(exact)}.$$

The reduced universal linking matrix $\hat{Q}_L^{red}$ gives a presentation matrix for $\hat{A}_L^{red}$ over $\hat{\Lambda}$.

Proof Since $\hat{A}_L^{\mathrm{red}} = \hat{A}_L \otimes_{\hat{\Lambda}_r} \hat{\Lambda}$, it suffices to show the assertion for $\hat{A}_L$. Let $\hat{F}(l)$ be the free pro-l group on $x_1, \ldots, x_r$ and let $\pi : \hat{F}(l) \to \hat{G}_L(l)$ be the natural homomorphism. By Corollary 10.3.6, we must show

$$(\hat{\psi} \circ \pi)\left(\frac{\partial[x_i, y_i]}{\partial x_j}\right) = \hat{Q}_L(ij), \quad 1 \leq i, j \leq r.$$

Firstly, by Proposition 9.3.3,

$$\frac{\partial[x_i, y_i]}{\partial x_j} = (1 - x_i y_i x_i^{-1})\delta_{ij} + (x_i - [x_i, y_i])\frac{\partial y_i}{\partial x_j}.$$

Next, note that the following diagram is commutative:

$$
\begin{array}{ccc}
\mathbb{Z}_l[[\hat{F}(l)]] & \xrightarrow{\ \pi\ } & \mathbb{Z}_l[[\hat{G}_L(l)]] \\
\hat{M}\downarrow & & \downarrow\hat{\psi} \\
\mathbb{Z}_l\langle\langle X_1, \ldots, X_r\rangle\rangle & \xrightarrow{\ \varphi\ } & \hat{\Lambda}_r = \mathbb{Z}_l[[T_1, \ldots, T_r]]
\end{array}
$$

where $\varphi : \mathbb{Z}_l\langle\langle X_1, \ldots, X_r\rangle\rangle \to \hat{\Lambda}_r = \mathbb{Z}_l[[T_1, \ldots, T_r]]; X_i \mapsto T_i$, is the Abelianization map and $\hat{M}$ is the pro-l Magnus isomorphism (Lemma 9.3.1). Since

$$\hat{M}(y_i) = 1 + \sum_{n \geq 1}\sum_{1 \leq i_1, \ldots, i_n \leq r} \hat{\mu}(i_1 \cdots i_n i)X_{i_1} \cdots X_{i_n},$$

$$\hat{M}\left(\frac{\partial y_i}{\partial x_j}\right) = \hat{\mu}(ji) + \sum_{n \geq 1}\sum_{1 \leq i_1, \ldots, i_n \leq r} \hat{\mu}(i_1 \cdots i_n ji)X_{i_1} \cdots X_{i_n},$$

we have

$$
(\hat{\psi} \circ \pi)\left(\frac{\partial[x_i, y_i]}{\partial x_j}\right)
$$

$$
= (\hat{\psi} \circ \pi)\left((1 - x_i y_i x_i^{-1})\delta_{ij} + (x_i - [x_i, y_i])\frac{\partial y_i}{\partial x_j}\right)
$$

$$
= (\varphi \circ \hat{M})\left((1 - x_i y_i x_i^{-1})\delta_{ij} + (x_i - [x_i, y_i])\frac{\partial y_i}{\partial x_j}\right)
$$

$$
= -\delta_{ij}\sum_{n \geq 1}\sum_{1 \leq i_1, \ldots, i_n \leq r} \hat{\mu}(i_1 \cdots i_n i)T_{i_1} \cdots T_{i_n}
$$

$$
+ \hat{\mu}(ji)T_i + \sum_{n \geq 1}\sum_{1 \leq i_1, \ldots, i_n \leq r} \hat{\mu}(i_1 \cdots i_n ji)T_i T_{i_1} \cdots T_{i_n}
$$

$$
= \begin{cases}
-\displaystyle\sum_{n \geq 1}\sum_{\substack{1 \leq i_1, \cdots, i_n \leq r \\ i_n \neq i}} \hat{\mu}(i_1 \cdots i_n i)T_{i_1} \cdots T_{i_n} & (i = j), \\
\hat{\mu}(ji)T_i + \displaystyle\sum_{r \geq 1}\sum_{1 \leq i_1, \cdots, i_n \leq r} \hat{\mu}(i_1 \cdots i_n ji)T_i T_{i_1} \cdots T_{i_n} & (i \neq j)
\end{cases}
$$

which yields the assertion. $\square$

Definition 11.1.3 For an integer $d \geq 2$, we define the *d-th truncated universal linking matrix* $\hat{Q}_L^{(d)} = (\hat{Q}_L^{(d)}(ij))$ of L by

$$
\hat{Q}_L^{(d)}(ij) =
\begin{cases}
-\displaystyle\sum_{n=1}^{d-1} \sum_{\substack{1 \leq i_1, \cdots, i_n \leq r \\ i_n \neq i}} \hat{\mu}(i_1 \cdots i_n i) T_{i_1} \cdots T_{i_n} & (i = j), \\[3mm]
\hat{\mu}(ji) T_i + \displaystyle\sum_{n=1}^{d-2} \sum_{1 \leq i_1, \cdots, i_n \leq r} \hat{\mu}(i_1 \cdots i_n ji) T_i T_{i_1} \cdots T_{i_n} & (i \neq j).
\end{cases}
$$

We also define the *d-th truncated reduced universal linking matrix* $\hat{Q}_L^{\mathrm{red},(d)}$ by $\eta(\hat{Q}_L^{(d)})$.

Example 11.1.4 For $d = 2$, one has $\hat{Q}_L^{\mathrm{red},(2)} = T \cdot C_L$ where $C_L = (C_L(ij))$ is the *linking matrix* of L defined by

$$
C_L(ij) =
\begin{cases}
-\displaystyle\sum_{j \neq i} \mathrm{lk}(K_j, K_i) & \text{if } i = j, \\[3mm]
\mathrm{lk}(K_j, K_i) & \text{if } i \neq j.
\end{cases}
$$

11.2 Higher-Order Genus Theory for a Link

Let $\psi_\infty : G_L \to \mathbb{Z}$ be the homomorphism sending all meridians of each component of L to 1 and let X_∞ be the total linking number covering of X_L corresponding to $\mathrm{Ker}(\psi_\infty)$. Let τ be the generator of $\mathrm{Gal}(X_\infty/X_L)$ corresponding to 1 and identify $\mathbb{Z}[\mathrm{Gal}(X_\infty/X_L)]$ with $\Lambda = \mathbb{Z}[t^{\pm 1}]$ by the correspondence $\tau \leftrightarrow t$. For $n \in \mathbb{N}$, let $\psi_n : G_L \to \mathbb{Z}/n\mathbb{Z}$ be the composite of ψ_∞ with the natural map $\mathbb{Z} \to \mathbb{Z}/n\mathbb{Z}$. Let X_n be the n-fold cyclic covering of X_L corresponding to $\mathrm{Ker}(\psi_n)$ and let M_n be the Fox completion of X_n (Example 2.1.15). We also denote by τ the generator of $\mathrm{Gal}(X_n/X_L)$ corresponding to 1 mod $n \in \mathbb{Z}/n\mathbb{Z}$. We set $\nu_n(t) := t^{n-1} + \cdots + t + 1$. Let $f : M_n \to S^3$ be the ramified covering map. Since the composite of $f_* : H_1(M_n) \to H_1(S^3)$ and the transfer $H_1(S^3) \to H_1(M_n)$ is $\nu_n(\tau)_*$ and $H_1(S^3) = 0$, we can regard $H_1(M_n)$ a module over $\mathcal{O}_n := \Lambda/(\nu_n(t))$.

Theorem 11.2.1 *We have the following isomorphisms of $\mathcal{O}_n$-modules:*

$$
H_1(X_\infty)/\nu_n(t) H_1(X_\infty) \simeq H_1(M_n), \quad A_L^{\mathrm{red}}/\nu_n(t) A_L^{\mathrm{red}} \simeq H_1(M_n) \oplus \Lambda/(\nu_n(t)).
$$

Proof Noting the Λ-isomorphism $I_\Lambda \simeq \Lambda$, the Crowell exact sequence (Theorem 10.2.1) splits and gives the Λ-isomorphism

$$
A_L^{\mathrm{red}} \simeq H_1(X_\infty) \oplus \Lambda.
$$

By tensoring $\Lambda/(\nu_n(t))$ with both sides over Λ, the second assertion follows from the first one. Therefore it suffices to show the first isomorphism. The exact sequence of coefficient modules over X_L

$$0 \longrightarrow \mathbb{Z} \xrightarrow{\times \nu_n(t)} \Lambda/(t^n - 1) \longrightarrow \Lambda/(\nu_n(t)) \longrightarrow 0$$

gives rise to the long exact sequence

$$\cdots \longrightarrow H_1(X_L) \xrightarrow{tr_1} H_1(X_n) \longrightarrow H_1(X_L, \Lambda/(\nu_n(t))) \longrightarrow H_0(X_L) \xrightarrow{tr_0} H_0(X_n).$$

Here tr_i ($i = 0, 1$) denotes the transfer. Since the image of tr_1 is generated by the classes of meridians of components of $f^{-1}(L)$, $\mathrm{Coker}(tr_1) \simeq H_1(M_n)$. And clearly $tr_0 : H_0(X_L) = \mathbb{Z} \to H_0(X_n) = \mathbb{Z}$ is the multiplication by n and so injective. Therefore

$$H_1(M_n) \simeq H_1(X_L, \Lambda/(\nu_n(t))). \tag{11.1}$$

On the other hand, the short exact sequence of chain complexes

$$0 \longrightarrow C_*(X_\infty) \xrightarrow{\nu_n(t)} C_*(X_\infty) \longrightarrow C_*(X_\infty) \otimes_\Lambda \Lambda/(\nu_n(t)) \longrightarrow 0$$

gives rise to the long exact sequence

$$\cdots \longrightarrow H_1(X_\infty) \xrightarrow{\nu_n(t)} H_1(X_\infty) \longrightarrow H_1(X_\infty, \Lambda/(\nu_n(t))) \longrightarrow H_0(X_\infty) \xrightarrow{\nu_n(t)} H_0(X_\infty).$$

Since $\nu_n(t)$ acts on $H_0(X_\infty) = \mathbb{Z}$ as the multiplication by n, we have

$$H_1(X_\infty, \Lambda/(\nu_n(t))) \simeq H_1(X_\infty)/\nu_n(t)H_1(X_\infty). \tag{11.2}$$

By (11.1) and (11.2), we obtain the first isomorphism. $\square$

In the following, let $n = l$ be a prime number and set $M := M_l$, $\mathcal{O} := \mathcal{O}_l$ for simplicity. Fix a primitive l-th root of unity ζ ($\in \overline{\mathbb{Q}}$) so that we identify $\mathcal{O}$ with the Dedekind ring $\mathbb{Z}[\zeta]$ by the correspondence $t \bmod (\nu_l(t)) \mapsto \zeta$. We assume that M is a rational homology 3-sphere. We let $H_1(M)(l)$ denote the l-Sylow subgroup of $H_1(M)$: $H_1(M)(l) = H_1(M) \otimes_\mathbb{Z} \mathbb{Z}_l = H_1(M, \mathbb{Z}_l)$. $H_1(M)(l)$ is regarded as a module over $\hat{\mathcal{O}} := \mathcal{O} \otimes_\mathbb{Z} \mathbb{Z}_l = \mathbb{Z}_l[\zeta]$. Here $\hat{\mathcal{O}}$ is a complete discrete valuation ring with the maximal ideal $\mathfrak{p} = (\varpi)$, $\varpi := \zeta - 1$, and the residue field $\hat{\mathcal{O}}/\mathfrak{p} = \mathbb{F}_l$. By the genus theory in Sect. 4.2, we have the following:

Lemma 11.2.2 $\dim_{\mathbb{F}_l} H_1(M)(l) \otimes_{\hat{\mathcal{O}}} \mathbb{F}_l = r - 1$.

Proof By Theorem 6.2.1,

$$H_1(M)/(\tau - 1)H_1(M) \simeq \mathbb{F}_l^{r-1}.$$

Since the left-hand side $= H_1(M) \otimes_{\mathcal{O}} \mathcal{O}/(\zeta - 1) \simeq H_1(M)(l) \otimes_{\hat{\mathcal{O}}} \mathbb{F}_l$, the assertion follows.   □

By Lemma 11.2.2, $H_1(M)(l)$ has the following form as an $\hat{\mathcal{O}}$-module

$$H_1(M)(l) = \bigoplus_{i=1}^{r-1} \hat{\mathcal{O}}/\mathfrak{p}^{a_i} \qquad (a_i \geq 1).$$

Hence the determination of the $\hat{\mathcal{O}}$-module structure of $H_1(M)(l)$ amounts to describing the $\mathfrak{p}^d$-rank

$$e_d := \#\{i \mid a_i \geq d\}$$

for each $d \geq 2$. In the following, we shall describe e_d in terms of the d-th truncated reduced universal linking matrix introduced in Sect. 8.1. We define the *reduced universal linking matrix* $\hat{Q}_L^{\mathrm{red}}(\varpi)$ of L over $\hat{\mathcal{O}}$ by $\hat{Q}_L^{\mathrm{red}}|_{T=\varpi}$ and define the *d-th truncated reduced universal linking matrix* $\hat{Q}_L^{\mathrm{red},(d)}(\varpi)$ of L over $\hat{\mathcal{O}}$ by $\hat{Q}_L^{\mathrm{red},(d)}|_{T=\varpi}$

Theorem 11.2.3 ([84]) *The matrix $\hat{Q}_L^{\mathrm{red}}(\varpi)$ gives a representation matrix for $H_1(M)(l) \oplus \hat{\mathcal{O}}$ over $\hat{\mathcal{O}}$. For each $d \geq 2$, the matrix $\hat{Q}_L^{\mathrm{red},(d)}(\varpi)$ gives a presentation matrix for $H_1(M)(l)/\mathfrak{p}^d \oplus \hat{\mathcal{O}}/\mathfrak{p}^d$ over $\hat{\mathcal{O}}/\mathfrak{p}^d$.*

Proof By Theorem 11.2.1, we have the following $\hat{\mathcal{O}}$-isomorphisms:

$$\begin{aligned}
H_1(M)(l) &\simeq (H_1(X_\infty)/\nu_l(t)H_1(X_\infty)) \otimes_{\mathbb{Z}} \mathbb{Z}_l \\
&\simeq H_1(X_\infty) \otimes_\Lambda ((\Lambda \otimes_{\mathbb{Z}} \mathbb{Z}_l)/(\nu_l(t))) \\
&\simeq H_1(X_\infty) \otimes_\Lambda \hat{\Lambda}/(\nu_l(1+T)) \\
&\simeq H_1(X_\infty) \otimes_\Lambda \hat{\mathcal{O}}.
\end{aligned}$$

Here the last isomorphism is given by the map sending T to ϖ. By tensoring $\hat{\mathcal{O}}$ with $A_L^{\mathrm{red}} \simeq H_1(X_\infty) \oplus \Lambda$ over Λ, we have an $\hat{\mathcal{O}}$-isomorphism $A_L^{\mathrm{red}} \otimes_\Lambda \hat{\mathcal{O}} \simeq H_1(M)(l) \oplus \hat{\mathcal{O}}$. Then, by Theorem 11.1.2, $\mathfrak{Q}_L^{\mathrm{red}}(\varpi)$ gives a representation matrix of $A_L^{\mathrm{red}} \otimes_\Lambda \hat{\mathcal{O}}$ over $\hat{\mathcal{O}}$ and hence the first assertion. The second assertion is obtained by taking mod $\mathfrak{p}^d$.   □

From Theorem 11.2.3, we have the following:

Theorem 11.2.4 ([84]) *For each $d \geq 2$, let $\varepsilon_1^{(d)}, \ldots, \varepsilon_{r-1}^{(d)}, \varepsilon_r^{(d)} = 0, \varepsilon_i^{(d)}|\varepsilon_{i+1}^{(d)}$, be the elementary divisors of $\hat{Q}_L^{\mathrm{red},(d)}(\varpi)$. Then*

$$e_d = \#\{i \mid 1 \leq i \leq r, \ \varepsilon_i^{(d)} \equiv 0 \bmod \mathfrak{p}^d\} - 1.$$

Corollary 11.2.5 *For $d = 2$, we have the following equality:*

$$e_2 = r - 1 - \mathrm{rank}_{\mathbb{F}_l}(C_L \bmod l),$$

where C_L is the linking matrix of L (Example 8.1.4).

Proof Since $\hat{Q}_L^{\mathrm{red},(2)}(\varpi) = \varpi C_L$, this follows from Theorem 11.2.4. □

For the case that $r = 2$,

$$H_1(M) = \hat{\mathcal{O}}/\mathfrak{p}^a \qquad (a \geq 1)$$

and so $e_d = 0$ or 1.

Corollary 11.2.6 *Let $r = 2$. Assume $e_d = 1$ for $d \geq 1$. Then*

$$e_{d+1} = 1 \Longleftrightarrow \begin{cases} \displaystyle\sum_{n=1}^{d} \sum_{i_1,\ldots,i_{n-1}=1,2} \hat{\mu}(i_1 \cdots i_{n-1}21)\varpi^n \equiv 0 \bmod \mathfrak{p}^{d+1}, \\[4mm] \displaystyle\sum_{n=1}^{d} \sum_{i_1,\ldots,i_{n-1}=1,2} \hat{\mu}(i_1 \cdots i_{n-1}12)\varpi^n \equiv 0 \bmod \mathfrak{p}^{d+1}. \end{cases}$$

Proof Noting $\hat{Q}_L^{\mathrm{red},(d)}(12)(\varpi) = -\hat{Q}_L^{\mathrm{red},(d)}(11)(\varpi)$ and $\hat{Q}_L^{\mathrm{red},(d)}(21)(\varpi) = -\hat{Q}_L^{\mathrm{red},(d)}(22)(\varpi)$ for $d \geq 1$,

$$\begin{aligned} e_{d+1} = 1 &\Longleftrightarrow \hat{Q}_L^{\mathrm{red},(d+1)}(\varpi) \equiv O_2 \bmod \mathfrak{p}^{d+1} \\ &\Longleftrightarrow \hat{Q}^{\mathrm{red},(d+1)}(12)(\varpi) \equiv 0 \bmod \mathfrak{p}^{d+1}, \\ &\qquad\quad \hat{Q}^{\mathrm{red},(d+1)}(21)(\varpi) \equiv 0 \bmod \mathfrak{p}^{d+1}. \end{aligned}$$

From Definition 11.1.3, the assertion follows. □

Example 11.2.7 Let $L = K_1 \cup K_2$ be the *Whitehead link* (Fig. 11.1):
Since $\mu(12) = \mu(21) = 0$, we have $e_2 = 1$ by Corollary 8.2.6. We note the
following formula for a 2-component link [170, Remark, p.100] and [52, (3.9),

Fig. 11.1 The Whitehead link

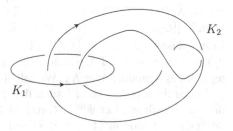

p.555]: For $(i, j) = (1, 2)$ or $(2, 1)$,

$$\bar{\mu}(\underbrace{i \cdots i}_{n \text{ times}} j) \equiv \binom{\text{lk}(K_1, K_2)}{n} \mod \Delta(\underbrace{i \cdots i}_{n \text{ times}} j), \tag{11.3}$$

Since $\mu(12) = \mu(21) = 0$, we have $\Delta(112) = \Delta(221) = 0$. So, by (11.3), $\mu(112) = \mu(221) = 0$. By the cyclic symmetry (Theorem 9.2.1 (4)), we have $\mu(121) = \mu(212) = 0$. Therefore, by Corollary 11.2.6, $e_3 = 1$. Since $\Delta(1112) = \Delta(2221) = 0$, by (11.3), $\mu(1112) = \mu(2221) = 0$. The cyclic symmetry yields $\mu(1121) = \mu(2212) = 0$. By the shuffle relation (Theorem 9.2.1 (3)), $\mu(1221) + \mu(2121) + \mu(2211) = 0$, $\mu(1212) + \mu(2112) + \mu(1122) = 0$. On the other hand, by Murasugi [171, Example 2, p131] or [170, Theorem 4.1], $\mu(1122) = \mu(2211) = 1$. Therefore, by Corollary 11.2.6, $e_4 = 0$. Hence we obtain, as an $\hat{\mathcal{O}}$-module,

$$H_1(M)(l) = \hat{\mathcal{O}}/\mathfrak{p}^3.$$

11.3 The Universal Linking Matrix for Primes

Let $S = \{p_1, \cdots, p_r\}$ be a set of r distinct prime numbers such that $p_i \equiv 1 \mod l$ $(1 \le i \le r)$. In the following, we shall use the same notation as in Sect. 8.2. Let $G_S(l) := \pi_1(\text{Spec}(\mathbb{Z}) \setminus S)(l) = \text{Gal}(\mathbb{Q}_{\overline{S}}(l)/\mathbb{Q})$, where $\mathbb{Q}_{\overline{S}}(l)$ is the maximal pro-l extension of $\mathbb{Q}$ unramified outside $\overline{S} := S \cup \{\infty\}$. By Theorem 8.2.1, the pro-l group $G_S(l)$ has the following presentation:

$$G_S(l) = \langle x_1, \ldots, x_r \mid x_1^{p_1-1}[x_1, y_1] = \cdots = x_r^{p_r-1}[x_r, y_r] = 1 \rangle,$$

where x_i and y_i represent respectively a monodromy and a Frobenius automorphism over p_i given in (8.4). Let $\psi : G_S(l) \to G_S(l)^{\text{ab}}$ be the Abelianization map and let $\mathfrak{A}_S$ be the complete ψ-differential module. Write $p_i - 1 = m_i q_i$, $(m_i = l^{e_i}, (l, q_i) = 1)$. Then $G_S(l)^{\text{ab}}$ is isomorphic to $\mathbb{Z}/m_1\mathbb{Z} \times \cdots \times \mathbb{Z}/m_r\mathbb{Z}$ by class field theory (Example 2.3.6). By the correspondence $1 \mod m_i \in \mathbb{Z}/m_i\mathbb{Z} \mapsto 1 + T_i$, the complete group algebra $\mathbb{Z}_l[[G_S(l)^{\text{ab}}]]$ is identified with $\Lambda_S := \hat{\Lambda}_r/((1+T_1)^{m_1} - 1, \ldots, (1+T_r)^{m_r} - 1)$ (Lemma 10.3.3). This also follows from $\mathbb{Z}_p[[G_S(l)^{\text{ab}}]] = \mathbb{Z}_p[G_S(l)^{\text{ab}}] = \mathbb{Z}_p[t_1, \ldots, t_r]/(t_1^{m_1} - 1, \ldots, t_r^{m_r} - 1)$ and the isomorphism $\mathbb{Z}_p[t_1, \ldots, t_r]/(t_1^{m_1} - 1, \ldots, t_r^{m_r} - 1) \simeq \hat{\Lambda}_r/((1+T_1)^{m_1} - 1, \ldots, (1+T_r)^{m_r} - 1)$ $(t_i \leftrightarrow 1 + T_i)$, which is shown by using Lemma 12.2.3 (1). Hence $\mathfrak{A}_S$ is regarded as a module over Λ_S. We call $\mathfrak{A}_S$ the *complete Alexander module* of S.

Let us fix a power m of l such that $p_i \equiv 1 \mod m$ $(1 \le i \le r)$. So m is a divisor of each m_i. Let $\psi^{\text{red}} : G_S(l) \to \mathbb{Z}/m\mathbb{Z}$ be the homomorphism defined by $\psi^{\text{red}}(x_i) = 1 \mod m$ $(1 \le i \le r)$ and let $\mathfrak{A}_S^{\text{red}}$ be the complete ψ^{red}-differential module. By the correspondence $1 \mod m \mapsto 1 + T$, the complete group algebra

$\mathbb{Z}_l[[\mathbb{Z}/m\mathbb{Z}]]$ is identified with $\Lambda_S^{\mathrm{red}} := \hat{\Lambda}/((1+T)^m - 1)$. Therefore $\mathfrak{A}_S^{\mathrm{red}}$ is regarded as a module over Λ_S^{red}. We call $\mathfrak{A}_S^{\mathrm{red}}$ the *reduced complete Alexander module of* S. Let $\eta : \Lambda_S \to \Lambda_S^{\mathrm{red}}$ be the homomorphism defined by $\eta(T_i) = T$ and we regard Λ_S^{red} as a Λ_S-module via η. Then one has $\mathfrak{A}_S^{\mathrm{red}} = \mathfrak{A}_S \otimes_{\Lambda_S} \Lambda_S^{\mathrm{red}}$.

Let $\hat{\mu}(I)$ denote an l-adic Milnor number of S in Sect. 9.4.

Definition 11.3.1 We define the *universal linking matrix* $\mathfrak{Q}_S = (\mathfrak{Q}_S(ij))_{1 \le i,j \le r}$ of S over Λ_S by

$$\mathfrak{Q}_S(ij)$$

$$:= \begin{cases} T_i^{-1}((1+T_i)^{p_i-1} - 1) - \displaystyle\sum_{n \ge 1} \sum_{\substack{1 \le i_1, \cdots, i_n \le r \\ i_n \ne i}} \hat{\mu}(i_1 \cdots i_n i) T_{i_1} \cdots T_{i_n} & (i = j), \\[2em] \hat{\mu}(ji) T_i + \displaystyle\sum_{n \ge 1} \sum_{1 \le i_1, \cdots, i_n \le r} \hat{\mu}(i_1 \cdots i_n ji) T_i T_{i_1} \cdots T_{i_n} & (i \ne j), \end{cases}$$

where the power series in the right-hand side are regarded as elements in Λ_S (namely, as the images under the natural map $\hat{\Lambda}_r \to \Lambda_S$). We also define the *reduced universal linking matrix* $\mathfrak{Q}_S^{\mathrm{red}}$ of S over Λ_S^{red} by $\eta(\mathfrak{Q}_S)$.

Theorem 11.3.2 ([163]) *The universal linking matrix* $\mathfrak{Q}_S$ *gives a presentation matrix for* $\mathfrak{A}_S$ *over* Λ_S:

$$(\Lambda_S)^r \xrightarrow{\mathfrak{Q}_S} (\Lambda_S)^r \longrightarrow \mathfrak{A}_S \longrightarrow 0 \quad \text{(exact)}.$$

The reduced universal linking matrix $\mathfrak{Q}_S^{\mathrm{red}}$ *gives a presentation matrix for* $\mathfrak{A}_S^{\mathrm{red}}$ *over* Λ_S^{red}.

Proof Since $\mathfrak{A}_S^{\mathrm{red}} = \mathfrak{A}_S \otimes_{\Lambda_S} \Lambda_S^{\mathrm{red}}$, it suffices to show the assertion for $\mathfrak{A}_S$. Let $\hat{F}(l)$ be the free pro-l group on $x_1, \dots, x_r$ and let $\pi : \hat{F}(l) \to G_S(l)$ be the natural homomorphism. By Corollary 7.3.7, we must show

$$(\psi \circ \pi)\left(\frac{\partial x_i^{p_i-1}[x_i, y_i]}{\partial x_j}\right) = \mathfrak{Q}_S(ij).$$

By Proposition 9.3.3 (3) and $(\psi \circ \pi)(x_i^{p_i-1}) = (1+T_i)^{p_i-1} = 1 \in \Lambda_S$, we have

$$(\psi \circ \pi)\left(\frac{\partial x_i^{p_i-1}[x_i, y_i]}{\partial x_j}\right) = (\psi \circ \pi)\left(\frac{\partial x_i^{p_i-1}}{\partial x_j}\right) + (\psi \circ \pi)\left(\frac{\partial [x_i, y_i]}{\partial x_j}\right). \quad (11.4)$$

For the first term in the right hand side, using

$$\frac{\partial x_i^{p_i-1}}{\partial x_j} = \frac{x_i^{p_i-1}-1}{x_i-1}\frac{\partial x_i}{\partial x_j},$$

we have

$$(\psi \circ \pi)\left(\frac{\partial x_i^{p_i-1}}{\partial x_j}\right) = T_i^{-1}((1+T_i)^{p_i-1}-1)\delta_{ij}. \tag{11.5}$$

For the second term, as in the proof of Theorem 11.1.2, we have

$$(\psi \circ \pi)\left(\frac{\partial [x_i, y_i]}{\partial x_j}\right) = -\delta_{ij}\sum_{n\geq 1}\sum_{1\leq i_1,\dots,i_n\leq r}\hat{\mu}(i_1\cdots i_n i)T_{i_1}\cdots T_{i_n}$$
$$+\hat{\mu}(ji)T_i + \sum_{n\geq 1}\sum_{1\leq i_1,\dots,i_n\leq r}\hat{\mu}(i_1\cdots i_n ji)T_i T_{i_1}\cdots T_{i_n}. \tag{11.6}$$

From (11.4)–(11.6), the assertion follows.   □

Definition 11.3.3 For an integer $d \geq 2$, we define the *d-th truncated universal linking matrix* $\mathfrak{Q}_S^{(d)} = (\mathfrak{Q}_S^{(d)}(ij))$ of S over Λ_S by the following:

$$\mathfrak{Q}_S^{(d)}(ij) = \begin{cases} -\sum_{n=1}^{d-1}\sum_{\substack{1\leq i_1,\cdots,i_n\leq r \\ i_n\neq i}}\hat{\mu}(i_1\cdots i_n i)T_{i_1}\cdots T_{i_n} & (i=j), \\ \hat{\mu}(ji)T_i + \sum_{n=1}^{d-2}\sum_{1\leq i_1,\cdots,i_n\leq r}\hat{\mu}(i_1\cdots i_n ji)T_i T_{i_1}\cdots T_{i_n} & (i\neq j). \end{cases}$$

We also define the *d-th truncated reduced universal linking matrix* $\mathfrak{Q}_S^{\text{red},(d)}$ of S over Λ_S^{red} by $\eta(\mathfrak{Q}_S^{(d)})$.

Example 11.3.4 For $d = 2$, one has $\mathfrak{Q}_S^{\text{red},(2)} = T \cdot C_S$, where $C_S = (C_L(ij))$ is the *l-adic linking matrix* of S defined by

$$C_S(ij) = \begin{cases} -\sum_{j\neq i}\hat{\mu}(ji) & \text{if } i = j, \\ \hat{\mu}(ji) & \text{if } i \neq j. \end{cases}$$

11.4   Higher Order Genus Theory for Primes

Let $\mathfrak{N}$ be the kernel of $\psi^{\mathrm{red}} : G_S(l) \to \mathbb{Z}/m\mathbb{Z}$ and let K be the subfield of $\mathbb{Q}_S(l)$ corresponding to $\mathfrak{N}$. Let τ be the generator of $\mathrm{Gal}(K/\mathbb{Q})$ corresponding to $1 \bmod m$. By sending τ to $1 + T$, we identify $\mathbb{Z}_l[\mathrm{Gal}(K/\mathbb{Q})]$ with Λ_S^{red}. Let k be the subfield of K of degree l over $\mathbb{Q}$, and we shall also write the same τ for the generator $\tau|_k$ of $\mathrm{Gal}(k/\mathbb{Q})$. Let M be the maximal Abelian subextension of $\mathbb{Q}_{\overline{S}}(l)/K$ so that $\mathfrak{N}^{\mathrm{ab}} = \mathrm{Gal}(M/K)$. The Galois group $\mathrm{Gal}(K/\mathbb{Q})$ acts on $\mathfrak{N}^{\mathrm{ab}}$ by $x^g := \tilde{g} x \tilde{g}^{-1}$ ($g \in \mathrm{Gal}(K/\mathbb{Q})$, $x \in \mathfrak{N}^{\mathrm{ab}}$), where $\tilde{g}$ denotes a lift of g to $\mathrm{Gal}(M/\mathbb{Q})$. Let L_k^+ be the narrow Hilbert l-class field of k (i.e., the maximal Abelian l-extension of k such that all finite primes of k are unramified) and let $H^+(k)(l)$ be the l-Sylow subgroup of the narrow ideal class group of k. By Artin's reciprocity, $\mathrm{Gal}(L_k^+/k) \simeq H^+(k)(l)$ (Example 2.3.4 and 6.1).

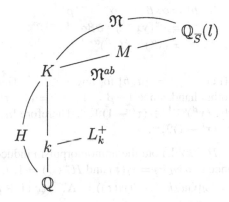

As in Sect. 8.2, we choose a prime $\mathfrak{p}_j$ in $\mathbb{Q}_{\overline{S}}(l)$ over p_j. Let $I_j = I_j(M/k)$ be the inertia group of $\mathfrak{p}_j|_M$ in M/k ($1 \le j \le r$). Let $s_j := \tau_j|_M$ so that I_j is generated by s_j^l. Since L_k^+ is the maximal Abelian subextension of M/k unramified over S, we have

$$
\begin{aligned}
H^+(k)(l) &\simeq \mathrm{Gal}(L_k^+/k) \\
&\simeq \mathrm{Gal}(M/k)/\langle \mathrm{Gal}(M/k)^{(2)}, I_j \ (1 \le j \le r)\rangle.
\end{aligned}
\tag{11.7}
$$

Since $s_j|_K = s_1|_K = \tau$, we may write $s_j = u_j s_1$, $u_j \in \mathfrak{N}^{\mathrm{ab}}$ ($1 \le j \le r$) where we put $u_1 = 1$. Let $\nu_l(t) := 1 + t + \cdots + t^{l-1}$.

Lemma 11.4.1 $\mathrm{Gal}(M/k) = \mathfrak{N}^{\mathrm{ab}} I_j$, $s_j^l = u_j^{\nu_l(\tau)} s_1^l$ ($1 \le j \le r$).

Proof Since $s_j|_K = \tau$ and $I_j = \langle s_j^l \rangle$, the composite map $I_j \hookrightarrow \mathrm{Gal}(M/k) \to \mathrm{Gal}(M/k)/\mathfrak{N}^{\mathrm{ab}} = \mathrm{Gal}(K/k) = \langle \tau^l \rangle$ is surjective. From this the first assertion follows. The latter half is verified as follows:

$$
\begin{aligned}
s_j^l &= (u_j s_1)^l \\
&= u_j s_1 u_j s_1^{-1} s_1^2 \cdots s_1^{l-1} u_j s_1^{-(l-1)} s_1^l \\
&= u_j u_j^{s_1} \cdots u_j^{s_1^{l-1}} s_1^l \\
&= u_j^{\nu_l(\tau)} s_1^l.
\end{aligned}
$$

□

Lemma 11.4.2 $\mathrm{Gal}(M/k)^{(2)} = (\tau^l - 1)\mathfrak{N}^{\mathrm{ab}}$.

Proof Let $a, b \in \mathrm{Gal}(M/k)$ and write $a = \alpha x, b = \beta y$ with $x, y \in \mathfrak{N}^{\mathrm{ab}}$ and $\alpha = \tilde{\tau}^{li}, \beta = \tilde{\tau}^{lj}$ where $\tilde{\tau}^l$ denotes an extension of τ^l to $\mathrm{Gal}(M/k)$. Then

$$
\begin{aligned}
[a, b] &= \alpha x \beta y x^{-1} \alpha^{-1} y^{-1} \beta^{-1} \\
&= x^\alpha (yx^{-1})^{\alpha\beta} \alpha\beta\alpha^{-1} y^{-1} \beta^{-1} \\
&= (x^\alpha)^{1-\beta} (y^\beta)^{\alpha-1}.
\end{aligned}
$$

Set $\beta = 1, \alpha = \tilde{\tau}^l$. Then $y^{\tilde{\tau}^l - 1} = [a, b]$ for any $y \in \mathfrak{N}^{\mathrm{ab}}$. Hence $(\tau^l - 1)\mathfrak{N}^{\mathrm{ab}} \subset \mathrm{Gal}(M/k)^{(2)}$. On the other hand, since $1 - \beta = 1 - \tilde{\tau}^{lj} = (1 - \tilde{\tau}^l)\nu_j(\tilde{\tau}^l), (x^\alpha)^{1-\beta} \in (\tau^l - 1)\mathfrak{N}^{\mathrm{ab}}$. Similarly, $(y^\beta)^{\alpha-1} \in (\tau^l - 1)\mathfrak{N}^{\mathrm{ab}}$. Therefore $[a, b] \in (\tau^l - 1)\mathfrak{N}^{\mathrm{ab}}$. Hence $\mathrm{Gal}(M/k)^{(2)} \subset (\tau^l - 1)\mathfrak{N}^{\mathrm{ab}}$. □

Let $tr : H^+(\mathbb{Q}) \to H^+(k)$ denote the homomorphism induced by the extension of ideals of $\mathbb{Q}$ to k. Since $tr \circ \mathrm{N}_{k/\mathbb{Q}} = \nu_l(\tau)$ and $H^+(\mathbb{Q}) = 1$, $H^+(k)(l)$ is regarded as a module over $\hat{\mathcal{O}} := \mathbb{Z}_l[\mathrm{Gal}(k/\mathbb{Q})]/(\nu_l(\tau)) = \Lambda_S^{\mathrm{red}}/(\nu_l(1 + T))$. Fix a primitive l-th root ζ of unity ($\in \overline{\mathbb{Q}}$). Then, by the correspondence $\tau \mapsto \zeta$, $\hat{\mathcal{O}}$ is identified with the complete discrete valuation ring $\mathbb{Z}_l[\zeta]$ whose maximal ideal $\mathfrak{p}$ is generated by $\varpi = \zeta - 1$ and the residue field is $\hat{\mathcal{O}}/\mathfrak{p} = \mathbb{F}_l$.

Theorem 11.4.3 ([163]) *We have the following isomorphism of $\hat{\mathcal{O}}$-modules:*

$$
\mathfrak{N}^{\mathrm{ab}}/\nu_l(\tau)\mathfrak{N}^{\mathrm{ab}} \simeq H^+(k)(l).
$$

Proof By (11.7), Lemmas 11.4.1 and 11.4.2, we get the following $\hat{\mathcal{O}}$-isomorphism:

$$
\begin{aligned}
\mathrm{Gal}(L_k^+/k) &\simeq \mathfrak{N}^{\mathrm{ab}} I_1(M/k)/\overline{\langle \tau^l - 1)\mathfrak{N}^{\mathrm{ab}}, u_j^{\nu_l(\tau)} s_1^l (1 \leq j \leq r)\rangle} \\
&\simeq \mathfrak{N}^{\mathrm{ab}}/\nu_l(\tau)\langle(\tau - 1)\mathfrak{N}^{\mathrm{ab}}, u_j (2 \leq j \leq r)\rangle.
\end{aligned}
$$

Similarly, replacing k by $\mathbb{Q}$,

$$
1 = H^+(\mathbb{Q})(l) = \mathfrak{N}^{\mathrm{ab}}/\langle(\tau - 1)\mathfrak{N}^{\mathrm{ab}}, u_j (2 \leq j \leq r)\rangle. \tag{11.8}
$$

By (11.4) and (11.8), we obtain the assertion. □

Theorem 11.4.4 *We have the following $\hat{\mathcal{O}}$-isomorphism:*

$$\mathfrak{A}_S^{\mathrm{red}}/v_l(\tau)\mathfrak{A}_S^{\mathrm{red}} = \mathfrak{A}_S^{\mathrm{red}} \otimes_{\Lambda_S^{\mathrm{red}}} \hat{\mathcal{O}} \simeq H^+(k)(l) \oplus \hat{\mathcal{O}}.$$

Proof The complete Crowell exact sequence (Theorem 10.4.1) yields the exact sequence of Λ_S^{red}-modules:

$$0 \longrightarrow \mathfrak{N}^{\mathrm{ab}} \xrightarrow{\theta_2} \mathfrak{A}_S^{\mathrm{red}} \xrightarrow{\theta_1} I_{\Lambda_S^{\mathrm{red}}} \longrightarrow 0.$$

Tensoring $\hat{\mathcal{O}}$ with the above sequence over Λ_S^{red}, we obtain, by Theorem 11.4.3, the following exact sequence of $\hat{\mathcal{O}}$-modules

$$\longrightarrow \mathrm{Tor}_1(I_{\Lambda_S^{\mathrm{red}}}, \hat{\mathcal{O}}) \xrightarrow{\delta} H^+(k)(l) \longrightarrow \mathfrak{A}_S^{\mathrm{red}} \otimes_{\Lambda_S^{\mathrm{red}}} \hat{\mathcal{O}} \longrightarrow I_{\Lambda_S^{\mathrm{red}}} \otimes_{\Lambda_S^{\mathrm{red}}} \hat{\mathcal{O}} \longrightarrow 0.$$

Let $\xi := (\tau^m - 1)/v_l(\tau) = ((1+T)^m - 1)/v_l(1+T)$ and consider a cyclic Λ_S^{red}-free resolution of $\hat{\mathcal{O}} = \Lambda_S^{\mathrm{red}}/(v_l(\tau))$:

$$\cdots \xrightarrow{\xi} \Lambda_S^{\mathrm{red}} \xrightarrow{v_l(\tau)} \Lambda_S^{\mathrm{red}} \xrightarrow{\xi} \Lambda_S^{\mathrm{red}} \xrightarrow{v_l(\tau)} \Lambda_S^{\mathrm{red}} \longrightarrow \hat{\mathcal{O}} \longrightarrow 0.$$

From this we get

$$I_{\Lambda_S^{\mathrm{red}}} \otimes_{\Lambda_S^{\mathrm{red}}} \hat{\mathcal{O}} = \hat{\mathcal{O}},$$
$$\mathrm{Tor}_1(I_{\Lambda_S^{\mathrm{red}}}, \hat{\mathcal{O}}) = \xi\Lambda_S^{\mathrm{red}}/\xi I_{\Lambda_S^{\mathrm{red}}} \simeq \Lambda_S^{\mathrm{red}}/(\tau - 1, v_l(\tau)) \simeq \mathbb{F}_l.$$

Here we note that $\xi \bmod \xi I_{\Lambda_S^{\mathrm{red}}}$ corresponds to $1 \bmod l$ in the second isomorphism. Since $\theta_1(\tau_1^m - 1) = v_l(\tau)\xi, \theta_2(s_1^m) = \tau_1^m - 1$ (where we identify $\mathfrak{A}_S^{\mathrm{red}}$ with $I_{\mathbb{Z}_l[[G_S(l)]]}/I_{\mathbb{Z}_l[[\mathfrak{N}]]}I_{\mathbb{Z}_l[[G_S(l)]]}$), the image of $1 \bmod l$ under the map

$$\delta : \mathbb{F}_l \simeq \mathrm{Tor}_1(I_{\Lambda_S^{\mathrm{red}}}, \hat{\mathcal{O}}) \longrightarrow \mathfrak{N}^{\mathrm{ab}}/v_l(\tau)\mathfrak{N}^{\mathrm{ab}} \simeq H^+(k)(l)$$

is $s_1^m \bmod v_l(\tau)\mathfrak{N}^{\mathrm{ab}}$. On the other hand, since the image of $s_1^m \bmod v_l(\tau)\mathfrak{N}^{\mathrm{ab}}$ under the isomorphism

$$\mathfrak{N}^{\mathrm{ab}}/v_l(\tau)\mathfrak{N}^{\mathrm{ab}} \xrightarrow{\sim} \mathrm{Gal}(M/k)/\langle(\tau^l - 1)\mathfrak{N}^{\mathrm{ab}}, u_j^{v_l(\tau)}s_1^l(1 \le j \le r)\rangle$$

in the proof of Theorem 11.4.3 is 0, we see that δ is the 0-map. Hence we have the exact sequence of $\hat{\mathcal{O}}$-modules

$$0 \longrightarrow H^+(k)(l) \longrightarrow \mathfrak{A}_S^{\mathrm{red}}/v_l(\tau)\mathfrak{A}_S^{\mathrm{red}} \longrightarrow \hat{\mathcal{O}} \longrightarrow 0,$$

which yields the assertion. $\square$

Now by genus theory in Sect. 6.3, we obtain the following:

Lemma 11.4.5 $\dim_{\mathbb{F}_l} H^+(k)(l) \otimes_{\hat{\mathcal{O}}} \mathbb{F}_l = r - 1$.

Proof By Theorem 6.3.1,

$$H^+(k)/(\tau - 1)H^+(k) \simeq \mathbb{F}_l^{r-1}.$$

Since the left-hand side $= H^+(k)(l) \otimes_{\hat{\mathcal{O}}} \hat{\mathcal{O}}/(\zeta - 1) \simeq H^+(k)(l) \otimes_{\hat{\mathcal{O}}} \mathbb{F}_l$, the assertion follows.   □

By Lemma 11.4.5, $H^+(k)(l)$ has the following form as an $\hat{\mathcal{O}}$-module:

$$H^+(k)(l) = \bigoplus_{i=1}^{r-1} \hat{\mathcal{O}}/\mathfrak{p}^{a_i} \quad (a_i \geq 1).$$

Hence the determination of the $\hat{\mathcal{O}}$-module structure of $H^+(k)(l)$ amounts to describing the $\mathfrak{p}^d$-rank

$$e_d := \#\{i \mid a_i \geq d\}$$

for each $d \geq 2$. We define the *reduced universal linking matrix* $\mathfrak{Q}_S^{\text{red}}(\varpi)$ of S over $\hat{\mathcal{O}}$ by $\mathfrak{Q}_S^{\text{red}}|_{T=\varpi}$ and define the *d-th truncated reduced universal linking matrix* $\mathfrak{Q}_S^{\text{red},(d)}(\varpi)$ of S over $\hat{\mathcal{O}}$ by $\mathfrak{Q}_S^{\text{red},(d)}|_{T=\varpi}$. Here we remark that the term $T^{-1}((1 + T)^{p_i - 1} - 1)$ in the definition of $\mathfrak{Q}_S^{\text{red}}(ii)$ becomes 0 if we set $T = \varpi$. Hence $\mathfrak{Q}_S^{\text{red}}(\varpi) = \lim_{d \to \infty} \mathfrak{Q}_S^{\text{red},(d)}(\varpi)$.

By Theorem 11.3.2 and Theorem 11.4.4, we can show the following Theorems 11.4.6, 11.4.7, and Corollary 11.4.8 as in proofs of Theorems 11.2.3, 11.2.4, and Corollary 11.2.5.

Theorem 11.4.6 ([163]) *The matrix* $\mathfrak{Q}_S^{\text{red}}(\varpi)$ *gives a presentation matrix for* $H^+(k)(l) \oplus \hat{\mathcal{O}}$ *over* $\hat{\mathcal{O}}$. *For* $d \geq 2$, *the matrix* $\mathfrak{Q}_S^{\text{red},(d)}(\varpi)$ *gives a presentation matrix for* $H^+(k)(l)/\mathfrak{p}^d \oplus \hat{\mathcal{O}}/\mathfrak{p}^d$ *over* $\hat{\mathcal{O}}/\mathfrak{p}^d$.

Theorem 11.4.7 ([163]) *For* $d \geq 2$, *let* $\varepsilon_1^{(d)}, \ldots, \varepsilon_{r-1}^{(d)}, \varepsilon_r^{(d)} = 0$ $(\varepsilon_i^{(d)} | \varepsilon_{i+1}^{(d)})$ *be the elementary divisors of* $\mathfrak{Q}_S^{\text{red},(d)}(\varpi)$. *Then*

$$e_d = \#\{i \mid 1 \leq i \leq r, \ \varepsilon_i^{(d)} \equiv 0 \bmod \mathfrak{p}^d\} - 1.$$

For $d = 2$, Theorem 11.4.7 yields the following theorem by Rédei ([191], [192], §4] deals with the case $l = 2$).

Corollary 11.4.8 *For $d = 2$, we have the following equality:*

$$e_2 = r - 1 - \mathrm{rank}_{\mathbb{F}_l}(C_S \bmod l),$$

where C_S is the linking matrix of S (Example 11.3.4).

For the case $r = 2$,

$$H^+(k)(l) = \hat{\mathcal{O}}/\mathfrak{p}^a \qquad (a \geq 1)$$

and so $e_d = 0$ or 1.

Corollary 11.4.9 *Let $r = 2$. Assume $e_d = 1$ for $d \geq 1$. Then we have the following:*

$$e_{d+1} = 1 \iff \begin{cases} \displaystyle\sum_{n=1}^{d} \sum_{i_1,\dots,i_{n-1}=1,2} \hat{\mu}(i_1 \cdots i_{n-1}21)\varpi^n \equiv 0 \bmod \mathfrak{p}^{d+1}, \\ \displaystyle\sum_{n=1}^{d} \sum_{i_1,\dots,i_{n-1}=1,2} \hat{\mu}(i_1 \cdots i_{n-1}12)\varpi^n \equiv 0 \bmod \mathfrak{p}^{d+1}. \end{cases}$$

Proof Noting $\mathfrak{Q}_S^{\mathrm{red},(d)}(12)(\varpi) = -\mathfrak{Q}_S^{\mathrm{red},(d)}(11)(\varpi)$, $\mathfrak{Q}_L^{\mathrm{red},(d)}(21)(\varpi) = -\mathfrak{Q}_L^{\mathrm{red},(d)}(22)(\varpi)$ for $d \geq 1$, the assertion is shown in the same way as in the proof of Corollary 11.2.6. □

Example 11.4.10 Let $l = 2$, $r = 3$, $(p_1, p_2, p_3) = (13, 41, 937)$, and $k = \mathbb{Q}(\sqrt{13 \cdot 41 \cdot 937})$. We then have by Example 9.4.4

$$\begin{cases} \mu_2(ij) = 0 \ (1 \leq i, j \leq 3), \\ \mu_2(ijk) = 1 \ (ijk \text{ is a permutation of } 123), \quad \mu_2(ijk) = 0 \ \text{(otherwise)}. \end{cases}$$

Furthermore, we see that $\mu_4(ij) = 0$ if $i \neq j$, since $p_j^{(p_i-1)/4} \equiv 1 \bmod p_i$. Therefore $\mathfrak{Q}_S^{\mathrm{red},(2)}(-2) \equiv O_3 \bmod 4 \ (\varpi = -2)$ and

$$\mathfrak{Q}_S^{\mathrm{red},(3)}(-2) \equiv \begin{pmatrix} 0 & 4 & 4 \\ 4 & 0 & 4 \\ 4 & 4 & 0 \end{pmatrix} \sim \begin{pmatrix} 4 & 0 & 0 \\ 0 & 4 & 0 \\ 0 & 0 & 0 \end{pmatrix} \bmod 8$$

and so $e_2 = 2$, $e_3 = 0$. Hence $H^+(k)(2) \simeq \mathbb{Z}/4\mathbb{Z} \oplus \mathbb{Z}/4\mathbb{Z}$.

Sections 11.1 and 11.2 are an application of number-theoretic methods to link theory, while Sects. 11.3 and 11.4 may be regarded as an application of the knot-theoretic idea to number theory.

Summary

Higher-order genus theory for a link	Higher-order genus theory for primes
$H_1(M)(l) = \bigoplus_{i=1}^{r-1} \hat{\mathcal{O}}/\mathfrak{p}^{a_i}$	$H^+(k)(l) = \bigoplus_{i=1}^{r-1} \hat{\mathcal{O}}/\mathfrak{p}^{a_i}$
Description of $\mathfrak{p}^d$-rank	Description of $\mathfrak{p}^d$-rank
by the Milnor numbers	by the l-adic Milnor numbers

Chapter 12
Homology Groups and Ideal Class Groups III: Asymptotic Formulas

As we discussed in Chap. 10, there is a group-theoretic analogy between the knot module associated to the infinite cyclic covering of a knot exterior and the Iwasawa module associated to the cyclotomic $\mathbb{Z}_p$-extension of number fields. Based on this analogy, there are found close parallels between the Alexander–Fox theory and Iwasawa theory. In this chapter, as a consequence of this analogy, we shall show asymptotic formulas on the orders of the homology groups (p-ideal class groups) of cyclic ramified coverings (extensions). Up to Chap. 8, we have dealt mainly with tame quotients of the Galois group $G_S(k) = \pi_1(\mathrm{Spec}(\mathcal{O}_k) \setminus S)$. In the rest of this book, we shall deal with quotient groups of $G_S(k)$ with wild ramification and investigate analogies with knot groups.

As for the basic materials on Iwasawa theory, we refer to [126], [241].

12.1 The Alexander Polynomial and Homology Groups

Let K be a knot in a rational homology 3-sphere M. Let α be a meridian of K, $X_K := M \setminus \mathrm{int}(V_K)$ the knot exterior and $G_K = \pi_1(X_K)$. In the following, we assume that K is null-homologous in M, namely, there is an oriented surface $\Sigma \subset M$ such that $\partial \Sigma = K$. Then

$$H_1(X_K) \simeq \langle [\alpha] \rangle \oplus H_1(M), \quad \langle [\alpha] \rangle \simeq \mathbb{Z}.$$

In fact, if we denote by $\varphi(c)$ the intersection number (with signature) of 1-cycle $c \in Z_1(X_K)$ with Σ, φ defines a surjective homomorphism $H_1(X_K) \to \mathbb{Z}$ with $\varphi(\alpha) = 1$. Therefore $H_1(X_K) = \langle [\alpha] \rangle \oplus \mathrm{Ker}(\varphi)$, $\langle [\alpha] \rangle \simeq \mathbb{Z}$. By the relative homology exact sequence and the excision, we have $H_1(X_K) \simeq \langle [\alpha] \rangle \oplus H_1(X_K, \partial V_K) \simeq \langle [\alpha] \rangle \oplus H_1(M, V_K) \simeq \langle [\alpha] \rangle \oplus H_1(M)$. Hence $\mathrm{Ker}(\varphi) \simeq H_1(M)$.

© The Author(s), under exclusive license to Springer Nature Singapore Pte Ltd. 2024   171
M. Morishita, *Knots and Primes*, Universitext,
https://doi.org/10.1007/978-981-99-9255-3_12

Let X_∞ be the infinite cyclic covering of X_K corresponding to the kernel of the natural projection $\psi : G_K \to H_1(X_K) \to \langle[\alpha]\rangle = \mathbb{Z}$, and let τ be a generator of $\mathrm{Gal}(X_\infty/X_K)$ corresponding to $1 \in \mathbb{Z}$. Let $\Lambda := \mathbb{Z}[t^{\pm 1}] = \mathbb{Z}[\mathrm{Gal}(X_\infty/X_K)]$ $(t \leftrightarrow \tau)$. For $n \in \mathbb{N}$, let X_n be the cyclic subcovering of $X_\infty \to X_K$ of degree n, and let M_n be the Fox completion of X_n.

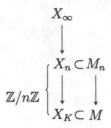

Proposition 12.1.1 *We have the following isomorphism:*

$$H_1(M_n) \simeq H_1(X_\infty)/(t^n - 1)H_1(X_\infty) \quad (n \geq 1).$$

Proof By the Wang exact sequence $H_1(X_\infty) \overset{t^n-1}{\to} H_1(X_\infty) \to H_1(X_n) \to \mathbb{Z} \to 0$,

$$H_1(X_n) \simeq H_1(X_\infty)/(t^n - 1)H_1(X_\infty) \oplus \mathbb{Z}.$$

Here $1 \in \mathbb{Z}$ corresponds to a lift $[\tilde{\alpha}^n]$ of $[\alpha^n]$ to X_n. (Since the image of α^n in $\mathrm{Gal}(X_n/X_K) \simeq \mathbb{Z}/n\mathbb{Z}$ is 0, α^n can be lifted to X_n.) Since $H_1(M_n) = H_1(X_n)/\langle[\tilde{\alpha}^n]\rangle$, we obtain our assertion.   □

Let $G_K = \langle x_1, \dots, x_m \mid R_1 = \cdots = R_{m-1} = 1 \rangle$ be a presentation of G_K (Example 2.1.6) and let $\pi : F(x_1, \dots, x_m) \to G_K$ be the natural homomorphism. The *Alexander module* of K is defined by the ψ-differential module and is denoted by A_K. Then $Q_K := ((\psi \circ \pi)(\partial R_i/\partial x_j))$ gives a presentation matrix of the Λ-module A_K (Corollary 10.1.6). Since $A_K \simeq H_1(X_\infty) \oplus \Lambda$ (Λ-isomorphism) by the Crowell exact sequence (Theorem 10.2.1), we may assume $Q_K = (Q_1|0)$ by some elementary operations if necessary. Here Q_1 gives a presentation square matrix of the Λ-module $H_1(X_\infty)$.

Proposition 12.1.2 $H_1(X_\infty)$ *is a finitely generated, torsion Λ-module, and we have* $E_0(H_1(X_\infty)) = (\det(Q_1)), \quad \Delta_0(H_1(X_\infty)) = \det(Q_1)(\neq 0).$

Proof Since $H_1(X_\infty) = \Lambda^{m-1}/Q_1(\Lambda^{m-1})$, $H_1(X_\infty)$ is finitely generated over Λ. If $\mathrm{rank}_\Lambda H_1(X_\infty) \geq 1$, Proposition 12.1.1 and $\Lambda/(t-1)\Lambda \simeq \mathbb{Z}$ imply $\mathrm{rank}_\mathbb{Z} H_1(M) \geq 1$. This is a contradiction. The latter part is obvious.   □

We let $\Delta_K(t) := \Delta_0(H_1(X_\infty)) = \det(Q_1)$ and call it the *Alexander polynomial* of K. $\Delta_K(t)$ is determined up to multiplication by an element of $\Lambda^\times$. Since $\Lambda_\mathbb{Q} := \Lambda \otimes_\mathbb{Z} \mathbb{Q} = \mathbb{Q}[t^{\pm 1}]$ is a principal ideal domain, we have a $\Lambda_\mathbb{Q}$-isomorphism

$$H_1(X_\infty) \otimes_\mathbb{Z} \mathbb{Q} \simeq \bigoplus_{i=1}^{s} \Lambda_\mathbb{Q}/(f_i), \quad f_i \in \Lambda_\mathbb{Q}.$$

Here noting that τ acts on the right-hand side as the multiplication by t, one has[1]

$$\Delta_K(t) = f_1 \cdots f_s = \det(t \cdot \mathrm{id} - \tau \mid H_1(X_\infty) \otimes_\mathbb{Z} \mathbb{Q}) \mod \Lambda_\mathbb{Q}^\times. \tag{12.1}$$

Now, suppose that M is a homology 3-sphere. Then Proposition 12.1.1 and Lemma 12.1.3 below imply that $\Delta_K(1) = \pm 1$, and hence $\Delta_K(t)$ coincides with the characteristic polynomial $\det(t\,\mathrm{id} - \tau \mid H_1(X_\infty) \otimes_\mathbb{Z} \mathbb{Q})$ up to multiplication of an element $\Lambda^\times = \{\pm t^n \mid n \in \mathbb{Z}\}$.

Next, we prepare an algebraic lemma. For the proof, we refer to [83, Theorem 3.13].

Lemma 12.1.3 *Let N be a finitely generated, torsion Λ-module and suppose that $E_0(N) = (\Delta)$. Then, for a non-constant $f(t) \in \mathbb{Z}[t]$, $N/f(t)N$ is a torsion Abelian group if and only if $\Delta(\xi) \neq 0$ for any root $\xi \in \overline{\mathbb{Q}}$ of $f(t) = 0$. Further, if $f(t)$ is decomposed into the form $\pm \prod_{j=1}^{n}(t - \xi_j)$, one has*

$$\#(N/f(t)N) = \prod_{j=1}^{n} |\Delta(\xi_j)|.$$

For $g(t) = c_0 t^d + \cdots + c_d \in \mathbb{Z}[t]$ $(c_0 \neq 0, d \geq 1)$, we define the *Mahler measure* $m(g)$ of $g(t)$ by

$$m(g) := \exp\left(\int_0^1 \log |g(e^{2\pi\sqrt{-1}x})| \, dx\right).$$

If $g(t) = c_0 \prod_{i=1}^{d}(t - \theta_i)$, by Jensen's formula [1, Ch.5, 3.1], we obtain $m(g) = c_0 \prod_{i=1}^{d} \max(|\theta_i|, 1)$.

Theorem 12.1.4 *Assume that there is no root of $\Delta_K(t) = 0$ which is an n-th root of unity for some n. Then M_n's are rational homology 3-spheres and we have*

$$\#H_1(M_n) = \prod_{j=0}^{n-1} |\Delta_K(e^{2\pi\sqrt{-1}j/n})|$$

---

[1] $a = b \mod R^\times$ means that $b = au$ for some $u \in R^\times$.

for any n, and

$$\lim_{n \to \infty} \frac{1}{n} \log \#H_1(M_n) = \log m(\Delta_K).$$

Proof By Proposition 12.1.1, Lemma 12.1.3 and the assumption, any $H_1(M_n)$ is finite for any n and we have

$$\#H_1(M_n) = \prod_{j=0}^{n-1} |\Delta_K(e^{2\pi\sqrt{-1}j/n})|.$$

Hence

$$\lim_{n \to \infty} \frac{1}{n} \log \#H_1(M_n) = \lim_{n \to \infty} \frac{1}{n} \sum_{j=0}^{n-1} \log |\Delta_K(e^{2\pi\sqrt{-1}j/n})|$$

$$= \int_0^1 \log |\Delta_K(e^{2\pi\sqrt{-1}x})| \, dx$$

$$= \log m(\Delta_K).$$

$\square$

Example 12.1.5 Let $K \subset S^3$ be the figure eight knot $B(5, 3)$ (cf. Fig. 7.1). Using the presentation $G_K = \langle x_1, x_2 \mid x_2 x_1^{-1} x_2 x_1 x_2^{-1} = x_1^{-1} x_2 x_1 x_2^{-1} x_1 \rangle$, we obtain $\Delta_K = t^2 - 3t + 1$. Hence

$$\lim_{n \to \infty} \frac{1}{n} \log \#H_1(M_n) = \log m(\Delta_K) = \log \frac{3 + \sqrt{5}}{2}.$$

Remark 12.1.6

(1) The above formula for $\#H_1(M_n)$ is due to Fox ([53]). The asymptotic formula in Theorem 12.1.5 was shown in [68], [183] when M is a homology 3-sphere. For some extensions to the case of cyclic covering ramified along a link, we refer to [210], [232].

(2) For Iwasawa-theoretic type formulas on the growth of the p-primary parts of the homology groups (cf. Sect. 12.2), we refer to [84], [95], where topological analogues for links of Iwasawa λ and μ-invariants were introduced. For further study of those Iwasawa-type invariants of links, we refer to [96], [229], [230], [231] and [49].

(3) The homology growth rate in Theorem 12.1.4 is also interpreted as the entropy of the natural $\mathbb{Z} = \{\tau^n | n \in \mathbb{Z}\}$-action on the compact Pontryagin dual $H_1(X_\infty, \mathbb{Z})^*$ (cf. [210], [198], [232]).

(4) A generalization of Fox's formula to the twisted homology for a knot group representation was obtained in [219]. For twisted Iwasawa invariants for knots, see [221].

12.2 The Iwasawa Polynomial and p-Ideal Class Groups

Let k be a finite algebraic number field and let p be a prime number. Let k_∞ be the cyclotomic $\mathbb{Z}_p$-extension of k (Example 2.3.6). In the following, we assume that only one prime ideal $\mathfrak{p}$ of k is ramified in k_∞/k and it is totally ramified. (This is an assumption analogous to the knot case.) We note that $\mathfrak{p}$ must be a prime ideal over p by class field theory. We denote by μ_{p^d} the group of p^d-th roots of unity. The fields $k = \mathbb{Q}(\mu_{p^d})$ or their subfields satisfy the assumption.

For an integer $n \geq 0$, let k_n be the cyclic subfield of k_∞/k of degree p^n, and let H_n be the p-Sylow subgroup of the ideal class group of k_n: $H_n := H(k_n)(p)$. Let L_n be the maximal unramified Abelian p-extension of k_n (Hilbert p-class field). By unramified class field theory (Example 2.3.4, 6.1), we obtain the following isomorphism for each n:

$$\varphi_n : H_n \simeq \mathrm{Gal}(L_n/k_n); \quad \varphi_n([\mathfrak{a}]) = \sigma_\mathfrak{a}.$$

For $n \geq m$, let $\mathrm{N}_{n/m} : H_n \to H_m$ be the norm map. Then we have the commutative diagram:

$$
\begin{array}{ccc}
H_n & \xrightarrow{\varphi_n} & \mathrm{Gal}(L_n/k_n) \\
\mathrm{N}_{n/m}\downarrow & & \downarrow \\
H_m & \xrightarrow{\varphi_m} & \mathrm{Gal}(L_m/k_m),
\end{array}
\tag{12.2}
$$

where the right vertical map is the restriction map. Note that $\{H_n\}_{n\geq 0}$ forms a projective system with respect to the norm maps, and let $H_\infty := \varprojlim_n H_n$. Let $L_\infty := \cup_{n\geq 0} L_n$. By (12.2), we have an isomorphism of pro-p Abelian groups:

$$\varphi_\infty : H_\infty \simeq \mathrm{Gal}(L_\infty/k_\infty).$$

Since L_n is the maximal unramified Abelian p-extension of k_n, we see that L_n is a Galois extension of k, and hence L_∞/k is a pro-p Galois extension.

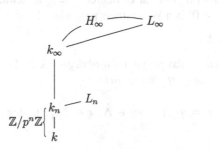

Let $\tilde{\mathfrak{p}}$ be a prime of L_∞ lying over $\mathfrak{p}$ and let $I_\mathfrak{p}$ be the inertia group of $\tilde{\mathfrak{p}}$. By the assumption, $I_\mathfrak{p} \simeq \mathrm{Gal}(L_\infty/k)/\mathrm{Gal}(L_\infty/k_\infty) \simeq \mathrm{Gal}(k_\infty/k)$. We fix a topological generator γ of $\mathrm{Gal}(k_\infty/k)$. By sending γ to $1 + T$, we identify

$\mathbb{Z}_p[[\mathrm{Gal}(k_\infty/k)]]$ with $\hat{\Lambda} := \mathbb{Z}_p[[T]]$ (pro-p Magnus isomorphism). Note that the Galois group $\mathrm{Gal}(L_\infty/k_\infty)$ is an *Iwasawa module* (ψ-Galois module) in the sense of Example 10.4.2 for the natural homomorphism ψ : $\mathrm{Gal}(L_\infty/k) \to \mathrm{Gal}(k_\infty/k) \simeq \mathbb{Z}_p$. We note that $g \in \mathrm{Gal}(k_\infty/k)$ acts on $\mathrm{Gal}(L_\infty/k_\infty)$ by the inner-automorphism $g(x) = \tilde{g} \circ x \circ \tilde{g}^{-1}$ ($\tilde{g}$ being an extension of g to $\mathrm{Gal}(L_\infty/k)$) and that $\hat{\Lambda}$ acts on H_∞ in the natural manner, and we see that the isomorphism φ_∞ commutes with these actions. The following proposition is an analogue of Proposition 12.1.1.

Proposition 12.2.1 *We have the following isomorphism:*

$$H_n \simeq H_\infty/((1+T)^{p^n} - 1)H_\infty \quad (n \geq 0).$$

Proof Note that $H_n \simeq \mathrm{Gal}(L_n/k_n) \simeq \mathrm{Gal}(L_\infty/k_n)/\mathrm{Gal}(L_\infty/L_n)$. Since $I_{\mathfrak{p}}^{p^n} \simeq \mathrm{Gal}(k_\infty/k_n)$,

$$\mathrm{Gal}(L_\infty/k_n) = \mathrm{Gal}(L_\infty/k_\infty) \cdot I_{\mathfrak{p}}^{p^n}, \ \ \mathrm{Gal}(L_\infty/L_n) = \langle \mathrm{Gal}(L_\infty/k_n)^{(2)}, I_{\mathfrak{p}}^{p^n} \rangle.$$

By an argument similar to the proof of Lemma 11.4.2, we can show

$$\mathrm{Gal}(L_\infty/k_n)^{(2)} = (\gamma^{p^n} - 1)\mathrm{Gal}(L_\infty/k_\infty).$$

Hence

$$\begin{aligned} H_n &\simeq \mathrm{Gal}(L_\infty/k_\infty)I_{\mathfrak{p}}^{p^n}/\langle(\gamma^{p^n} - 1)\mathrm{Gal}(L_\infty/k_\infty), I_{\mathfrak{p}}^{p^n}\rangle \\ &\simeq \mathrm{Gal}(L_\infty/k_\infty)/(\gamma^{p^n} - 1)\mathrm{Gal}(L_\infty/k_\infty) \\ &\simeq H_\infty/((1+T)^{p^n} - 1)H_\infty. \end{aligned}$$

$\square$

Next, we prepare algebraic lemmas concerning the structures of the Iwasawa algebra $\hat{\Lambda}$ and $\hat{\Lambda}$-modules. A polynomial $g(T) \in \mathbb{Z}_p[T]$ is called a *Weierstrass polynomial* if it is of the form $g(T) = T^\lambda + c_1 T^{\lambda-1} + \cdots + c_\lambda, c_1, \ldots, c_\lambda \equiv 0 \bmod p$. (The Weierstrass polynomial of degree $\lambda = 0$ is defined to be 1.) For example, $(1+T)^{p^n} - 1$ ($n \geq 0$) is a Weierstrass polynomial.

Proposition 12.2.2

(1) *Let g be a Weierstrass polynomial of degree $\lambda(\geq 1)$. Then any element $f \in \hat{\Lambda}$ can be written uniquely in the form*

$$f = qg + r, \ \ q \in \hat{\Lambda}, \ r \in \mathbb{Z}_p[T], \ \deg(r) \leq \lambda - 1.$$

(2) (*p-adic Weierstrass preparation theorem*) *Any* $f(T)(\neq 0) \in \hat{\Lambda}$ *can be written uniquely in the form*

$$
\begin{cases}
f(T) = p^{\mu} g(T) u(T), \\
\mu \in \mathbb{Z}_{\geq 0}, \ g(T) \ \text{is a Weierstrass polynomial,} \ u(T) \in \hat{\Lambda}^{\times}.
\end{cases}
$$

The quantities $\mu = \mu(f)$, $\lambda = \lambda(f) := \deg(g)$ are called the *μ-invariant* and the *λ-invariant* of f respectively.

For the proof of Lemma 12.2.3, we refer to [176, 5.3.1, 5.3.4].

Two $\hat{\Lambda}$-modules $\mathfrak{N}$, $\mathfrak{N}'$ are said to be *pseudo-isomorphic*, written as $\mathfrak{N} \sim \mathfrak{N}'$, if there is a $\hat{\Lambda}$-homomorphism $\varphi : \mathfrak{N} \to \mathfrak{N}'$ such that $\mathrm{Ker}(\varphi)$ and $\mathrm{Coker}(\varphi)$ are finite.

Lemma 12.2.3 *Let $\mathfrak{N}$ be a compact $\hat{\Lambda}$-module.*

(1) (*Nakayama's lemma*) $\mathfrak{N}$ *is a finitely generated $\hat{\Lambda}$-module if and only if $\mathfrak{N}/(p, T)\mathfrak{N}$ is finite.*
(2) Suppose that $\mathfrak{N}$ is a finitely generated $\hat{\Lambda}$-module. Then

$$
\mathfrak{N} \sim \hat{\Lambda}^r \oplus \bigoplus_{i=1}^{s} \hat{\Lambda}/(p^{m_i}) \oplus \bigoplus_{i=1}^{t} \hat{\Lambda}/(f_i^{e_i}),
$$

where r is an integer (≥ 0), $m_i, e_i \in \mathbb{N}$ and f_i is an irreducible Weierstrass polynomial.

For the proof of Lemma 9.2.4, we refer to [176, 5.2.8, 5.3.8].

Let $\mathfrak{N}$ be a finitely generated, torsion $\hat{\Lambda}$-module. By Lemma 12.2.3 (2),

$$
\mathfrak{N} \sim \bigoplus_{i=1}^{s} \hat{\Lambda}/(p^{m_i}) \oplus \bigoplus_{i=1}^{t} \hat{\Lambda}/(f_i^{e_i}).
$$

The ideal generated by $f := \prod_{i=1}^{s} p^{m_i} \prod_{i=1}^{t} f_i^{e_i}$, which is determined by the $\hat{\Lambda}$-module $\mathfrak{N}$, is called the *characteristic ideal* of $\mathfrak{N}$. f is determined up to multiplication of an element of $\hat{\Lambda}^{\times}$ and is called the *Iwasawa polynomial* of $\mathfrak{N}$. When $\mathfrak{N}$ has no non-trivial finite $\hat{\Lambda}$-submodule, it can be shown that the characteristic ideal of $\mathfrak{N}$ coincides with the 0-th elementary ideal $E_0(\mathfrak{N})$ and $f = \Delta_0(\mathfrak{N})$ [241, p.299, Ex.(3)] and [141, Appendix]. Since $\hat{\Lambda}_{\mathbb{Q}_p} := \hat{\Lambda} \otimes_{\mathbb{Z}_p} \mathbb{Q}_p = \mathbb{Q}_p[[T]]$ is a principal ideal domain, we have a $\hat{\Lambda}_{\mathbb{Q}_p}$-isomorphism

$$
\mathfrak{N} \otimes_{\mathbb{Z}_p} \mathbb{Q}_p \simeq \bigoplus_{i=1}^{t} \hat{\Lambda}_{\mathbb{Q}_p}/(f_i^{e_i}).
$$

Noting that $\gamma - 1$ acts on the right-hand side by the multiplication by T, we obtain

$$f(T) = \det(T \cdot \mathrm{id} - (\gamma - 1) \mid \mathfrak{N} \otimes_{\mathbb{Z}_p} \mathbb{Q}_p) \quad \mathrm{mod}\ (\hat{\Lambda}_{\mathbb{Q}_p})^{\times}. \tag{12.3}$$

The quantities $\mu(f) := \sum_{i=1}^{t} m_i$, $\lambda(f) := \sum_{i=1}^{s} \deg(f_i^{e_i})$, which are determined uniquely by $\mathfrak{N}$, are called the *Iwasawa μ-invariant, Iwasawa λ-invariant* of $\mathfrak{N}$, written as $\mu(\mathfrak{N})$, $\lambda(\mathfrak{N})$, respectively. If $\mu(\mathfrak{N}) = 0$, f equals the characteristic polynomial $\det(T\,\mathrm{id} - (\gamma - 1) \mid \mathfrak{N} \otimes_{\mathbb{Z}_p} \mathbb{Q}_p) \bmod \hat{\Lambda}^{\times}$.

Proposition 12.2.4 H_{∞} *is a finitely generated, torsion $\hat{\Lambda}$-module.*

Proof Since H_n is finite (2.2.14), by Proposition 12.2.1 and Proposition 12.2.4 (1), H_{∞} is a finitely generated $\hat{\Lambda}$-module. If H_{∞} is not a torsion over $\hat{\Lambda}$, $H_{\infty} \approx \hat{\Lambda}^r \oplus \cdots$, $r \geq 1$. Since $\hat{\Lambda}/T\hat{\Lambda} \simeq \mathbb{Z}_p$ is infinite, this contradicts Proposition 12.2.1.   □

We have the following asymptotic formula for $\#H_n$ which may be regarded as an arithmetic analogue of Theorem 12.1.4 for a knot.

Theorem 12.2.5 (Iwasawa's Class Number Formula) *Let* $\mu = \mu(H_{\infty})$, $\lambda = \lambda(H_{\infty})$. *For sufficiently large n,*

$$\log_p \#H_n = \mu p^n + \lambda n + \nu,$$

where ν is a constant independent of n.

Proof By Lemma 12.2.3 (2) and Proposition 12.2.5,

$$H_{\infty} \sim E := \bigoplus_{i=1}^{s} \hat{\Lambda}/(p^{m_i}) \oplus \bigoplus_{i=1}^{t} \hat{\Lambda}/(f_i^{e_i}),$$

where $m_i, e_i \in \mathbb{N}$, f_i is an irreducible Weierstrass polynomial. From this, it can be shown [241, p.284] that there is a constant c independent of n such that

$$\#(H_{\infty}/((1 + T)^{p^n} - 1)H_{\infty}) = p^c \#(E/((1 + T)^{p^n} - 1)E). \tag{12.4}$$

When $E = \hat{\Lambda}/(p^m)$:

$$\begin{aligned}
\#(E/((1 + T)^{p^n} - 1)E) &= \#(\mathbb{Z}/p^m\mathbb{Z}[T]/((1 + T)^{p^n} - 1)) \\
&= (p^m)^{\deg((1+T)^{p^n}-1)} \\
&= p^{mp^n}.
\end{aligned} \tag{12.5}$$

When $E = \hat{\Lambda}/(g)$ (g being a Weierstrass polynomial of $\deg(g) \geq 1$):

$$\#(E/((1+T)^{p^n} - 1)E) = \#(\mathbb{Z}_p[T]/(g, (1+T)^{p^n} - 1))$$
$$= \prod_{\zeta^{p^n}=1} |g(\zeta - 1)|_p^{-1}.$$

Here $|\cdot|_p$ stands for the p-adic multiplicative valuation with $|p|_p^{-1} = p$ (Note that $g(\zeta - 1) \neq 0$ for any p^n-th root ζ of unity, as $E/((1+T)^{p^n} - 1)E$ is finite). Let v_p be the p-adic additive valuation ($|x|_p = p^{-v_p(x)}$) and suppose $g(T) = T^d + a_1 T^{d-1} + \cdots + a_d$, $a_i \equiv 0 \bmod p$. If n is sufficiently large, we have $v_p((\zeta - 1)^d) < v_p(p)$ for a primitive p^n-th root ζ of unity and hence $v_p(g(\zeta - 1)) = v_p((\zeta - 1)^d)$. Therefore, when n is sufficiently large,

$$v_p\left(\prod_{\zeta^{p^n}=1} g(\zeta - 1)\right) = v_p\left(\prod_{\substack{\zeta^{p^n}=1 \\ \zeta \neq 1}} (\zeta - 1)^d\right) + C$$
$$= v_p(p^{nd}) + C$$
$$= nd + C,$$

where C is a constant independent of n. Hence

$$\#(E/((1+T)^{p^n} - 1)E) = p^{nd+C}. \tag{12.6}$$

By (12.4)–(12.6) and $\mu = \sum_{i=1}^{s} m_i$, $\lambda = \sum_{i=1}^{t} \deg(f_i^{e_i})$, we obtain the desired formula. $\square$

Remark 12.2.6

(1)   Note that the $\hat{\Lambda}$-module H_∞ and hence the Iwasawa invariants $\mu(H_\infty)$, $\lambda(H_\infty)$ depend only on k and p and are called the *Iwawasa μ-invariant, Iwasawa λ-invariant* of (k, p). It is known that $\mu(H_\infty) = 0$ if $k/\mathbb{Q}$ is an Abelian extension [241, 7.5], and conjectured that this is always the case. In [151], Iwasawa invariants were studied for certain (k, p) in terms of mod p arithmetic linking numbers of primes ramified in $k/\mathbb{Q}$.

(2)   As in the case for knots (Remark 12.1.6 (2)), it would be interesting to study some arithmetic meaning of the entropy of the $\mathbb{Z} = \{\gamma^n | n \in \mathbb{Z}\}$-action on the Iwasawa module H_∞.

Example 12.2.7 Let $k = \mathbb{Q}(\mu_p)$. Then $k_\infty = \mathbb{Q}(\mu_{p^\infty})$, $\mu_{p^\infty} = \cup_{d \geq 1} \mu_{p^d}$, and our assumption is satisfied. By Remark 12.2.6 (1),

$$\log_p \#H_n = \lambda n + v \quad (n \gg 0).$$

If we assume the Vandiver conjecture which asserts that the class number of $\mathbb{Q}(\zeta + \zeta^{-1})$ is not divisible by p, it is known that the following formula

$$\#H_n = \prod_{\zeta^{p^n}=1} |f(\zeta - 1)|_p^{-1}$$

holds for any $n \geq 1$ ([241, Theorem 10.16]). Here $f(T)$ stands for the Iwasawa polynomial of H_∞.

Summary

Infinite cyclic covering $X_\infty \to X_K$	Cyclotomic $\mathbb{Z}_p$-extension k_∞/k
$\mathrm{Gal}(X_\infty/X_K) = \langle \tau \rangle \simeq \mathbb{Z}$	$\mathrm{Gal}(k_\infty/k) = \langle \gamma \rangle \simeq \mathbb{Z}_p$
Knot module $H_1(X_\infty)$	Iwasawa module H_∞
Alexander polynomial	Iwasawa polynomial
$\det(t \cdot \mathrm{id} - \tau \mid H_1(X_\infty) \otimes_{\mathbb{Z}} \mathbb{Q})$	$\det(T \cdot \mathrm{id} - (\gamma - 1) \mid H_\infty \otimes_{\mathbb{Z}_p} \mathbb{Q}_p)$
Asymptotic formula for $\#H_1(M_n)$	Asymptotic formula for $\#H(k_n)(p)$

Chapter 13
Torsions and the Iwasawa Main Conjecture

The Iwasawa main conjecture asserts that the Iwasawa polynomial coincides essentially with the Kubota–Leopoldt p-adic analytic zeta function. This can also be regarded as a determinant expression of the p-adic zeta function, which was originally conjectured by Iwasawa as an analogue of the determinant expression, due to Weil and Grothendieck, of the congruence zeta function. According to the analogy between the Iwasawa polynomial and the Alexander polynomial discussed in Chaps. 7 and 9, an analogue of the Iwasawa main conjecture in knot theory may be a connection between the Alexander polynomial and a certain analytically defined zeta function. As Milnor [149] already pointed out, such a connection is given as the relation between the Reidemeister–Milnor torsion and the Lefschetz zeta function associated to the infinite cyclic covering of a knot exterior. If one takes the Ray–Singer spectral zeta function as an analytic zeta function, such a connection is also given as the relation between the Reidemeister torsion and the analytic torsion (J. Cheeger, W. Müller and W. Lück).

13.1 Torsions and Zeta Functions

Let V be an n-dimensional vector space over a field F. For two (ordered) bases $b = (b_1, \ldots, b_n)$ and $c = (c_1, \ldots, c_n)$, we let $[b/c] := \det(a_{ij}) \in F^\times$, where $b_i = \sum_{j=1}^{n} a_{ij} c_j$. Let

$$C : \quad 0 \longrightarrow C_m \xrightarrow{\partial_m} C_{m-1} \xrightarrow{\partial_{m-1}} \cdots \xrightarrow{\partial_2} C_1 \xrightarrow{\partial_1} C_0 \longrightarrow 0$$

be an acyclic complex (i.e., exact sequence) of finite-dimensional based vector spaces C_i over F and let c_i be a given basis of C_i (a based vector space means

© The Author(s), under exclusive license to Springer Nature Singapore Pte Ltd. 2024
M. Morishita, *Knots and Primes*, Universitext,
https://doi.org/10.1007/978-981-99-9255-3_13

a vector space with a distinguished basis). Choose a basis b_i of $B_i := \mathrm{Im}(\partial_{i+1}) = \mathrm{Ker}(\partial_i)$. Consider the short exact sequence

$$0 \longrightarrow B_i \longrightarrow C_i \xrightarrow{\partial_i} B_{i-1} \longrightarrow 0.$$

Choosing a lift $\tilde{b}_{i-1}$ of b_{i-1} in C_i, the pair $(b_i, \tilde{b}_{i-1})$ forms a basis of C_i (b_{-1} is defined to be empty). We then define the *torsion* of C by

$$\tau(C) := \prod_{i=0}^{m} [(b_i, \tilde{b}_{i-1})/c_i]^{(-1)^{i+1}}.$$

One easily see that $\tau(C)$ depends on C_i, c_i, but not on the choices of b_i, $\tilde{b}_i$. Next, let R be a Noetherian unique factorization domain and let

$$D: \quad 0 \longrightarrow D_m \xrightarrow{\partial_m} D_{m-1} \xrightarrow{\partial_{m-1}} \cdots \xrightarrow{\partial_2} D_1 \xrightarrow{\partial_1} D_0 \longrightarrow 0$$

be a complex of free R-modules with finite rank such that the homology group $H_i(D)$ is a torsion R-module for any i. Let $\Delta_i := \Delta_0(H_i(D))$, and let F be the quotient field of R. We then define the *homology torsion* of D by

$$\tau^h(D) := \prod_{i=0}^{m} \Delta_i^{(-1)^{i+1}} \quad (\in F),$$

which is determined up to multiplication of a unit in $R^\times$. Let us choose an R-basis d_i of each D_i and let $C := D \otimes_R F$. Since C is then an acyclic complex of finite-dimensional based vector spaces over F, the torsion $\tau(C)$ is defined. Then we have the following lemma.

Lemma 13.1.1 $\tau(C) = \tau^h(D) \bmod R^\times$.

Proof We follow [83, Theorem 3.15]. We shall prove the assertion by induction on the length m of D. The assertion is clearly true for $m = 1$ and so we assume that $m > 1$ and that the assertion holds for all such complexes over R of length less than m.

Let $Z_{m-2} = \mathrm{Ker}(\partial_{m-2})$ have rank r, and choose $d'_1, \ldots, d'_r \in D_{m-1}$ whose images under ∂_{m-1} generate a rank r free submodule D'_{m-1} of D_{m-1}. Let $j : D'_{m-1} \hookrightarrow D_{m-1}$ be the inclusion. Let D' be the subcomplex of D such that $(D')_i = D_i$ if $i < m-1$, $(D')_{m-1} = D'_{m-1}$ and $(D')_i = 0$ if $i \geq m$. Let E be the complex $\cdots D'_{m-1} \xrightarrow{\mathrm{id}} D'_{m-1} \cdots$ concentrated in degrees m and $m-1$, and let $D'' := D \oplus E$. Define a chain homomorphism $\alpha : D' \to D''$ by $\alpha_i := \mathrm{id}_{D_i}$ if $i < m-1$ and $\alpha_{m-1} := j \oplus \mathrm{id}_{D'_{m-1}}$. Then α is injective, and the cokernel A of α is the complex $\cdots D_m \oplus D'_{m-1} \xrightarrow{\partial_m + j} D_{m-1} \cdots$ concentrated in degrees m and $m-1$.

Let $C' := D' \otimes_R F$, $C'' := D'' \otimes_R F$ and $B := A \otimes_R F$. Note that the complexes D', D'' and A are all free over R and that C', C'' and B are acyclic.

Equip each of D', D'' and A with obvious bases. Since torsions are multiplicative with respect to an exact sequence of complexes, we have $\tau(C) = \tau(C'') = \tau(C')\tau(B)$. The inclusion of D into D'' is a chain homotopy equivalence, and so $H_i(D) \simeq H_i(D'')$ for all i. The long exact sequence $0 \to D' \xrightarrow{\alpha} D'' \to A \to 0$ breaks up into isomorphisms $H_i(D') = H_i(D)$ if $i < m - 2$ and an exact sequence $0 \to H_{m-1}(D) \to H_{m-1}(A) \to H_{m-2}(D') \to H_{m-2}(D) \to 0$. These modules are all R-torsion and so $\Delta_0(H_{m-1}(A))\Delta_0(H_{m-2}(D'))^{-1} = \Delta_0(H_{m-1}(D))\Delta_0(H_{m-2}(D))^{-1}$. As D' and A are each of length $\leq m - 1$, the result follows from the induction hypothesis. $\square$

Now let $K \subset S^3$ be a knot, $X_K = S^3 \setminus \text{int}(V_K)$ the link exterior and $G_K = \pi_1(X_K)$ the knot group. Let $\psi : G_K \to G_K^{ab} = \langle \alpha \rangle \simeq \mathbb{Z}$ be the Abelianization map and let X_∞ be the infinite cyclic covering of X_K corresponding to $\text{Ker}(\psi)$. Set $\Lambda := \mathbb{Z}[t^{\pm 1}] = \mathbb{Z}[\langle \alpha \rangle]$ ($t \leftrightarrow \alpha$). Let $G_K = \langle x_1, \ldots, x_n \mid R_1 = \cdots = R_{n-1} = 1 \rangle$ be a Wirtinger presentation (Example 2.1.6) and let $\pi : F = \langle x_1, \ldots, x_n \rangle \to G_K$ be the natural homomorphism. We associate to the presentation of G_K a complex of Λ-modules

$$D : \quad 0 \longrightarrow D_2 = \Lambda^{n-1} \xrightarrow{\partial_2} D_1 = \Lambda^n \xrightarrow{\partial_1} D_0 = \Lambda \longrightarrow 0, \quad (13.1)$$

where $\partial_2 = ((\psi \circ \pi)(\partial R_i/\partial x_j))$ is the Alexander matrix and $\partial_1 = ((\psi \circ \pi)(x_i - 1))$. Since X_K collapses to a 2-dimensional complex obtained from the presentation of G_K, we may regard the complex D of (13.1) as $C_*(X_\infty) = C_*(X_K, \Lambda)$ and so the homology group $H_i(X_\infty) = H_i(X_K, \Lambda)$ is given by $H_i(D)$.

Proposition 13.1.2 For any $i \geq 0$, $H_i(X_\infty)$ is a finitely generated, torsion Λ-module. Furthermore we have $H_i(X_\infty) = 0$ ($i \geq 2$), $E_0(H_1(X_\infty)) = (\Delta_K(t))$, $E_0(H_0(X_\infty)) = (t - 1)$, where $\Delta_K(t)$ is the Alexander polynomial of K.

Proof It is obvious that $H_i(X_\infty) = 0$ ($i \geq 3$) and $E_0(H_0(X_\infty)) = (t - 1)$. The assertion about $H_1(X_\infty)$ follows from Proposition 12.1.2. Thus it suffices to show $H_2(X_\infty) = 0$. We associate to (13.1) an exact sequence of Λ-modules

$$0 \longrightarrow H_2(X_\infty) \longrightarrow \Lambda^{n-1} \xrightarrow{\partial_2} \Lambda^n \longrightarrow A_K \longrightarrow 0,$$

where A_K is the Alexander module of K. Since $H_2(X_\infty)$ is a Λ-submodule of the free Λ-module Λ^{n-1}, it has no torsion. On the other hand, as A_K has Λ-rank 1, the Λ-rank of $H_2(X_\infty)$ is 0. Hence $H_2(X_\infty) = 0$. $\square$

Let k be an extension of $\mathbb{Q}$ and let $F = k(t)$. Choose the standard basis of D_i in (13.1). Then, by Proposition 13.1.2, the complex $C_*(X_K, F) = C_*(X_K, \Lambda) \otimes_\Lambda F = C_*(X_\infty) \otimes_\Lambda F$ becomes an acyclic complex of finite-dimensional based vector spaces over F. Therefore, by Lemma 13.1.1 and Proposition 13.1.2, we obtain the following:

Proposition 13.1.3 $\tau(C_*(X_K, F)) = \tau^h(C_*(X_\infty)) = \dfrac{\Delta_K(t)}{t-1} \mod \Lambda^\times.$

$\tau(C_*(X_K, F)) = \tau^h(C_*(X_\infty))$ is called the *Reidemeister–Milnor torsion* of X_K, which we denote by $\tau(X_K, \Lambda)$.

Remark 13.1.4

(1) The torsion $\tau(X_K, \Lambda)$ can be described in terms of the determinant module [110] as follows: Tensoring Λ over $\mathbb{Z}$ with the exact sequence

$$0 \longrightarrow C_*(X_\infty) \xrightarrow{\alpha-1} C_*(X_\infty) \longrightarrow C_*(X_K) \longrightarrow 0,$$

we have the exact sequence of Λ-modules

$$0 \longrightarrow C_*(X_\infty, \Lambda) \xrightarrow{\alpha-1} C_*(X_\infty, \Lambda) \longrightarrow C_*(X_K, \Lambda) \longrightarrow 0.$$

From this, we obtain the following Λ-isomorphisms:

$$\begin{aligned}
\Lambda &\simeq \det{}_\Lambda C_*(X_\infty, \Lambda) \otimes_\Lambda \det{}_\Lambda C_*(X_\infty, \Lambda)^{-1} \\
&\simeq \det{}_\Lambda C_*(X_K, \Lambda) \\
&\simeq \det{}_\Lambda H_*(X_K, \Lambda) \text{ (Euler isomorphism)}.
\end{aligned}$$

Now let $\zeta(X_K, \Lambda)$ be the image of $1 \in \Lambda$ in $\det_\Lambda H_*(X_K, \Lambda)$ under the above isomorphism. The torsion $\tau(X_K, \Lambda)$ is then nothing but the image of $\zeta(X_K, \Lambda)$ in F under the isomorphism $\det_\Lambda H_*(X_K, \Lambda) \otimes_\Lambda F \simeq \det_F H_*(X_K, F) = \det_F(0) = F$. $\zeta(X_K, \Lambda)$ is an analogue in knot theory of K. Kato's *zeta element* [100].

(2) More generally, we may consider the Reidemeister–Milnor torsion associated to a representation of a knot group $\rho : G_K \to GL(V)$ (V being a finite-dimensional vector space over k) (cf. [106, 107]). Let $\Lambda_k := \Lambda \otimes_\mathbb{Z} k$, $V[t^{\pm 1}] := V \otimes_k \Lambda_k$ and $V(t) := V \otimes_k F$. We regard $V[t^{\pm 1}]$ as a left $k[G_K]$-module via the representation $\rho \otimes \psi : G_K \to GL(V[t^{\pm 1}])$ and consider the complex $C_*(X_K, V[t^{\pm 1}]) = C_*(\tilde{X}_K, k) \otimes_{k[G_K]} V[t^{\pm 1}]$. Since $C_*(X_K, V[t^{\pm 1}]) = C_*(X_K, \Lambda_k) \otimes_k V$, by Lemma 13.1.3, $H_i(X_K, V[t^{\pm 1}]) = H_i(X_K, \Lambda_k) \otimes_k V$ is a finitely generated, torsion Λ_k-module for any i. So, choosing a basis of V over k, $C_*(X_K, V(t)) := C_*(X_K, \Lambda_k) \otimes_{\Lambda_k} F$ becomes an acyclic complex of finite-dimensional based vector spaces over F. Letting $\Delta_{i,\rho}(t) := \Delta_0(H_i(X_K, V[t^{\pm 1}]))$, we have by Lemma 13.1.1,

$$\tau(C_*(X_K, V(t))) = \tau^h(C_*(X_K, V[t^{\pm 1}])) = \frac{\Delta_{1,\rho}(t)}{\Delta_{0,\rho}(t)} \mod (\Lambda_k)^\times.$$

Here $\Delta_{1,\rho}$ is called the *twisted Alexander polynomial* of K associated to the representation ρ.

The torsion $\tau(X_K, \Lambda_k)$ has the following dynamical interpretation [149]. The monodromy (meridian) action $\alpha : X_\infty \to X_\infty$ defines a discrete dynamical system on X_∞. The torsion $\tau(X_K, \Lambda_k)$ is then expressed by the Lefschetz zeta function of this dynamical system. For $n \in \mathbb{N}$, let $L(\alpha^n)$ be the Lefschetz number of α^n defined by

$$L(\alpha^n) := \sum_{i=0}^{1} (-1)^i \mathrm{Tr}((\alpha_*)^n \mid H_i(X_\infty, k)).$$

The *Lefschetz zeta function* is then defined by

$$\zeta_K(t) := \exp\left(\sum_{n=1}^{\infty} L(\alpha^n) \frac{t^n}{n}\right) \quad (\in k[[t]]).$$

Theorem 13.1.5 ([149], [182]) $\zeta_K(t) = \tau(X_K, \Lambda_k) \bmod (\Lambda_k)^\times$.

Proof Note that for a matrix $A \in M_N(k)$, the following equality holds in $k[[t]]$:

$$\det(I - tA)^{-1} = \exp\left(\sum_{n=1}^{\infty} \mathrm{Tr}(A^n) \frac{t^n}{n}\right).$$

Therefore

$$\begin{aligned}
\zeta_K(t) &= \prod_{i=1}^{1} \det(I - t\alpha_* \mid H_i(X_\infty, k))^{(-1)^{i+1}} \\
&= \tau(X_K, \Lambda_k) \bmod (\Lambda_k)^\times.
\end{aligned}$$

$\square$

The Reidemeister torsion is also expressed by the spectral zeta function (Hodge theoretic interpretation). Let $\rho : G_K \to O(V)$ be an orthogonal representation where V is a finite-dimensional vector space over $\mathbb{R}$ equipped with an inner product. We consider the relative chain complex $C_*(X_K, \partial X_K, V) := C_*(X_K, \partial X_K) \otimes_{\mathbb{Z}[G_K]} V$. Choosing a cell decomposition of X_K and an orthonormal basis of V, the Reidemeister torsion $\tau(X_K, \partial X_K, V)$ is defined up to ± 1 [107, Lemma 5.2.5–5.2.7]. On the other hand, we give a Riemannian metric on X_K. The metric is assumed to be a product near ∂X_K. Then the space of V-valued i-forms $\Omega^i(X_K, V)$ is equipped with an inner product, and the Hodge star operator $* : \Omega^i(X_K, V) \to$

$\Omega^{3-i}(X_K, V)$ and the adjoint of the differential operator $\delta := - * d^{3-i} * :$ $\Omega^i(X_K, V) \to \Omega^{i-1}(X_K, V)$ are defined. We set

$$\Omega^i(X_K, \partial X_K, V) := \{\omega \in \Omega^i(X_K, V) \mid \omega|_{\partial X_K} = \delta\omega|_{\partial X_K} = 0\}$$

on which the Laplace operator $\Delta^i := d^{i-1} \circ \delta^i + \delta^{i+1} \circ d^i$ acts as a self-adjoint operator. The *Ray–Singer spectral zeta function* is then defined by

$$\zeta_{\Delta^i}(s) := \sum_{\lambda > 0} \lambda^{-s}, \quad \zeta_{\Delta}(s) := \sum_{i=0}^{3} (-1)^i i \zeta_{\Delta^i}(s),$$

where λ ranges over positive eigenvalues of Δ^i. It is shown that $\zeta_{\Delta}(s)$ is continued analytically to the whole complex plane and is analytic at $s = 0$. The connection with the Reidemeister torsion is given as follows.

Theorem 13.1.6 ([130]) $\exp(\zeta'_{\Delta}(0)) = \pm\tau(X_K, \partial X_K, V)$.

Here $\exp(\zeta'_{\Delta}(0))$ is called the *analytic torsion* of $(X_K, \partial X_K, V)$. This is independent of the choice of Riemannian metric.

Remark 13.1.7 When M is a complete hyperbolic 3-manifold which is closed or of finite volume, one has a closer analogy with number theory. For an orthonormal representation $\rho : \pi_1(M) \to O(V)$, consider the following zeta function of M:

$$Z(s) := \prod_{\mathfrak{p}} \det(I - \rho(\mathfrak{p})e^{-sl(\mathfrak{p})})^{-1},$$

where $\mathfrak{p}$ runs over prime closed geodesics of M and $l(\mathfrak{p})$ stands for the length of $\mathfrak{p}$. $Z(s)$ is a geometric analogue of the zeta function of a number field (Artin motive). The analytic torsion $\exp(\zeta'_{\Delta}(0))$ is then shown to coincide with the coefficient $Z_M(0)^*$ of the main part of $Z_M(s)$ at $s = 0$ up to an elementary factor ($\in \mathbb{Q}^{\times}$). From this, $Z_M(0)^*$ is expressed essentially by the Reidemeister torsion $\tau(M, V)$ ([56], [213], [214], [215]). This formula may be seen as an analogue of the analytic class number formula of a number field. Furthermore, if M admits an infinite cyclic covering, the relation between the order of the Milnor torsion at $t = 1$ and the order of $Z(s)$ at $s = 0$ is also shown ([213], [214], [215]). See also [44] for a formulation using a dynamical analogue of the Lichtenbaum's Weil-étale cohomology.

13.2   The Iwasawa Main Conjecture

Let p be an odd prime number, $X_{\{p\}} := \mathrm{Spec}(\mathbb{Z}[1/p])$ and $G_{\{p\}} := \pi_1(X_{\{p\}})$ the prime group. Let $\psi : G_{\{p\}} \to G_{\{p\}}^{\mathrm{ab}}$ be the Abelianization map and let $\mathfrak{X}_{\infty}$ be the pro-étale covering of $X_{\{p\}}$ corresponding to $\mathrm{Ker}(\psi)$. By class field theory, we have

$G_{\{p\}}^{ab} = \mathrm{Gal}(\mathbb{Q}(\mu_{p^\infty})/\mathbb{Q}) \simeq \mathbb{Z}_p^\times$, where $\mu_{p^\infty} := \bigcup_{n\geq 1}\mu_{p^n}$, μ_{p^n} being the group of p^n-th roots of unity. According to the decomposition $\mathbb{Z}_p^\times = \mathbb{F}_p^\times \times (1 + p\mathbb{Z}_p)$, one has the decomposition of $G_{\{p\}}^{ab}$: $G_{\{p\}}^{ab} = H \times \Gamma$, $H = \mathrm{Gal}(\mathbb{Q}(\mu_{p^\infty})/\mathbb{Q}_\infty) = \mathrm{Gal}(\mathbb{Q}(\mu_p)/\mathbb{Q}) = \mathbb{F}_p^\times$, $\Gamma = \mathrm{Gal}(\mathbb{Q}(\mu_{p^\infty})/\mathbb{Q}(\mu_p)) = \mathrm{Gal}(\mathbb{Q}_\infty/\mathbb{Q}) = 1 + p\mathbb{Z}_p \simeq \mathbb{Z}_p$. Here $\mathbb{Q}_\infty$ stands for the cyclotomic $\mathbb{Z}_p$-extension of $\mathbb{Q}$ (Example 2.3.6). $\mathfrak{X}_\infty$ is the integral closure of $X_{\{p\}}$ in $\mathbb{Q}(\mu_{p^\infty})$. Let $\mathfrak{X}_p := \mathrm{Spec}(\mathbb{Z}[\mu_p, 1/p])$ and let X_∞ be the integral closure of $X_{\{p\}}$ in $\mathbb{Q}_\infty$.

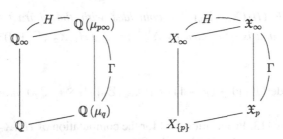

For $g \in \mathrm{Gal}(\mathbb{Q}(\mu_{p^\infty})/\mathbb{Q})$, we define $\kappa(g) \in \mathbb{Z}_p^\times$ by $g(\zeta) = \zeta^{\kappa(g)}$ ($\zeta \in \mu_{p^\infty}$). κ gives the isomorphism $\mathrm{Gal}(\mathbb{Q}(\mu_{p^\infty})/\mathbb{Q}) \simeq \mathbb{Z}_p^\times$ and is called the *cyclotomic character*. We set $\omega := \kappa|_H : H \hookrightarrow \mathbb{Z}_p^\times$. Fix a generator δ of H and a topological generator γ of Γ and let $\Lambda := \mathbb{Z}_p[[T]] = \mathbb{Z}_p[[\Gamma]]$ ($1 + T \leftrightarrow \gamma$) and $\tilde{\Lambda} := \mathbb{Z}_p[[G_{\{p\}}^{ab}]] = \mathbb{Z}_p[H][[\Gamma]]$.

For each $j \bmod p - 1$, we define the H-module $\mathbb{Z}/p^n\mathbb{Z}[j]$ ($n \geq 1$) by $\mathbb{Z}/p^n\mathbb{Z}$ as an Abelian group on which $\delta \in H$ acts as multiplication by $\omega(\delta)^j$, and let $\mathbb{Z}_p[j] := \varprojlim_n \mathbb{Z}/p^n\mathbb{Z}[j]$. Then $\mathbb{Z}_p[H] = \bigoplus_j \mathbb{Z}_p[j]$. Let $\tilde{\Lambda}^{(j)} := \mathbb{Z}_p[j][[\Gamma]]$ so that $\tilde{\Lambda} = \bigoplus_j \tilde{\Lambda}^{(j)}$. For a $\tilde{\Lambda}$-module M, we let $M^{(j)} := M \otimes_{\tilde{\Lambda}} \tilde{\Lambda}^{(j)}$. $M^{(j)}$ is the maximal quotient of M on which $\delta \in H$ acts as multiplication by $\omega(\delta)^j$. For $n \in \mathbb{Z}$, we define the $\hat{\Lambda}$-module $\mathbb{Z}_p(n)$ by $\mathbb{Z}_p$ as an Abelian group on which $\gamma \in \Gamma$ acts as multiplication by $\kappa(\gamma)^n$, and let $A(n) := A \otimes_{\mathbb{Z}_p} \mathbb{Z}_p(n)$ (Tate twist) for a $\hat{\Lambda}$-module A.

We identify an H-module with a finite étale sheaf on $X_{\{p\}}$ which becomes a constant sheaf on $\mathfrak{X}_p$. We define $H_i(X_\infty, \mathbb{Z}_p[j])$ (which is also denoted by $H_i(X_{\{p\}}, \tilde{\Lambda}^{(j)})$) by

$$H_i(X_\infty, \mathbb{Z}_p[j]) := (\varinjlim_n H^i(X_\infty, \mathbb{Z}/p^n\mathbb{Z}[j]))^*,$$

where $*$ means the Pontryagin dual. (Although the étale sheaf $\mathbb{Z}/p^n\mathbb{Z}[j]$ in the right-hand side should be $\mathbb{Z}/p^n\mathbb{Z}[-j]$, we adopt the above definition to simplify the notations below.) Finally we let $\mathfrak{M} := \pi_1^{ab}(\mathfrak{X}_\infty)(p) = \mathrm{Gal}(M/\mathbb{Q}(\mu_{p^\infty}))$, where M denotes the maximal Abelian extension of $\mathbb{Q}(\mu_{p^\infty})$ unramified outside p. The Galois group $G_{\{p\}}^{ab}$ acts on $\mathfrak{M}$ by the inner automorphism and $\mathfrak{M}$ is then regarded as a compact $\tilde{\Lambda}$-module.

Proposition 13.2.1 *For any i and even j, $H_i(X_\infty, \mathbb{Z}_p[j])$ is a finitely generated, torsion $\hat{\Lambda}$-module. More precisely,*

$$H_i(X_\infty, \mathbb{Z}_p[j]) = 0 \quad (i \geq 2, \ j : even),$$
$$H_1(X_\infty, \mathbb{Z}_p[j]) = \mathfrak{M}^{(j)} \quad (j : even),$$
$$H_0(X_\infty, \mathbb{Z}_p[j]) = \begin{cases} \mathbb{Z}_p & (j = 0), \\ 0 & (j \neq 0), \end{cases} \quad E_0(H_0(X_\infty, \mathbb{Z}_p[j])) = \begin{cases} (T) & (j = 0), \\ (1) & (j \neq 0). \end{cases}$$

If j is even, $E_0(H_1(X_\infty, \mathbb{Z}_p[j]))$ coincides with the characteristic ideal of $H_1(X_\infty, \mathbb{Z}_p[j])$ and is generated by $\Delta_p^{(j)} := \Delta_0(H_1(X_\infty, \mathbb{Z}_p[j]))$. Furthermore $\Delta_p^{(j)}$ satisfies

$$\Delta_p^{(j)} = \det(T \cdot \mathrm{id} - (\gamma - 1) \mid H_1(X_\infty, \mathbb{Z}_p[j]) \otimes_{\mathbb{Z}_p} \mathbb{Q}_p) \ mod \ \hat{\Lambda}^\times.$$

Proof We follow [13, Proposition 5.5] for the computation of $H^i(X_\infty, \mathbb{Z}/p^n\mathbb{Z}[j])$. Since $\#H$ is prime to p, the Hochschild–Serre spectral sequence

$$H^k(H, H^i(\mathfrak{X}_\infty, \mathbb{Z}/p^n\mathbb{Z}[j])) \Rightarrow H^{k+i}(X_\infty, \mathbb{Z}/p^n\mathbb{Z}[j])$$

degenerates and yields

$$\begin{aligned} H^i(X_\infty, \mathbb{Z}/p^n\mathbb{Z}[j]) &= H^0(H, H^i(\mathfrak{X}_\infty, \mathbb{Z}/p^n\mathbb{Z}[j])) \\ &= H^0(H, H^i(\mathfrak{X}_\infty, \mathbb{Z}/p^n\mathbb{Z})^{(j)}) \\ &= H^i(\mathfrak{X}_\infty, \mathbb{Z}/p^n\mathbb{Z})^{(-j)}. \end{aligned}$$

Since $\mathbb{Z}[\mu_{p^\infty}, 1/p]$ is the Dedekind domain containing μ_{p^∞}, the cohomological p-dimension of $\mathfrak{X}_\infty = \mathrm{Spec}(\mathbb{Z}[\mu_{p^\infty}, 1/p])$ is 2. Therefore $H^i(\mathfrak{X}_\infty, \mathbb{Z}/p^n\mathbb{Z}[j]) = 0$ $(i \geq 3)$ and so $H_i(X_\infty, \mathbb{Z}_p[j]) = 0$ $(i \geq 3)$.

To see $H_2(X_\infty, \mathbb{Z}_p[j]) = 0$, we note first that

$$\begin{aligned} H^2(\mathfrak{X}_\infty, \mathbb{Z}/p^n\mathbb{Z})^{(-j)} &= (H^2(\mathfrak{X}_\infty, \mu_{p^n})(-1))^{(-j)} \\ &= ((Cl(\mathfrak{X}_\infty) \otimes_{\mathbb{Z}} \mathbb{Z}/p^n\mathbb{Z})(-1))^{(-j)} \\ &= (Cl(\mathfrak{X}_\infty) \otimes_{\mathbb{Z}} \mathbb{Z}/p^n\mathbb{Z})^{(1-j)}(-1), \end{aligned}$$

where $Cl(\mathfrak{X}_\infty)$ denotes the ideal class group of $\mathbb{Z}[\mu_{p^\infty}, 1/p]$. Let $\mathfrak{X}_k := \mathrm{Spec}(\mathbb{Z}[\mu_{p^k}, 1/p])$. Since $Cl(\mathfrak{X}_k)$ is a finite Abelian group, $Cl(\mathfrak{X}_\infty) = \varinjlim_k Cl(\mathfrak{X}_k)$ is a torsion Abelian group. Therefore $\varinjlim_n H^2(\mathfrak{X}_\infty, \mathbb{Z}/p^n)^{(-j)} = (Cl(\mathfrak{X}_\infty) \otimes_{\mathbb{Z}} \mathbb{Q}_p/\mathbb{Z}_p)^{(1-j)}(-1) = 0$. Hence $H_2(X_\infty, \mathbb{Z}_p[j]) = 0$.

Next, we have

$$\begin{aligned} H^1(\mathfrak{X}_\infty, \mathbb{Z}/p^n\mathbb{Z})^{(-j)} &= \mathrm{Hom}_c(\mathfrak{M}, \mathbb{Z}/p^n\mathbb{Z})^{(-j)} \\ &= \mathrm{Hom}_c(\mathfrak{M}^{(j)}, \mathbb{Z}/p^n\mathbb{Z}), \end{aligned}$$

where Hom_c stands for the group of continuous homomorphisms. Therefore $H_1(X_\infty, \mathbb{Z}_p[j]) = \mathrm{Hom}_c(\mathfrak{M}^{(j)}, \mathbb{Q}_p/\mathbb{Z}_p)^* = \mathfrak{M}^{(j)}$. If j is even, it is known ([13, Lemma 5.3], [241, Proposition 15.36] that $\mathfrak{M}^{(j)}$ is a finitely generated, torsion $\hat{\Lambda}$-module and has no non-trivial finite $\hat{\Lambda}$-submodule and $\mu(\mathfrak{M}^{(j)}) = 0$. So $E_0(H_1(X_\infty, \mathbb{Z}_p[j]))$ coincides with the characteristic ideal of $H_1(X_\infty, \mathbb{Z}_p[j])$ and $\Delta_p^{(j)}$ equals, mod $\hat{\Lambda}^\times$, the characteristic polynomial in the statement. Finally, we see obviously

$$H^0(\mathfrak{X}_\infty, \mathbb{Z}/p^n\mathbb{Z})^{(-j)} = (\mathbb{Z}/p^n\mathbb{Z})^{(-j)} = \begin{cases} \mathbb{Z}/p^n\mathbb{Z} & (j = 0) \\ 0 & (j \neq 0) \end{cases}$$

and so $H_0(X_\infty, \mathbb{Z}_p[j]) = \mathbb{Z}_p$ ($j = 0$), $= 0$ ($j \neq 0$). The assertion for $E_0(H_0(X_\infty, \mathbb{Z}_p[j]))$ is also obvious. $\qquad\square$

By Proposition 13.2.1, for an even integer j mod $p - 1$, we define the *homology torsion* of $H_*(X_\infty, \mathbb{Z}_p[j])$ by

$$\tau(X_{\{p\}}, \tilde{\Lambda}^{(j)}) := \begin{cases} \Delta_p^{(j)}/T & (j = 0), \\ \Delta_p^{(j)} & (j \neq 0). \end{cases}$$

This is determined up to multiplication by an element in $\hat{\Lambda}^\times$.

The torsion $\tau(X_{\{p\}}, \tilde{\Lambda}^{(j)})$ is expressed by the Kubota–Leopoldt p-adic analytic zeta function. Let $\zeta_\mathbb{Q}(s) = \sum_{n=1}^\infty n^{-s}$ be the Riemann zeta function. It is known that $\zeta_\mathbb{Q}(s)$ is continued analytically to the whole complex plane and $\zeta_\mathbb{Q}(-n) \in \mathbb{Q}$ for $n \in \mathbb{N}$. Further, one has $\zeta_\mathbb{Q}(-n) \neq 0$ if and only if n is odd. T. Kubota and H. Leopoldt constructed a p-adic analytic function which interpolates the values $\zeta_\mathbb{Q}(-n)$ for odd n, called the *p-adic zeta function*.

Theorem 13.2.2 ([120]) *For each even j mod $p - 1$, there is a p-adic analytic function*

$$\zeta_p(\omega^j, \ ) : \mathbb{Z}_p \setminus \{1\} \longrightarrow \mathbb{Q}_p$$

such that for any $n \equiv j - 1$ mod $p - 1$, one has the equality

$$\zeta_p(\omega^j, -n) = (1 - p^n)\zeta_\mathbb{Q}(-n).$$

For the proof of Theorem 13.2.2, we refer to [33], [120], [241, Ch.5, Ch.7].

The following theorem, due to B. Mazur and A. Wiles, gives the relation between the torsion $\tau(X_{\{p\}}, \tilde{\Lambda}^{(j)})$ and the p-adic zeta function $\zeta_p(\omega^j, s)$.

Theorem 13.2.3 (The Iwasawa Main Conjecture [141]) *For even j, there is a generator $\Delta_p^{(j)}$ of the ideal $E_0(H_1(X_\infty, \mathbb{Z}_p^{(j)}))$ such that one has*

$$\zeta_p(\omega^j, s) = \tau(X_{\{p\}}, \tilde{\Lambda}^{(j)})|_{T=q^{1-s}-1},$$

where $q := \kappa(\gamma)$.

For the proof of Theorem 13.2.3, we refer to [33], [126, Appendix], [141], [241, Ch.15].

Remark 13.2.4 As in the case of knots, we can introduce a generalization of the Iwasawa module associated to a certain p-adic representation $\rho : G_{\{p\}} \to \text{Aut}(L)$ (L being a free $\mathbb{Z}_p$-module of finite rank). It is in fact defined as the Pontryagin dual of a certain subgroup $\text{Sel}(X_\infty, \rho)$, called the *Selmer group*, of $H^1(X_\infty, V/L)$ ($V = L \otimes_{\mathbb{Z}_p} \mathbb{Q}_p$) [70]. A generator of the characteristic ideal of $\text{Sel}(X_\infty, \rho)^*$, called the *twisted Iwasawa polynomial*, is an analogue of the twisted Alexander polynomial. When ρ is coming from a motive $H^m(Y)(n)$ (Y being a smooth projective variety over $\mathbb{Q}$), it is conjectured that an associated p-adic analytic L-function is defined and coincides essentially with the twisted Iwasawa polynomial (*generalized Iwasawa main conjecture*, [32], [100]).

Summary

Relation between Alexander polynomial and Lefschetz or spectral zeta function	Relation between Iwasawa polynomial and p-adic analytic zeta function

Chapter 14
Moduli Spaces of Representations of Knot and Prime Groups

In view of the analogy between a knot group $G_K = \pi_1(S^3 \setminus K)$ and a prime group $G_{\{p\}} = \pi_1(\mathrm{Spec}(\mathbb{Z}) \setminus \{p\})$, we expect some analogies between the moduli spaces of representations of knot and prime groups [139]. In particular, the Alexander–Fox theory and Iwasawa theory are regarded as the theories on the moduli spaces of 1-dimensional representations of a knot and prime groups and associated topological and arithmetic invariants respectively.

14.1 Character Varieties of Complex Representations of a Knot Group

For a knot $K \subset S^3$ and $N \in \mathbb{N}$, let $\mathcal{R}_{K,N}$ be the set of N-dimensional complex representations of the knot group G_K:

$$\mathcal{R}_{K,N} := \mathrm{Hom}(G_K, GL_N(\mathbb{C}))$$
$$:= \{\rho : G_K \longrightarrow GL_N(\mathbb{C}) \mid \rho \text{ is a homomorphism}\}.$$

Let $G_K = \langle x_1, \ldots, x_n \mid r_1 = \cdots = r_{n-1} = 1 \rangle$ be a Wirtinger presentation for the knot group, as in Example 2.1.6. Then by the correspondence $\rho \mapsto (\rho(x_1), \ldots, \rho(x_n))$, $\mathcal{R}_{K,N}$ is identified with the affine algebraic set in $GL_N(\mathbb{C})^n$ defined by $r_1(X_1, \ldots, X_n) = \cdots = r_{n-1}(X_1, \ldots, X_n) = I$:

$$\mathcal{R}_{K,N} = \{(X_1, \ldots, X_n) \in GL_N(\mathbb{C})^n \mid$$
$$r_1(X_1, \ldots, X_n) = \cdots = r_{n-1}(X_1, \ldots, X_n) = I\},$$

© The Author(s), under exclusive license to Springer Nature Singapore Pte Ltd. 2024   191
M. Morishita, *Knots and Primes*, Universitext,
https://doi.org/10.1007/978-981-99-9255-3_14

Let $R_{K,N}$ be the coordinate ring of $\mathcal{R}_{K,N}$ and consider the tautological representation

$$\rho_{K,N} : G_K \longrightarrow GL_N(R_{K,N}); \quad x_k \mapsto X_k \ (1 \le k \le n),$$

which has the following universal property: For any representation $\rho : G_K \to GL_N(A)$ with a $\mathbb{C}$-algebra A, there exists uniquely a $\mathbb{C}$-algebra homomorphism $\varphi : R_{K,N} \to A$ such that $\varphi \circ \rho_{K,N} = \rho$.

The group $GL_N(\mathbb{C})$ acts on the ring $R_{K,N}$ by $(g, X_k) \mapsto gX_kg^{-1}$ ($1 \le k \le n$). Let $R_{K,N}^{GL_N(\mathbb{C})}$ be the invariant ring of this conjugate action. Then we define the *character variety* $\mathcal{X}_{K,N} = \mathcal{R}_{K,N}//GL_N(\mathbb{C})$ of N-dimensional complex representations of G_K by the affine algebraic set whose coordinate ring is $R_{K,N}^{GL_N(\mathbb{C})}$:

$$\mathcal{X}_{K,N} := (\mathrm{Spec}(R_{K,N}^{GL_N(\mathbb{C})}))(\mathbb{C}) = \mathrm{Hom}_{\mathbb{C}\text{-alg}}(R_{K,N}^{GL_N(\mathbb{C})}, \mathbb{C}), \tag{14.1}$$

The inclusion $R_{K,N}^{GL_N(\mathbb{C})} \hookrightarrow R_{K,N}$ induces the morphism $\mathcal{R}_{K,N} \to \mathcal{X}_{K,N}$. We write $[\rho] \in \mathcal{X}_{K,N}$ for the image of $\rho \in \mathcal{R}_{K,N}$ under this morphism. Then, for $\rho, \rho' \in \mathcal{R}_{K,N}$, one has $[\rho] = [\rho'] \Leftrightarrow \mathrm{Tr}(\rho) = \mathrm{Tr}(\rho')$ [37].

14.2 The Character Variety of Complex 1-Dimensional Representations of a Knot Group and Alexander Ideals

We fix a meridian α of K. Note that a 1-dimensional representation of G_K factors through the Abelianization G_K^{ab}. Since G_K^{ab} is an infinite cyclic group generated by the class of α, we have the following theorem on $\mathcal{X}_{K,1}$ and $\rho_{K,1}$. We set $\Lambda := \mathbb{Z}[t^{\pm 1}] = \mathbb{Z}[G_K^{\mathrm{ab}}]$ ($[\alpha] \leftrightarrow t$).

Theorem 14.2.1 *The correspondence $\rho \mapsto \rho(\alpha)$ gives an isomorphism*

$$\mathcal{X}_{K,1} \simeq \mathbb{C}^\times.$$

Hence the coordinate ring $R_{K,1}$ of $\mathcal{X}_{K,1}$ is the Laurent polynomial ring $\Lambda_\mathbb{C} := \Lambda \otimes_\mathbb{Z} \mathbb{C} = \mathbb{C}[t^{\pm 1}]$ over $\mathbb{C}$ and the tautological representation $\rho_{K,1} : G_K \to \Lambda_\mathbb{C}^\times$ is given by the composite of the Abelianization map $G_K \to G_K^{\mathrm{ab}}$ with the inclusion $G_K^{\mathrm{ab}} \subset \Lambda_\mathbb{C}^\times$.

Let A_K be the Alexander module of K and let $E_d(A_K)$ be the d-th Alexander ideal for each $d \in \mathbb{N}$. The ideal $E_d(A_K)$ coincides with $E_{d-1}(H_1(X_\infty))$ where X_∞ is the infinite cyclic covering of the knot exterior $X_K = S^3 \setminus \mathrm{int}(V_K)$ (Example 10.2.2), and $E_0(H_1(X_\infty))$ is generated by the Alexander polynomial $\Delta_K(t)$ (Proposition 13.1.2). We then define the *d-th Alexander set* in $\mathcal{X}_{K,1}$ by

$$\mathcal{A}_K(d) := \{\rho \in \mathcal{X}_{K,1} \mid f(\rho(\alpha)) = 0 \text{ for any } f \in E_d(A_K)\}.$$

Thus we have a descending series $\mathcal{X}_{K,1} \supset \mathcal{A}_K(1) \supset \cdots \supset \mathcal{A}_K(d) \supset \cdots$. On the other hand, for $\rho \in \mathcal{X}_{K,1}$, we define the G_K-module $\mathbb{C}(\rho)$ by the additive group $\mathbb{C}$ equipped with G_K-action given by $g.z = \rho(g)z$ ($g \in G_K$, $z \in \mathbb{C}$). We then define the *d-th cohomology jumping set* in $\mathcal{X}_{K,1}$ by

$$\mathcal{C}_K(d) := \{\rho \in \mathcal{X}_{K,1} \mid \dim_{\mathbb{C}} H^1(G_K, \mathbb{C}(\rho)) \geq d\}.$$

Thus we have another descending series $\mathcal{X}_{K,1} \supset \mathcal{C}_K(1) \supset \cdots \supset \mathcal{C}_K(d) \supset \cdots$.

Theorem 14.2.2 ([89], [128]) *One has*

$$\mathcal{A}_K(d) = \mathcal{C}_K(d) \; (d > 1), \quad \mathcal{A}_K(1) \cup \{\mathbf{1}\} = \mathcal{C}_K(1),$$

where $\mathbf{1}$ stands for the trivial representation of G_K.

For the proof of Theorem 14.2.2, we prepare one lemma. We regard $\mathbb{C}(\rho)$ as a Λ-module by $t.z = \rho(\alpha)z$ ($t \in \Lambda$, $z \in \mathbb{C}(\rho)$) and set $A_K(\rho) := A_K \otimes_\Lambda \mathbb{C}(\rho)$.

Lemma 14.2.3

(1) *We have the following isomorphism:*

$$\mathrm{Hom}_{\mathbb{C}}(A_K(\rho), \mathbb{C}) \simeq Z^1(G_K, \mathbb{C}(\rho)),$$

where $Z^1(G_K, \mathbb{C}(\rho))$ stands for the group of 1-cocycles.
(2) *We have the following:*

$$\dim_{\mathbb{C}} H^1(G_K, \mathbb{C}(\rho)) = \begin{cases} \dim A_K(\rho) - 1, & \rho \neq \mathbf{1}, \\ 1, & \rho = \mathbf{1}. \end{cases}$$

Proof

(1) By Definition 10.1.1 and Example 10.1.7,

$$A_K(\rho) = \left(\bigoplus_{g \in G_K} \mathbb{C}(\rho) \, dg \right) / \langle d(g_1 g_2) - dg_1 - \rho(g_1) \, dg_2 \rangle_{\mathbb{C}}.$$

Hence

$$\begin{aligned} \mathrm{Hom}_{\mathbb{C}}(A_K(\rho), \mathbb{C}) &\simeq \{c : G_K \longrightarrow \mathbb{C} \mid c(g_1 g_2) - c(g_1) - \rho(g_1)c(g_2) = 0\} \\ &= Z^1(G_K, \mathbb{C}(\rho)). \end{aligned}$$

(2) Noting that $\rho(g)z = z \Leftrightarrow \rho(g) = 1$ or $z = 0$, the group of 1-coboundaries $B^1(G_K, \mathbb{C}(\rho))$ is given by

$$B^1(G_K, \mathbb{C}(\rho)) = \begin{cases} \mathbb{C}, & \rho \neq 1, \\ \{0\}, & \rho = 1. \end{cases}$$

Together with (1), we get the assertion.

$\square$

Proof of Theorem 14.2.4 Let Q_K be a representation matrix for the Λ-module A_K:

$$\Lambda^{n-1} \xrightarrow{Q_K} \Lambda^n \longrightarrow A_K \longrightarrow 0 \quad \text{(exact)}. \qquad (14.2)$$

The ideal $E_d(A_K)$ is generated by $(n - d)$-minors of Q_K if $d < n$, and is Λ if $d \geq n$. Tensoring $\mathbb{C}(\rho)$ with (14.2) over Λ, we have

$$\mathbb{C}(\rho)^{n-1} \xrightarrow{\rho(Q_K)} \mathbb{C}(\rho)^n \longrightarrow A_K(\rho) \longrightarrow 0 \quad \text{(exact)},$$

where $\rho(Q_K)$ is the matrix over $\mathbb{C}$ obtained by the evaluation $\Lambda \ni t \mapsto \rho(\alpha) \in \mathbb{C}$. Therefore $\dim A_K(\rho) = n - (\text{rank of } \rho(Q_K))$. Hence, for $d > 1$,

$$\begin{aligned}
\dim_{\mathbb{C}} H^1(G_K, \mathbb{C}(\rho)) \geq d &\Longleftrightarrow \dim_{\mathbb{C}} A_K(\rho) \geq d + 1 \quad \text{(Lemma 14.2.3, (2))} \\
&\Longleftrightarrow \text{rank of } \rho(Q_K) \leq n - (d + 1) \\
&\Longleftrightarrow \text{any } (n - d)\text{-minor of } \rho(Q_K) = 0 \\
&\Longleftrightarrow f(\rho(\alpha)) = 0 \text{ for any } f \in E_d(A_K),
\end{aligned}$$

and, for $d = 1$,

$$\begin{aligned}
&\dim_{\mathbb{C}} H^1(G_K, \mathbb{C}(\rho)) \geq 1 \\
&\Longleftrightarrow \rho \neq 1 \text{ and } \dim_{\mathbb{C}} A_K(\rho) \geq 2, \text{ or } \rho = 1 \quad \text{(Lemma 14.2.3, (2))} \\
&\Longleftrightarrow \rho \neq 1 \text{ and } f(\rho(\alpha)) = 0 \text{ for any } f \in E_1(A_K), \text{ or } \rho = 1.
\end{aligned}$$

The proof is done.

Corollary 14.2.5 *Assume that* $\rho \neq 1$. *Then*

$$\Delta_K(\rho(\alpha)) = 0 \Longleftrightarrow H^1(G_K, \mathbb{C}(\rho)) \neq 0.$$

Proof By $E_1(A_K) = (\Delta_K(t))$ and $\Delta_K(1) = \pm 1$, our assertion follows from Theorem 14.2.2.

$\square$

14.3 Universal Deformation Spaces of p-Adic Representations of a Prime Group

Let p be a prime number and let $G_{\{p\}} = \pi_1(\mathrm{Spec}(\mathbb{Z}[1/p])) = \mathrm{Gal}(\mathbb{Q}_{\{p,\infty\}}/\mathbb{Q})$ be the prime group. Here $\mathbb{Q}_{\{p,\infty\}}$ is the maximal Galois extension of $\mathbb{Q}$ unramified outside $\{p, \infty\}$. In the following, we assume $p > 2$ and consider representations of $G_{\{p\}}$. Since $G_{\{p\}}$ is a huge pro-finite group (it is not known if $G_{\{p\}}$ is finitely generated or not), we do not have a good moduli space by considering naively all N-dimensional representations of $G_{\{p\}}$ over a ring. So, following B. Mazur [137], we consider the set of p-adic deformations of a given residual representation. Let

$$\overline{\rho} : G_{\{p\}} \longrightarrow GL_N(\mathbb{F}_p)$$

be a given N-dimensional continuous residual (mod p) representation of $G_{\{p\}}$. A pair (A, ρ) is called a *deformation* of $\overline{\rho}$ if:

$$\begin{cases} \text{1) } A \text{ is a complete Noetherian local } \mathbb{Z}_p\text{-algebra with residue field} \\ \quad A/\mathfrak{m}_A = \mathbb{F}_p, \\ \text{2) } \rho : G_p \to GL_N(A) \text{ is a continuous representation such that} \\ \quad \rho \bmod \mathfrak{m}_A = \overline{\rho}. \end{cases}$$

If the composite of $\overline{\rho}$ with the inclusion $GL_N(\mathbb{F}_p) \subset GL_N(\overline{\mathbb{F}}_p)$ for an algebraic closure $\overline{\mathbb{F}}_p$ of $\mathbb{F}_p$ is an irreducible representation over $\overline{\mathbb{F}}_p$, $\overline{\rho}$ is said to be absolutely irreducible. (This is independent of the choice of $\overline{\mathbb{F}}_p$.) We say that two deformations (A, ρ) and (A, ρ') are strictly equivalent, written as $\rho \approx \rho'$, if there is a $P \in I + M_N(\mathfrak{m}_A)$ such that one has $\rho'(g) = P\rho(g)P^{-1}$ for all $g \in G_{\{p\}}$. The following theorem due to Mazur is fundamental.

Theorem 14.3.1 ([137, 1.2]) *Assume that $\overline{\rho}$ is absolutely irreducible. There is a deformation of $\overline{\rho}$*

$$\rho_{p,N} : G_{\{p\}} \longrightarrow GL_N(R_{p,N})$$

having the following universal property: For any deformation $\rho : G_{\{(p)\}} \to GL_N(A)$ of $\overline{\rho}$, there exists uniquely a $\mathbb{Z}_p$-algebra homomorphism $\varphi : R_{p,N} \to A$ such that $\varphi \circ \rho_{p,N} \approx \rho$. (Although it is more precise to write $R_{p,N}(\bar{\rho})$ since $R_{p,N}$ depends on $\bar{\rho}$, we abbreviate it to $R_{p,N}$ for simplicity.)

If two deformations (A, ρ) and (A', ρ') of $\overline{\rho}$ satisfy the above universal property, we have a $\mathbb{Z}_p$-algebra isomorphism $\varphi : A \xrightarrow{\sim} A'$ such that $\varphi \circ \rho \approx \rho'$. Therefore the pair $(R_{p,N}, \rho_{p,N})$ is unique in this sense (up to equivalence) and is called the *universal deformation* of $\overline{\rho}$. We define the *universal deformation space* of $\overline{\rho}$ by

$$\mathcal{X}_{p,N}(\overline{\rho}) := (\mathrm{Spec}(R_{p,N}))(\overline{\mathbb{Q}}_p) = \mathrm{Hom}_{\mathbb{Z}_p\text{-alg}}(R_{p,N}, \overline{\mathbb{Q}}_p), \tag{14.3}$$

where $\overline{\mathbb{Q}}_p$ is an algebraic closure of $\mathbb{Q}_p$ and $\mathcal{X}_{p,N}(\overline{\rho})$ is regarded as a p-adic analytic space. For $\varphi \in \mathcal{X}_{p,N}(\overline{\rho})$, we denote $\varphi \circ \rho_{p,N}$ by ρ_φ.

14.4   The Universal Deformation Space of p-Adic 1-Dimensional Representations of a Prime Group and Iwasawa Ideals

Assume that $p > 2$ for simplicity. Note that a 1-dimensional representation of $G_{\{p\}}$ factors through the Abelianization $G_{\{p\}}^{\mathrm{ab}}$ of $G_{\{p\}}$. By class field theory, we have $G_{\{p\}}^{\mathrm{ab}} = \mathrm{Gal}(\mathbb{Q}(\mu_{p^\infty})/\mathbb{Q}) = H \times \Gamma$, $H := \mathrm{Gal}(\mathbb{Q}(\mu_p)/\mathbb{Q}) = \mathbb{F}_p^\times$, $\Gamma := \mathrm{Gal}(\mathbb{Q}_\infty/\mathbb{Q}) = 1 + p\mathbb{Z}_p$. (We keep the same notations as in Sect. 10.2.) We denote by $[g] = (g^{(p)}, g_p)$ the image of $g \in G_{\{p\}}$ under the Abelianization $G_{\{p\}} \to G_{\{p\}}^{\mathrm{ab}} = \mathbb{F}_p^\times \times (1 + p\mathbb{Z}_p)$. Fixing a topological generator γ of $\Gamma = 1 + p\mathbb{Z}_p$, we identify $\mathbb{Z}_p[[\Gamma]]$ with $\hat{\Lambda} := \mathbb{Z}_p[[T]]$ by the correspondence $\gamma \leftrightarrow 1 + T$.

Let $\overline{\rho} : G_{\{p\}} \to \mathbb{F}_p^\times$ be a 1-dimensional continuous residual representation. We identify $\mathbb{F}_p^\times$ with the group of $(p-1)$-th roots of 1 in $\mathbb{Z}_p^\times$. Then the *Teichmüller lift* $\tilde{\rho}$ of $\overline{\rho}$ is defined by the composite of $\overline{\rho}$ with the inclusion $\mathbb{F}_p^\times \hookrightarrow \mathbb{Z}_p^\times$. The following theorem may be regarded as an arithmetic analogue of Theorem 11.2.1.

Theorem 14.4.1 ([137, 1.4]) *Define $\rho_{p,1} : G_{\{p\}} \to \hat{\Lambda}^\times$ by*

$$\rho_{p,1}(g) := \tilde{\rho}(g)g_p.$$

Then the pair $(\hat{\Lambda}, \rho_{p,1})$ is the universal deformation of $\overline{\rho}$. Hence $\mathcal{X}_{p,1}(\overline{\rho})$ is identified with the p-adic unit disk $\mathcal{D}_p := \{x \in \overline{\mathbb{Q}}_p \mid |x|_p < 1\}$:

$$\mathcal{X}_{p,1}(\overline{\rho}) \xrightarrow{\sim} \mathcal{D}_p; \quad \varphi \longmapsto \varphi(T).$$

Proof Since $\rho_{p,1}(g) \bmod \mathfrak{m}_{\hat{\Lambda}} = \tilde{\rho}(g) \bmod p = \overline{\rho}(g)$, $\rho_{p,1}$ is a deformation of $\overline{\rho}$. Let (A, ρ) be any deformation of $\overline{\rho}$: $A/\mathfrak{m}_A = \mathbb{F}_p, \rho(g) \bmod \mathfrak{m}_A = \overline{\rho}(g)$. Define the $\mathbb{Z}_p$-algebra homomorphism $\varphi : \hat{\Lambda} \to A$ by $\varphi(g_p) := \rho((1, g_p))$, $g_p \in \mathrm{Gal}(\mathbb{Q}_\infty/\mathbb{Q})$. For any $g \in G_{\{p\}}$,

$$\begin{aligned}
(\varphi \circ \rho_{p,1})(g) &= \varphi(\tilde{\rho}(g)g_p) \\
&= \tilde{\rho}(g)\rho((1, g_p)) \\
&= \rho(g) \quad (\tilde{\rho}(g) = \rho((g^{(p)}, 1))).
\end{aligned}$$

The uniqueness of φ is clear and hence $(\hat{\Lambda}, \rho_{p,1})$ is the universal deformation of $\overline{\rho}$. The assertion about $\mathcal{X}_{p,1}(\overline{\rho})$ is obvious.   □

As in the case of the Alexander polynomial, the zeros of the Iwasawa polynomial are described by the variation of Galois cohomology of $G_{\{p\}}$ on the universal deformation space $\mathfrak{X}_p(\bar{\rho})$. In the following, we show it for the Iwasawa polynomial $\Delta_p^{(j)}(T)$ in Sect. 13.2. We keep the same notations as in Sect. 13.2.

Let $\bar{\omega} : G_{\{p\}} \to H = \mathrm{Gal}(\mathbb{Q}(\mu_p)/\mathbb{Q}) \simeq \mathbb{F}_p^\times$ be the natural homomorphism. We denote the composite of the natural map $G_{\{p\}} \to H$ with $\omega : H \hookrightarrow \mathbb{Z}_p^\times$ in Sect. 10.2 by the same ω. This ω is the Teichmüller lift of $\bar{\omega}$. For each $j \mod p - 1$, let $\rho^{(j)} : G_{\{p\}} \to \hat{\Lambda}^\times$ be the universal deformation of $\bar{\omega}^j$ and let $\mathcal{X}_p^{(j)} := \mathcal{X}_{p,1}(\bar{\omega}^j)$ be the universal deformation space of $\bar{\omega}^j$. For $\varphi \in \mathcal{X}_p^{(j)}$, we set $\rho_\varphi^{(j)} := \varphi \circ \rho^{(j)}$.

Let $\psi : G_{\{p\}} \to \mathrm{Gal}(\mathbb{Q}(\mu_{p^\infty})/\mathbb{Q}) = H \times \Gamma$ be the Abelianization map and let $\mathfrak{A}_p$ be the complete ψ-differential module. Set $\tilde{\Lambda} := \mathbb{Z}_p[[H \times \Gamma]] = \mathbb{Z}_p[H][[\Gamma]]$. Then $\mathfrak{A}_p$ is a $\tilde{\Lambda}$-module. Let $\mathfrak{A}_p^{(j)} := \mathfrak{A}_p \otimes_{\tilde{\Lambda}} \tilde{\Lambda}^{(j)}$. By tensoring $\tilde{\Lambda}^{(j)}$ with the complete Crowell exact sequence (Theorem 10.4.1)

$$0 \longrightarrow \mathfrak{M} \longrightarrow \mathfrak{A}_p \longrightarrow I_{\tilde{\Lambda}} \longrightarrow 0$$

over $\tilde{\Lambda}$, we obtain the exact sequence of $\tilde{\Lambda}^{(j)}$-modules

$$0 \longrightarrow \mathfrak{M}^{(j)} \longrightarrow \mathfrak{A}_p^{(j)} \longrightarrow I_{\tilde{\Lambda}^{(j)}} \longrightarrow 0$$

for each $j \mod p - 1$. From this, we have $E_{d-1}(\mathfrak{M}^{(j)}) = E_d(\mathfrak{A}_p^{(j)})$ for $d \geq 1$. When j is even, $E_0(\mathfrak{M}^{(j)}) = E_1(\mathfrak{A}_p^{(j)})$ is generated by the Iwasawa polynomial $\Delta_p^{(j)}(T)$ (Proposition 13.2.1). We then define the d-th Iwasawa set in $\mathcal{X}_p^{(j)}$ by

$$\mathcal{A}_p^{(j)}(d) := \{\varphi \in \mathcal{X}_p^{(j)} \mid f(\varphi(\gamma) - 1) = 0 \text{ for all } f \in E_d(\mathfrak{A}_p^{(j)})\}.$$

Thus we have a descending series $\mathcal{X}_p^{(j)} \supset \mathcal{A}_p^{(j)}(1) \supset \cdots \supset \mathcal{A}_p^{(j)}(d) \supset \cdots$. On the other hand, for $\varphi \in \mathcal{X}_p^{(j)}$, we define the $G_{\{p\}}$-module $\overline{\mathbb{Q}}_p(\rho_\varphi^{(j)})$ by the additive group $\overline{\mathbb{Q}}_p$ equipped with $G_{\{p\}}$-action given by $g.z = \rho_\varphi^{(j)}(g)z$, $(g \in G_{\{p\}}, z \in \overline{\mathbb{Q}}_p)$. We then define the d-th cohomology jumping set in $\mathcal{X}_p^{(j)}$ by

$$\mathcal{C}_p^{(j)}(d) := \{\rho \in \mathcal{X}_p^{(j)} \mid \dim_{\overline{\mathbb{Q}}_p} H^1(G_K, \overline{\mathbb{Q}}_p(\rho_\varphi^{(j)})) \geq d\}.$$

Thus we have another descending series $\mathcal{X}_p^{(j)} \supset \mathcal{C}_p^{(j)}(1) \supset \cdots \supset \mathcal{C}_p^{(j)}(d) \supset \cdots$.

Theorem 14.4.2 ([162]) *For each even j (mod $p - 1$), one has*

$$\mathcal{A}_p^{(j)}(d) = \mathcal{C}_p^{(j)}(d) \ (d > 1), \quad \mathcal{A}_p^{(j)}(1) = \mathcal{C}_p^{(j)}(1) \ (j \neq 0), \quad \mathcal{A}_p^{(0)}(1) \cup \{\mathbf{1}\} = \mathcal{C}_p^{(0)}(1),$$

where $\mathbf{1}$ stands for the trivial representation of $G_{\{p\}}$.

We regard $\overline{\mathbb{Q}}_p(\rho_\varphi^{(j)})$ as a $\tilde{\Lambda}$-module and set $\mathfrak{A}_p(\rho_\varphi^{(j)}) := \mathfrak{A}_p \otimes_{\tilde{\Lambda}} \overline{\mathbb{Q}}_p(\rho_\varphi^{(j)})$. If we denote by $\overline{\mathbb{Q}}_p(\varphi)$ the additive group $\overline{\mathbb{Q}}_p$ with Γ-action defined by $\gamma.z := \varphi(\gamma)z$, then $\mathfrak{A}_p(\rho_\varphi^{(j)}) = \mathfrak{A}_p^{(j)} \otimes_{\tilde{\Lambda}} \overline{\mathbb{Q}}_p(\varphi)$.

Lemma 14.4.3

(1) We have the following isomorphism:

$$\mathrm{Hom}_{\overline{\mathbb{Q}}_p}(\mathfrak{A}_p(\rho_\varphi^{(j)}), \overline{\mathbb{Q}}_p) \simeq Z^1(G_{\{p\}}, \overline{\mathbb{Q}}_p(\rho_\varphi^{(j)})),$$

where $Z^1(G_{\{p\}}, \overline{\mathbb{Q}}_p(\rho_\varphi^{(j)}))$ stands for the group of continuous 1-cocycles.
(2) We have the following:

$$\dim_{\overline{\mathbb{Q}}_p} H^1(G_{\{p\}}, \overline{\mathbb{Q}}_p(\rho_\varphi^{(j)})) = \begin{cases} \dim \mathfrak{A}_p(\rho_\varphi^{(j)}) - 1, & (j, \varphi) \neq (0, \mathbf{1}), \\ 1, & (j, \varphi) = (0, \mathbf{1}). \end{cases}$$

Proof

(1) By Definition 10.3.1,

$$\mathfrak{A}_p(\rho_\varphi^{(j)}) = \left(\bigoplus_{g \in G_{\{p\}}} \overline{\mathbb{Q}}_p(\rho_\varphi^{(j)}) \, dg \right) / \langle d(g_1 g_2) - dg_1 - \rho_\varphi^{(j)}(g_1) \, dg_2 \rangle_{\overline{\mathbb{Q}}_p}.$$

Hence

$$\begin{aligned} &\mathrm{Hom}_{\overline{\mathbb{Q}}_p}(\mathfrak{A}_p(\rho_\varphi^{(j)}), \overline{\mathbb{Q}}_p) \\ &\simeq \{c : G_{\{p\}} \to \overline{\mathbb{Q}}_p \mid c(g_1 g_2) - c(g_1) - \rho_\varphi^{(j)}(g_1)c(g_2) = 0\} \\ &= Z^1(G_{\{p\}}, \overline{\mathbb{Q}}_p(\rho_\varphi^{(j)})). \end{aligned}$$

(2) By definition of $\rho^{(j)}$, $\rho_\varphi^{(j)} = \mathbf{1} \Leftrightarrow (i, \varphi) = (0, \mathbf{1})$. Therefore the group of coboundaries $B^1(G_{\{p\}}, \overline{\mathbb{Q}}_p(\rho_\varphi^{(j)}))$ is given by

$$B^1(G_{\{p\}}, \overline{\mathbb{Q}}_p(\rho_\varphi^{(j)})) = \begin{cases} \overline{\mathbb{Q}}_p, & (j, \varphi) \neq (0, \mathbf{1}), \\ \{0\}, & (j, \varphi) = (0, \mathbf{1}). \end{cases}$$

Together with (1), we get the assertion.

$\square$

Proof of Theorem 14.4.4 Let j be an even integer. Then $\mathfrak{M}^{(j)}$ is a finitely generated, torsion $\hat{\Lambda}$-module (Proposition 13.2.1). Let $\mathfrak{Q}_p^{(j)}$ be a representation matrix of the $\hat{\Lambda}$-module $\mathfrak{A}_p^{(j)}$:

$$\hat{\Lambda}^m \xrightarrow{\mathfrak{Q}_p^{(j)}} \hat{\Lambda}^n \longrightarrow \mathfrak{A}_p^{(j)} \longrightarrow 0 \quad \text{(exact).} \tag{14.4}$$

The ideal $E_d(\mathfrak{A}_p^{(j)})$ is generated by $(n-d)$-minors of $\mathfrak{Q}_p^{(j)}$ if $d < n$, and is $\hat{\Lambda}$ if $d \geq n$. By tensoring $\overline{\mathbb{Q}}_p(\varphi)$ with (14.4.3.1) over $\hat{\Lambda}$, we obtain the exact sequence

$$\overline{\mathbb{Q}}_p(\varphi)^m \xrightarrow{\varphi(\mathfrak{Q}_p^{(j)})} \overline{\mathbb{Q}}_p(\varphi)^n \longrightarrow \mathfrak{A}_p(\rho_\varphi^{(j)}) \longrightarrow 0 \quad \text{(exact),}$$

where $\varphi(\mathfrak{Q}_p^{(j)})$ is the matrix over $\overline{\mathbb{Q}}_p$ obtained by the evaluation $\hat{\Lambda} \ni T \mapsto \varphi(T) \in \overline{\mathbb{Q}}_p$. Therefore $\dim_{\overline{\mathbb{Q}}_p} \mathfrak{A}_p(\rho_\varphi^{(j)}) = n - (\text{rank of } \varphi(\mathfrak{Q}^{(j)}))$. Hence, for $d > 1$,

$$\dim_{\overline{\mathbb{Q}}_p} H^1(G_{\{p\}}, \overline{\mathbb{Q}}_p(\rho_\varphi^{(j)})) \geq d \iff \dim \mathfrak{A}_p(\rho_\varphi^{(j)}) \geq d + 1 \quad \text{(Lemma 14.4.3, (2))}$$
$$\iff \text{rank of } \varphi(\mathfrak{Q}_p^{(j)}) \leq n - (d+1)$$
$$\iff \text{any } (n-d)\text{-minor of } \varphi(\mathfrak{Q}_p^{(j)}) = 0$$
$$\iff \text{for all } f \in E_d(\mathfrak{A}_p(\rho_\varphi^{(j)})),$$
$$f(\varphi(\gamma) - 1) = 0,$$

and, for $d = 1$,

$$\dim_{\overline{\mathbb{Q}}_p} H^1(G_{\{p\}}, \overline{\mathbb{Q}}_p(\rho_\varphi^{(j)})) \geq 1$$
$$\iff \rho_\varphi^{(j)} \neq \mathbf{1} \text{ and } \dim \mathfrak{A}_p(\rho_\varphi^{(i)}) \geq 2, \text{ or } \rho_\varphi^{(j)} = \mathbf{1} \quad \text{(Lemma 14.4.3, (2))}$$
$$\iff (j, \varphi) \neq (0, \mathbf{1}) \text{ and } f(\varphi(\gamma) - 1) = 0 \text{ for all } f \in E_1(\mathfrak{A}_p(\rho_\varphi^{(j)})),$$
$$\text{or } (j, \varphi) = (0, \mathbf{1}).$$

The proof is done.

Corollary 14.4.5 *For each even j (mod $p - 1$),*

$$\Delta_p^{(j)}(\varphi(\gamma) - 1) = 0 \iff H^1(G_{\{p\}}, \overline{\mathbb{Q}}_p(\rho_\varphi^{(j)})) \neq 0.$$

Proof This follows from $E_1(\mathfrak{A}_p^{(j)}) = (\Delta_p^{(j)})$ and Theorem 14.4.2. □

Remark 14.4.6 It is known that the Alexander module A_K (or the knot module $H_1(X_\infty)$) has a standard resolution by the Seifert matrix [101, 5.4]. As for the Iwasawa module $H_\infty^{(i)}$, M. Kurihara gave a resolution using a certain Euler–Kolyvagin system and showed a refined form of the main conjecture of Iwasawa [121], [122].

Summary

Character variety of N-dim. complex representations of G_K $\mathcal{X}_{K,N}$	Deformation space of N-dim. p-adic representations of $G_{\{p\}}$ $\mathcal{X}_{p,N}(\overline{\rho})$
$\mathcal{X}_{K,1} = \mathrm{Hom}_{\mathbb{C}-\mathrm{alg.}}(\Lambda, \mathbb{C}) = \mathbb{C}^{\times}$ $(\Lambda = \mathbb{Z}[t^{\pm 1}])$	$\mathcal{X}_{p,1}(\overline{\rho}) = \mathrm{Hom}_{\mathbb{Z}_p-\mathrm{alg.}}(\hat{\Lambda}, \overline{\mathbb{Q}}_p) = \mathcal{D}_p$ $(\hat{\Lambda} = \mathbb{Z}_p[[T]])$
d-th Alexander set (zeros of Alexander polynomial) and cohomology jump set	d-th Iwasawa set (zeros of Iwasawa polynomial) and cohomology jump sets

Chapter 15
Deformations of Hyperbolic Structures and p-Adic Ordinary Modular Forms

As we have seen in Chap. 14, the Alexander–Fox theory and Iwasawa theory may be regarded as theories on the moduli spaces of 1-dimensional representations. H. Hida [78], [79] and B. Mazur [139] generalized the Iwasawa theory to a non-Abelian theory from the viewpoint of the deformation theory of higher-dimensional representations. A similar theory on the moduli of higher-dimensional representations of a knot group would provide a natural non-Abelian generalization of the Alexander–Fox theory. In particular, for 2-dimensional representations, we find intriguing analogies between the family of holonomy representations associated to deformations of hyperbolic structures and the family of Galois representations associated to deformations of p-adic ordinary modular forms.

We refer to [224] and [80] for 3-dimensional hyperbolic geometry and modular forms respectively.

15.1 Deformation of Hyperbolic Structures

Let K be a hyperbolic knot in S^3. Namely, $S^3 \setminus K$ is a complete hyperbolic 3-manifold of finite volume with a cusp. Thus the knot group G_K is a discrete subgroup of $PSL_2(\mathbb{C}) = \mathrm{Aut}(\mathbb{H}^3)$, where $\mathbb{H}^3$ denotes the hyperbolic 3-space, and $S^3 \setminus K$ is written as the quotient space $\mathbb{H}^3/G_K$. Let V_K be a tubular neighborhood of K and $X_K := S^3 \setminus \mathrm{int}(V_K)$, and fix a meridian α and a longitude β of K ($\alpha, \beta \subset \partial X_K$).

Let $\bar{\rho}_h : G_K \to PSL_2(\mathbb{C})$ be the holonomy representation associated to the complete hyperbolic structure on $S^3 \setminus K$. Since $H^2(G_K, \mathbb{F}_2) = H^2(X_K, \mathbb{F}_2) = 0$, we note that $\bar{\rho}_h \in H^1(G_K, PSL_2(\mathbb{C}))$ can be lifted to $\rho_h \in H^1(G_K, SL_2(\mathbb{C}))$ by considering the exact sequence of group cohomology with non-commutative coefficients ([205, Ch. VII, Appendix]) attached to the exact sequence $1 \to \{\pm I\} \to SL_2(\mathbb{C}) \to PSL_2(\mathbb{C}) \to 1$. In this chapter, we shall be concerned

© The Author(s), under exclusive license to Springer Nature Singapore Pte Ltd. 2024   201
M. Morishita, *Knots and Primes*, Universitext,
https://doi.org/10.1007/978-981-99-9255-3_15

with representations of G_K associated to (incomplete) hyperbolic structures and therefore we shall consider $SL_2(\mathbb{C})$-representations of G_K in the following, and thus set

$$\mathcal{R}_K := \mathrm{Hom}(G_K, SL_2(\mathbb{C})), \quad \mathcal{X}_K := \mathcal{R}_K /\!/ GL_2(\mathbb{C}).$$

Since $D_K = \pi_1(\partial X_K)$ is Abelian, the restriction of $\rho \in \mathrm{Hom}(G_K, SL_2(\mathbb{C}))$ to D_K is equivalent to an upper triangular representation:

$$\rho|_{D_K} \simeq \begin{pmatrix} \chi_\rho & * \\ 0 & \chi_\rho^{-1} \end{pmatrix}, \tag{15.1}$$

where $\chi_\rho : G_K \rightarrow \mathbb{C}^\times$ is a 1-dimensional representation. Let $\mathcal{X}_K^\circ(\rho_h)$ be the connected component of $\mathcal{X}_K$ containing ρ_h. Then we have the following theorem due to W. Thurston. (For the proof, we refer to [224], [18, Appendix B].)

Theorem 15.1.1 *The mapping*

$$\Psi_K : \mathcal{X}_K^\circ(\rho_h) \longrightarrow \mathbb{C}; \ \Psi_K([\rho]) := \mathrm{Tr}(\rho(\alpha))$$

is biholomorphic in a neighborhood W of $[\rho_h]$. In particular, $\mathcal{X}_K^\circ(\rho_h)$ is a complex algebraic curve.

The character curve $\mathcal{X}_K^\circ(\rho_h)$ is related to the deformation space of hyperbolic structures on $S^3 \setminus K$ as follows. Let $S(z)$ denote the ideal tetrahedron in $\mathbb{H}^3$ with vertices $0, 1, z$ and ∞ ($z \in \mathbb{C} \setminus \{0, 1\}$) (Fig. 15.1).

Fix an ideal triangulation of $S^3 \setminus K$

$$S^3 \setminus K = S(z_1^\circ) \cup \cdots \cup S(z_n^\circ).$$

Fig. 15.1 An the ideal tetrahedron in the hyperbolic 3-space

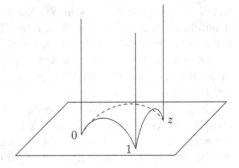

At each edge, the sum of dihedral angles of the tetrahedron around the edge is equal to 2π. The parameters $z_1^\circ, \dots, z_n^\circ$ satisfy a system of equation of the form

$$\prod_{i=1}^n z_i^{r'_{ij}} (1 - z_i)^{r''_{ij}} = \pm 1 \quad (j = 1, \dots, n). \tag{15.2}$$

Let $\mathcal{Y}$ be the affine algebraic set in $(\mathbb{C} \setminus \{0, 1\})^n$ defined by (15.1.3). W. Neumann and D. Zagier [177] showed that the irreducible component $\mathcal{X}_K^H(z^\circ)$ of $\mathcal{Y}$ containing $z^\circ := (z_1^\circ, \dots, z_n^\circ)$ is a complex algebraic curve, called the *deformation curve of hyperbolic structures*. Then $\mathcal{X}_K^H(z^\circ)$ is a double covering over $W \subset \mathcal{X}_K^\circ(\rho_h)$. Namely, there are a neighborhood U of z° in $\mathcal{X}_K^H(z^\circ)$ and a holomorphic local coordinate x_K on U with $x_K(z^\circ) = 0$ such that we have the following commutative diagram:

$$
\begin{array}{ccc}
U & \xrightarrow{x_K} & \mathbb{C} \\
\pi \downarrow & & \downarrow h \\
W & \xrightarrow{\psi_K} & \mathbb{C},
\end{array}
$$

where $h(x) := e^{x/2} + e^{-x/2}$ and π is a double covering ramified over z°. In fact, for each $z \in U$, we can construct the developing mapping $\widetilde{S^3 \setminus K} \to \mathbb{H}^3$ with holonomy representation $\bar{\rho}_z$, and express a lift ρ_z, $SL_2(\mathbb{C})$-representation, of $\bar{\rho}_z$ as $\rho_z = t(\pi(z))$ by a local section $t : W \to \mathcal{R}_K$. Then

$$\rho_z(\alpha) \simeq \begin{pmatrix} e^{x_K(z)/2} & * \\ 0 & e^{-x_K(z)/2} \end{pmatrix}, \quad \rho_z(\beta) \simeq \begin{pmatrix} e^{y_K(z)/2} & * \\ 0 & e^{-y_K(z)/2} \end{pmatrix}.$$

Here, using χ_ρ of (15.1),

$$x_K(z) = 2 \log \chi_{\rho_z}(\alpha), \quad y_K(z) = 2 \log \chi_{\rho_z}(\beta) \quad (z \in U).$$

For each $z \in U$, we define $(q, p) \in \mathbb{R}^2 \cup \{\infty\} = S^2$ by $(q, p) := \infty$ if $z = z^\circ$ and by $q x_K(z) + p y_K(z) = 2\pi \sqrt{-1}$ if $z \neq z^\circ$. Then the correspondence $z \mapsto (q, p)$ gives an homeomorphism from U to a neighborhood of ∞ in S^2. When p, q are coprime integers, we obtain a closed hyperbolic 3-manifold $M(q, p)$ by the *Dehn surgery* along K with surgery coefficient q/p. (For a solid torus V and a meridian $m \subset \partial V$, $M(q, p)$ is obtained by gluing X_K and V by a homeomorphism $f : \partial V \xrightarrow{\approx} \partial V_K = \partial X_K$ such that $[f_*(m)] = q[\alpha] + p[\beta]$.) A point $z \in U$ corresponding to such a (q, p) or z° is called a *Dehn surgery point with integral coefficient*.

Since x_K is a local coordinate on U, y_K can be regarded as a function of x_K. Then it was shown by Neumann–Zagier [177] that there is a holomorphic function $\tau(x_K)$ such that $y_K = x_K \tau(x_K)$ and $\tau(0)$ is a modulus of the 2-dimensional torus

∂X_K whose complex structure is induced by the complete hyperbolic structure on $S^3 \setminus K$.

Theorem 15.1.2 ([177, Lemma 4.1]) *Let* $q_K := \exp(2\pi\sqrt{-1}\tau(0))$. *Then* q_K *gives a multiplicative period of* ∂X_K: $\partial X_K = \mathbb{C}^\times/(q_K)^{\mathbb{Z}}$, *and*

$$\left.\frac{dy_K}{dx_K}\right|_{x_K=0} = \frac{1}{2\pi\sqrt{-1}}\log q_K.$$

The integral of y_K with respect to x_K is related with the $SL_2(\mathbb{C})$-Chern–Simons functional as follows. For an $sl_2(\mathbb{C})$-valued 1-form A on X_K, the $SL_2(\mathbb{C})$-*Chern–Simons functional* is defined by

$$CS(A) := \frac{1}{8\pi^2}\int_{X_K}\mathrm{Tr}\left(dA \wedge A + \frac{2}{3}A \wedge A \wedge A\right).$$

For $z \in \mathcal{X}_K^{\mathrm{H}}(z^\circ)$, we define $CS(\rho_z)$ by $CS(A_{\rho_z})$, where A_{ρ_z} is the connection 1-form of the flat connection corresponding to the representation ρ_z.

Theorem 15.1.3 ([168]) *Notation being as above, we have*

$$\int_0^{x_K(z)} y_K\, dx_K - \frac{1}{2}x_K y_K = 8\pi^2 CS(\rho_z) \qquad (z \in U).$$

Remark 15.1.4 Following the deformation theory for Galois representations in Sect. 14.3, in [169], we introduced the universal deformation of a SL_2-knot group representation, which may be regarded as an infinitesimal deformation of the SL_2-character scheme (Sect. 14.1). Motivated by the problems posed by Mazur ([139]), [108], [109] and [222] investigated the algebraic L-functions (Fitting ideals) associated to the twisted knot modules or Selmer modules for the universal representations together with various examples.

15.2 Deformation of p-Adic Ordinary Modular Galois Representations

Let p be an odd prime number, $G_{\{p\}}$ the prime group, and $D_{\{p\}}$, $I_{\{p\}}$ the decomposition, inertia groups over p respectively (Example 2.2.24). We shall be concerned with 2-dimensional p-adic representations of $G_{\{p\}}$. Firstly, in order to obtain an arithmetic analogue of the boundary condition (15.1), which is automatically satisfied for the knot case, we consider the "boundary" condition, called p-ordinaliness, on representations $G_{\{p\}}$, following after the analogy (3.9).

Namely, a representation $\rho : G_{\{p\}} \to GL_2(A)$ is said to be *p-ordinary*, if $\rho|_{D_{\{p\}}}$ is equivalent to an upper triangular representation of the following type:

$$\rho|_{D_{\{p\}}} \simeq \begin{pmatrix} \chi_1 & * \\ 0 & \chi_2 \end{pmatrix}, \quad \chi_2|_{I_{\{p\}}} = \mathbf{1},$$

where $\chi_i : D_{\{p\}} \to A^\times$ is a 1-dimensional representation ($i = 1, 2$).

Now we fix an absolutely irreducible and p-ordinary continuous residual representation $\bar\rho : G_{\{p\}} \to GL_2(\mathbb{F}_p)$. Then we have the following theorem for *p-ordinary deformation* of $\bar\rho$:

Theorem 15.2.1 ([137, 1.7]) *There is a p-ordinary deformation of $\bar\rho$*

$$\rho_p^\circ : G_{\{p\}} \longrightarrow GL_2(R_p^\circ)$$

such that for any p-ordinary deformation $\rho : G_{\{p\}} \to GL_N(A)$ of $\bar\rho$, there exists uniquely a $\mathbb{Z}_p$-algebra homomorphism $\varphi : R_p^\circ \to A$ with $\varphi \circ \rho_p^\circ \approx \rho$.

The pair $(R_p^\circ, \rho_p^\circ)$ is unique up to equivalence and called the *universal p-ordinary deformation* of $\bar\rho$. We define the *universal p-ordinary deformation space* of $\bar\rho$ by

$$\mathcal{X}_p^\circ(\bar\rho) := (\mathrm{Spec}(R_p^\circ))(\overline{\mathbb{Q}}_p) = \mathrm{Hom}_{\mathbb{Z}_p\text{-alg}}(R_p^\circ, \overline{\mathbb{Q}}_p).$$

We regard $\mathcal{X}_p^\circ(\bar\rho)$ as an arithmetic analogue of $\mathcal{X}_K^\circ$.

Since $\det \rho_p^\circ : G_{\{p\}} \to (R_p^\circ)^\times$ is a deformation of $\det \bar\rho : G_{\{p\}} \to \mathbb{F}_p^\times$, by Theorem 14.3.1, there exists a unique $\mathbb{Z}_p$-algebra homomorphism $\iota^\circ : \hat\Lambda \to R_p^\circ$ such that $\iota^\circ \circ \rho_{p,1} = \det \rho_p^\circ$. We regard R_p° as a $\hat\Lambda$-algebra via ι°.

The deformation space $\mathcal{X}_p^\circ(\bar\rho)$ is related with the family of Galois representations in Hida's theory on the universal ordinary p-adic Hecke algebra. Firstly, we recall some facts on Galois representations associated to modular forms. Let f be a cuspidal eigenform of level p-power on $\Gamma_0(p^m)$ ($m \geq 1$) of weight $w_f \geq 2$ and character ε_f. Here we mean by f being an eigenform that f is an eigenfunction of the Hecke operator T_l for each prime number $l \neq p$ and an eigenfunction of the Atkin operator U_p. For the Fourier expansion $f = \sum_{n \geq 1} a_n(f) e^{2\pi\sqrt{-1}nz}$ at ∞, the Fourier coefficients $a_l(f)$ ($l \neq p$) and $a_p(f)$ are eigenvalues of T_l and U_p respectively. Let $\mathcal{O}_f$ be the integral closure of the ring $\mathbb{Z}[a_n(f)|n \geq 1]$ in $\mathbb{C}$. Then $\mathcal{O}_f$ is the ring of integers in a finite algebraic number field k_f. For a prime ideal $\mathfrak{p}$ of $\mathcal{O}_f$ over p, let $k_{f,\mathfrak{p}}$ be the $\mathfrak{p}$-adic completion of k_f and $\mathcal{O}_{f,\mathfrak{p}}$ the ring of integers of $k_{f,\mathfrak{p}}$.

Theorem 15.2.2

(1) [38] There is an absolutely irreducible representation

$$\rho_{f,\mathfrak{p}} : G_{\{p\}} \longrightarrow GL_2(\mathcal{O}_{f,\mathfrak{p}})$$

such that for any prime number $l \neq p$, one has

$$\mathrm{Tr}(\rho_{f,\mathfrak{p}}(\sigma_l)) = a_l(f), \quad \det(\rho_{f,\mathfrak{p}}(\sigma_l)) = \varepsilon_f(l) l^{w_f - 1},$$

and such a representation is unique up to equivalence over $k_{f,\mathfrak{p}}$. Here $\sigma_l \in G_{\{p\}}$ stands for the Frobenius automorphism over l.

(2) *[142] Assume that f is p-ordinary, namely, $a_p(f) \in \mathcal{O}_{f,\mathfrak{p}}^{\times}$ (This condition is independent of the choice of $\mathfrak{p}$.) Then $\rho_{f,\mathfrak{p}}$ is also p-ordinary:*

$$\rho_{f,\mathfrak{p}}|_{D_{\{p\}}} \simeq \begin{pmatrix} \chi_1 & * \\ 0 & \chi_2 \end{pmatrix}, \quad \chi_2|_{I_{\{p\}}} = 1.$$

In the rest of this chapter, we fix an absolutely irreducible representation

$$\bar{\rho} = \rho_{f^{\circ},\mathfrak{p}} \bmod \mathfrak{p} : G_{\{p\}} \longrightarrow GL_2(\mathbb{F}_p)$$

associated to a p-ordinary cuspidal eigenform f° with $\mathcal{O}_{f^{\circ}}/\mathfrak{p} = \mathbb{F}_p$. Hida [78], [79] constructed a big Galois representation from a certain completion of the Hecke algebra which has the following universal property.

Theorem 15.2.3 ([78, 79]) *There is a p-ordinary deformation of $\bar{\rho}$*

$$\rho_p^{\mathrm{H}} : G_{\{p\}} \longrightarrow GL_2(R_p^{\mathrm{H}}),$$

which has the following property: If $\rho_{f,\mathfrak{p}} : G_{\{p\}} \to GL_2(\mathcal{O}_{f,\mathfrak{p}})$ is a deformation of $\bar{\rho}$ associated to a p-ordinary cuspidal eigenform f of level p-power, there exists a unique $\mathbb{Z}_p$-algebra homomorphism $\varphi : R_p^{\mathrm{H}} \to \mathcal{O}_{f,\mathfrak{p}}$ such that $\varphi \circ \rho_2^{\mathrm{H}} \approx \rho_{f,\mathfrak{p}}$.

Such a pair $(R_p^{\mathrm{H}}, \rho_p^{\mathrm{H}})$ is unique up to equivalence and is called the *universal p-ordinary modular deformation* of $\bar{\rho}$. We define the *universal p-ordinary modular deformation space* by

$$\mathcal{X}_p^{\mathrm{H}}(\bar{\rho}) := \mathrm{Spec}(R_p^{\mathrm{H}})(\overline{\mathbb{Q}}_p) = \mathrm{Hom}_{\mathbb{Z}_p\text{-algebra}}(R_p^{\mathrm{H}}, \overline{\mathbb{Q}}_p).$$

For $\varphi \in \mathcal{X}_p^{\mathrm{H}}(\bar{\rho})$, we write $\rho_{\varphi}^{\mathrm{H}}$ for $\varphi \circ \rho_p^{\mathrm{H}}$.

Let $(R_{p,1} = \hat{\Lambda}, \rho_{p,1})$ be the universal deformation of $\det \bar{\rho}$. Then there exists uniquely a $\mathbb{Z}_p$-algebra homomorphism $\iota^{\mathrm{H}} : \hat{\Lambda} \to R_p^{\mathrm{H}}$ such that $\iota^{\mathrm{H}} \circ \rho_{p,1} = \det \rho_p^{\mathrm{H}}$. Via ι^{H}, we regard R_p^{H} as a $\hat{\Lambda}$-algebra. ι^{H} induces a p-adic analytic mapping

$$x_p : \mathcal{X}_p^{\mathrm{H}}(\bar{\rho}) \longrightarrow \mathcal{D}_p; \quad \varphi \longmapsto (\varphi \circ \iota^{\mathrm{H}})(T).$$

A prime ideal $\mathfrak{P} \in \mathrm{Spec}(R_p^{\mathrm{H}})$ of height 1 is called an *arithmetic point*, if there are integers $w \geq 0$ and a Dirichlet character ε mod p-power such that the image of $1 + T$ under the natural homomorphism $R_{p,2}^{\mathrm{H}} \to R_{p,2}^{\mathrm{H}}/\mathfrak{P}$ equals $\varepsilon(1 + p)(1 +$

$p)^{w-2}$. Then the point $\mathfrak{P}$ corresponds to a cuspidal eigenform of weight w and character ε. A point $\varphi \in \mathcal{X}_p^H(\bar{\rho})$ is called an *arithmetic point*, if $\mathfrak{P} = \mathrm{Ker}(\varphi)$ is an arithmetic point. Arithmetic points in $\mathcal{X}_p^H(\bar{\rho})$ may be viewed as analogues of Dehn surgery points of integral coefficients in the deformation space $\mathcal{X}_K^H(z^\circ)$ of hyperbolic structures. As for the $\hat{\Lambda}$-algebra structure of R_p^H, the following theorem, due to Hida, is fundamental for us.

Theorem 15.2.4 ([78, Corollary 1.4], [79]) R_p^H *is a finite and flat algebra over* $\hat{\Lambda}$*. Furthermore,* $(R_p^H)_{\mathfrak{P}}$ *is an étale algebra over* $\hat{\Lambda}_{\mathfrak{p}}$ *for any arithmetic point* $\mathfrak{P} \in$ $\mathrm{Spec}(R_p^H)$*, where* $\mathfrak{p} := (\iota^H)^{-1}(\mathfrak{P})$*. In particular,* x_p *gives a* p*-adic analytic local coordinate around* φ*.*

Let $\mathbb{Q}_\infty$ be the cyclotomic $\mathbb{Z}_p$-extension of $\mathbb{Q}$ (Example 2.3.6). We fix a topological generator γ of $\mathrm{Gal}(\mathbb{Q}_\infty/\mathbb{Q}) = 1 + p\mathbb{Z}_p$ and identify $\mathbb{Z}_p[[\mathrm{Gal}(\mathbb{Q}_\infty/\mathbb{Q})]]$ with $\hat{\Lambda}$ by the correspondence $\gamma \leftrightarrow 1 + T$. We choose $\tau \in G_{\{p\}}$ which is sent to γ under the natural homomorphism $G_{\{p\}} \to \mathrm{Gal}(\mathbb{Q}_\infty/\mathbb{Q})$. The following theorem may be regarded as an arithmetic analogue for $\mathcal{X}_p^H(\bar{\rho})$ of Theorem 15.1.1.[1]

Theorem 15.2.5 ([167]) *The map*

$$\Psi_p : \mathcal{X}_p^H(\bar{\rho}) \longrightarrow \overline{\mathbb{Q}}_p; \ \Psi_p(\varphi) := \mathrm{Tr}(\rho_\varphi^H(\tau))$$

is p*-adically bianalytic in a neighborhood of* φ°*. Here* φ° *is defined by* $\rho_{\varphi^\circ}^H = \varphi^\circ \circ \rho_p^H \approx \rho_{f^\circ, p}$*.*

Proof Since ρ_φ^H is p-ordinary for $\varphi \in \mathcal{X}_p^H(\bar{\rho})$, we have

$$\rho_\varphi^H \simeq \begin{pmatrix} \chi_{1,\varphi} & * \\ 0 & \chi_{2,\varphi} \end{pmatrix}, \ \chi_{2,\varphi}|_{I_{\{p\}}} = \mathbf{1}.$$

Therefore

$$\begin{aligned}
\chi_{1,\varphi}(\tau) &= \det(\rho_\varphi^H(\tau)) \\
&= \varphi(\det(\rho_p^H(\tau))) \\
&= \varphi(\iota^H \circ \rho_{p,1}(\tau)) \\
&= \varphi(\omega(\tau)\iota^H(\gamma)) \\
&= \omega(\tau)\varphi(\iota^H(\gamma)) \\
&= \omega(\tau)(1 + x_p(\varphi)) \quad (\gamma = 1 + T),
\end{aligned}$$

---

[1] The analogy between two deformation spaces $\mathcal{X}_K^\circ(\rho_h)$ and $\mathcal{X}_p^H(\bar{\rho})$ was first pointed out by K. Fujiwara [58].

where ω denotes the Teichmüller lift of det $\bar{\rho}$. Hence

$$
\begin{aligned}
\Psi_p^H(\rho_\varphi^H(\tau)) &= \operatorname{Tr}(\rho_\varphi^H(\tau)) \\
&= \chi_{1,\varphi}(\tau) + 1 \\
&= \omega(\tau)(1 + x_p(\varphi)) + 1.
\end{aligned}
$$

Since x_p is a p-adically bianalytic function on φ° by Theorem 15.2.4, so is Ψ_p^H.

By the universality of $(R_p^\circ, \rho_p^\circ)$, there exists a unique $\mathbb{Z}_p$-algebra homomorphism $\psi : R_p^\circ \to R_p^H$ such that $\psi \circ \rho_p^\circ \approx \rho_p^H$. Since $\iota^H \circ \psi = \iota^\circ$, we see that ψ is a $\hat{\Lambda}$-algebra homomorphism. Then Mazur [138] conjectured that ψ is an isomorphism, which was proved by A. Wiles.

Theorem 15.2.6 ([223], [245]) *Let $k = \mathbb{Q}(\sqrt{p^*})$, $p^* = (-1)^{(p-1)/2}p$, and assume that the level of f° and p and that $\bar{\rho}|_{\operatorname{Gal}(\overline{\mathbb{Q}}/k)}$ is absolutely irreducible. Then, ψ is an isomorphism. Hence $\mathcal{X}_p^H(\bar{\rho}) = \mathcal{X}_p^\circ(\bar{\rho})$.*

Remark 15.2.7 Theorem 15.2.6 may be regarded as a non-Abelian generalization of class field theory in Sects. 2.3.3 and 7.3. Let $\mathbb{A}_\mathbb{Q} = \prod_p \mathbb{Q}_p \times \mathbb{R}$ be the adèle ring of $\mathbb{Q}$, where $\prod_p$ means the restricted product of $\mathbb{Q}_p$'s with respect to $\mathbb{Z}_p$'s. Note that the idèle group $J_\mathbb{Q}$ and the idèle class group $C_\mathbb{Q} = J_\mathbb{Q}/\mathbb{Q}^\times$ (2.7) are written as $J_\mathbb{Q} = \mathbb{A}_\mathbb{Q}^\times = GL_1(\mathbb{A}_\mathbb{Q})$ and $C_\mathbb{Q} = \mathbb{A}_\mathbb{Q}^\times/\mathbb{Q}^\times = GL_1(\mathbb{A}_\mathbb{Q})/GL_1(\mathbb{Q})$ respectively. Then, via the reciprocity homomorphism $\rho_k : C_\mathbb{Q} \to \operatorname{Gal}(\mathbb{Q}^{\mathrm{ab}}/\mathbb{Q})$ of class field theory (2.8), one has a bijection between the sets of certain 1-dimensional representations of $\operatorname{Gal}(\mathbb{Q}^{\mathrm{ab}}/\mathbb{Q})$ and $GL_1(\mathbb{A}_\mathbb{Q})/GL_1(\mathbb{Q})$. On the other hand, noting that an cuspidal eigenform may be regarded as a function on $GL_2(\mathbb{A}_\mathbb{Q})/GL_2(\mathbb{Q})$, Theorem 15.2.6 tells us that there is a bijection, called the *Langlands correspondence*, between the set of certain 2-dimensional representations of $\operatorname{Gal}(\overline{\mathbb{Q}}/\mathbb{Q})$ and the set of certain functions on $GL_2(\mathbb{A}_\mathbb{Q})/GL_2(\mathbb{Q})$. Therefore Theorem 15.2.6 may be viewed as a GL_2-version of class field theory. For more precise account of this issue, we refer to [80, 1].

Now we consider the case that $\rho_{f^\circ,\mathfrak{p}}$ is coming from an elliptic curve E defined over $\mathbb{Q}$: $\rho_{f^\circ,\mathfrak{p}} = \rho_E$. Here ρ_E is the representation arising from the action of $\operatorname{Gal}(\overline{\mathbb{Q}}/\mathbb{Q})$ on the Tate module $\varprojlim_n E[p^n] = \mathbb{Z}_p^{\oplus 2}$ of E and is assumed to be unramified outside $\{p, \infty\}$. In particular, the weight w_{f° of f° must be 2 and $\mathcal{O}_{f^\circ,\mathfrak{p}} = \mathbb{Z}_p$. Furthermore, we assume that E is split-multiplicative at p, namely, E is extended to a smooth group scheme $\mathcal{E}$ over $\mathbb{Z}_p$ whose special fiber $\mathcal{E} \otimes_{\mathbb{Z}_p} \mathbb{F}_p$ is isomorphic to $\mathbb{G}_m$ over $\mathbb{F}_p$, where $\mathbb{G}_m = \operatorname{Spec}(\mathbb{F}_p[X, Y]/(XY - 1))$ is the multiplicative group over $\mathbb{F}_p$.

We let

$$
\rho_p^H \simeq \begin{pmatrix} \chi_1 & * \\ 0 & \chi_2 \end{pmatrix}, \quad \chi_2|_{I_{\{p\}}} = 1
$$

and set $y_p := \chi_2(\sigma_p)$. Here σ_p stands for the Frobenius automorphism over p. Let $\varphi^\circ \in \mathcal{X}^H_{p,2}(\bar{\rho})$ be the arithmetic point corresponding to $\rho_E = \rho_{f^\circ, \mathfrak{p}}$. Since x_p is p-adically bianalytic in a neighborhood of φ° and $\varphi^\circ \circ \iota^H(T) = 0$, y_p is regarded as a p-adic analytic function of x_p around $0 \in \mathcal{D}_p$. Then we have the following theorem due to R. Greenberg and G. Stevens.

Theorem 15.2.8 ([71], [81, 1.5]) *Let q_E be the Tate period of E: $E(\overline{\mathbb{Q}}_p) = \overline{\mathbb{Q}}_p^\times / (q_E)^{\mathbb{Z}}$. Then we have the following formula:*

$$\left. \frac{dy_p}{dx_p} \right|_{x_p=0} = -\frac{1}{2\log_p(\gamma)} \frac{\log_p(q_E)}{\mathrm{ord}_p(q_E)},$$

where $\log_p$ stands for the p-adic logarithm. The quantity $\log_p(q_E)/\mathrm{ord}_p(q_E)$ is called the $\mathfrak{L}$-*invariant* of the elliptic curve E.

Remark 15.2.9

(1) In view of Theorem 15.1.3 and our analogy, it would be an interesting problem to study an arithmetic meaning of the p-adic integral of y_p with respect to x_p, $\int_0^{x_p} y_p \, dx_p$. A result by R. Coleman [34] on the relation between p-adic Dirichlet L-values and p-adic polylogarithms may suggest that there would be some relation between this integral and the p-adic modular L-value $L_p(f, 2)$ when x_p is an arithmetic point corresponding to a modular form f.

(2) In [187], Pappas introduced and studied an analogue for l-adic local systems on an algebraic curve of $CS(\rho)$ for representations ρ of the fundamental group of a 3-manifold which is fibered over S^1.

Summary

Hyperbolic structures and associated holonomy representations	*p*-Adic ordinary modular forms and associated Galois representations
Deformation space $\mathcal{X}^H_K(z^\circ)$	Deformation space $\mathcal{X}^H_p(\bar{\rho})$
Dehn surgery points with integral coefficient	Arithmetic points (modular forms of integral weight)
Meridian function x_K	Monodromy function x_p
Longitude function y_K	Frobenius function y_p
Modulus of ∂X_K	$\mathfrak{L}$-invariant

Chapter 16
Dijkgraaf–Witten Theory
for 3-Manifolds and Number Rings

Dijkgraaf–Witten theory [47] is Chern-Simons theory with finite gauge groups, which is an example of (2+1)-dimensional topological quantum field theory, TQFT for short. Here TQFT is a framework to produce topological invariants of manifolds [11], [12], [244]. So Dijkgraaf–Witten theory provides 3-manifold invariants, which are defined rigorously as partition functions. In recent years, Minhyong Kim and his collaborators [31], [103] has been studying an arithmetic analogue for number rings of Dijkgraaf–Witten theory for 3-manifolds, based on the analogies between 3-manifolds and number rings, knots and primes discussed in this book. So arithmetic Dijkgraaf–Witten theory provides arithmetic invariants of number rings, which are defined as analogues of Dijkgraaf–Witte partition functions. In this final chapter, we treat their work and some subsequent works as well as the original Dijkgraaf–Witten theory for 3-manifolds.

16.1   Chern–Simons Functionals and Dijkgraaf–Witten Invariants for Closed 3-Manifolds

Let M be an oriented, connected, closed 3-manifold. We fix, once and for all, a finite group G and a 3-cocycle c of G with coefficients in $\mathbb{R}/\mathbb{Z}$, $c \in Z^3(G, \mathbb{R}/\mathbb{Z})$. These are given data.

Let $\mathcal{F}_M := \mathrm{Hom}(\pi_1(M), G)$ be the set of homomorphisms $\pi_1(M) \to G$, which may be regarded as an analogue of the gauge fields on M for a finite group G. Note that $\mathcal{F}_M$ is a finite set, since $\pi_1(M)$ is a finitely generated group. We shall construct the Chern–Simons functional on $\mathcal{F}_M$ as follows.

By the Hochschild–Serre spectral sequence $H^i(\pi_1(M), H^j(\tilde{M}, \mathbb{R}/\mathbb{Z})) \Rightarrow H^{i+j}(M, \mathbb{R}/\mathbb{Z})$ associated with the universal covering $\tilde{M} \to M$, we have the edge homomorphism

$$j^3 : H^3(\pi_1(M), \mathbb{R}/\mathbb{Z}) \longrightarrow H^3(M, \mathbb{R}/\mathbb{Z}).$$

The natural pairing $\langle \ , \ \rangle : H^3(M, \mathbb{R}/\mathbb{Z}) \times H_3(M, \mathbb{Z}) \to \mathbb{R}/\mathbb{Z}$ gives us the homomorphism

$$\langle \ , [M] \rangle : H^3(M, \mathbb{R}/\mathbb{Z}) \longrightarrow \mathbb{R}/\mathbb{Z},$$

where $[M] \in H_3(M, \mathbb{Z})$ represents the fundamental homology class. Thus, for $\rho \in \mathcal{F}_M$, we obtain the homomorphism f_M^ρ as the composite of maps

$$f_M^\rho : H^3(G, \mathbb{R}/\mathbb{Z}) \xrightarrow{\rho^*} H^3(\pi_1(M), \mathbb{R}/\mathbb{Z}) \xrightarrow{j^3} H^3(M, \mathbb{R}/\mathbb{Z}) \xrightarrow{\langle \ , [M] \rangle} \mathbb{R}/\mathbb{Z}.$$

We then define the *Chern–Simons functional* over $\mathcal{F}_M$ associated to c, CS_M : $\mathcal{F}_M \longrightarrow \mathbb{R}/\mathbb{Z}$, by

$$CS_M(\rho) := f_M^\rho([c]). \tag{16.1}$$

It is easy to see that the value $CS_M(\rho)$, called the *Chern–Simons invariant*, depends on the cohomology class $[c] \in H^3(G, \mathbb{R}/\mathbb{Z})$ and also on the conjugacy class of ρ under the conjugate action of G on $\mathcal{F}_M$.

Now the *Dijkgraaf–Witten invariant* $Z(M)$ of the closed 3-manifold M is then defined by

$$Z(M) := \frac{1}{\#G} \sum_{\rho \in \mathcal{F}_M} e^{2\pi\sqrt{-1}CS_M(\rho)}. \tag{16.2}$$

Remark 16.1.1

(1) A cochain $c \in C^n(G, A)$ for a G-module A is said to be *normalized* if $c(g_1, \ldots, g_n) = 0$ whenever one of g_i's is 1, and it is known [51, Lemma 6.1], [176, Chapter I, §2, Exercise 4] that any cocycle is cohomologous to a normalized one, namely, any cohomology class can be represented by a normalized cocycle. Therefore we may assume that our 3-cocycle c is normalized.

(2) The Chern–Simons functional is originally defined for a compact Lie gauge group and the associated Witten invariant is given by a path integral, which is not defined rigorously, over the infinite-dimensional space of gauge fields. Witten interpreted the *Jones polynomial* of a knot in this manner (cf. [244], [115, Chap. 2]). Although the above Dijkgraaf-Witten invariant for a finite gauge group is a kind of toy model, it is rigorously defined as a finite sum.

Example 16.1.2

(1) Suppose $[c]$ is the trivial class in $H^3(G, \mathbb{R}/\mathbb{Z})$. Then, by (16.1) and (16.2), we have $CS_M(\rho) = 0$ for any $\rho \in \mathcal{F}_M$ and hence the Dijkgraaf–Witten invariant is given by the (averaged) number of all homomorphisms $\pi_1(M) \to G$

$$Z(M) = \frac{1}{\#G} \#\mathrm{Hom}(\pi_1(M), G),$$

which is the classical topological invariant of a 3-manifold M.

(2) Let $G = \mathbb{Z}/N\mathbb{Z}$ with $N \geq 2$. Note that $\mathbb{Z}/N\mathbb{Z}$ is regarded as a subgroup of $\mathbb{R}/\mathbb{Z}$ by identifying $a \bmod N\mathbb{Z}$ with $\dfrac{a}{N} \bmod \mathbb{Z}$. For $g \in \mathbb{Z}/N\mathbb{Z}$, we denote by $\overline{g}$ the unique element $\overline{g} \in \{0, 1, \ldots, N-1\}$ such that $\overline{g} \bmod N = g$. Then the map $c : (\mathbb{Z}/N\mathbb{Z})^3 \to \mathbb{R}/\mathbb{Z}$ defined by

$$c(g_1, g_2, g_3) := \frac{1}{N^2} \overline{g_1}(\overline{g_2} + \overline{g_3} - \overline{g_2 + g_3}) \bmod \mathbb{Z}$$

gives a normalized 3-cocycle of G with coefficient $\mathbb{R}/\mathbb{Z}$. In fact, it can be shown that the class $[c]$ generates $H^3(\mathbb{Z}/N\mathbb{Z}, \mathbb{R}/\mathbb{Z}) \simeq \mathbb{Z}/N\mathbb{Z}$.

Let us consider the case that $G = \mathbb{Z}/2\mathbb{Z}$. Let $L = K_1 \cup \cdots \cup K_r$ be a link of r components in S^3 ($r \leq 2$) and let M be the double covering of S^3 ramified along L. In order to describe the Dijkgraaf-Witten invariant $Z(M)$, we introduce the mod 2 *linking diagram* D_L of L as follows. The diagram D_L consists of r vertices and edges. Each vertex represents each component knot K_i and two vertices K_i and K_j are adjacent by an edge if and only if the linking number $\mathrm{lk}(K_i, K_j) \equiv 1 \bmod 2$. The diagram is called a *circuit* (or *closed trail*) if it can be written in one stroke. A graph consisting of a single vertex is considered to be a circuit. The following formula can be proved by using genus theory for M, Corollary 6.2.2.

Theorem 16.1.2.1 ([39], [85]) *Notations being as above, we have*

$$Z(M) = \begin{cases} 2^{r-2} & \text{if any connected component of } D_L \text{ is a circuit,} \\ 0 & \text{otherwise.} \end{cases}$$

Example 16.1.2.2 Let $L = K_1 \cup K_2 \cup K_3 \cup K_4$ be the following link (left figure) in S^3 so that the mod 2 linking diagram D_L is given by the right figure (Fig. 16.1). Let M be the double covering of S^3 ramified along L.

By Theorem 16.1.2.1, we have $Z(M) = 2^2 = 4$.

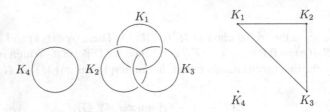

Fig. 16.1 The mod 2 linking diagram of a link

16.2   Dijkgraaf–Witten TQFT for 3-Manifolds with Boundaries

We start to recall the definition of topological quantum field theory, due to Atiyah, for the $(2 + 1)$-dimensional case.

Definition 16.2.1 ([11], [12]) A $(2 + 1)$-dimensional *topological quantum field theory*, called *TQFT* for short, consists of the following functor from the cobordism category of surfaces to the category of $\mathbb{C}$-vector spaces:

$$\text{oriented closed surface } \Sigma \qquad \rightsquigarrow \quad \mathbb{C}\text{-vector space } \mathcal{H}_\Sigma,$$
$$\text{oriented 3-manifold } M \text{ (with boundary)} \rightsquigarrow \text{ vector } Z_M \in \mathcal{H}_{\partial M},$$

where $\mathcal{H}_\Sigma$ is called the *state space* and Z_M is called the *partition function*. For some physical examples, $\mathcal{H}_\Sigma$ has the structure of a Hilbert space and is called the *quantum Hilbert space* [115]. The above correspondences must satisfy the following axioms.

(A1) *Functoriality.* An orientation-preserving homeomorphism $f : \Sigma \xrightarrow{\approx} \Sigma'$ induces an isomorphism $\mathcal{H}_\Sigma \xrightarrow{\sim} \mathcal{H}_{\Sigma'}$ of Hilbert quantum spaces. Moreover, if f extends to an orientation-preserving homeomorphism $M \xrightarrow{\approx} M'$, with $\partial M = \Sigma$, $\partial M' = \Sigma'$, then Z_M is sent to $Z_{M'}$ under the induced isomorphism $\mathcal{H}_{\partial M} \xrightarrow{\sim} \mathcal{H}_{\partial M'}$.

(A2) *Multiplicativity.* For disjoint surfaces Σ_1, Σ_2, we require

$$\mathcal{H}_{\Sigma_1 \sqcup \Sigma_2} = \mathcal{H}_{\Sigma_1} \otimes \mathcal{H}_{\Sigma_2}.$$

This multiplicative property is indicative of the quantum feature of the theory.

(A3) *Involutority.* For a surface Σ^*, which is Σ with opposite orientation, we require

$$\mathcal{H}_{\Sigma^*} = (\mathcal{H}_\Sigma)^*,$$

where $(\mathcal{H}_\Sigma)^*$ is the dual vector space of $\mathcal{H}_\Sigma$.

(A4) *Gluing formula.* If $\partial M_1 = \Sigma_1 \sqcup \Sigma_2$, $\partial M_2 = \Sigma_2^* \sqcup \Sigma_3$ and M is the 3-manifold obtained by gluing M_1 and M_2 along Σ_2, then we require the following gluing formula for the partition functions

$$\langle Z_{M_1}, Z_{M_2} \rangle = Z_M,$$

where $\langle \cdot, \cdot \rangle : \mathcal{H}_{\Sigma_1 \sqcup \Sigma_2} \times \mathcal{H}_{\Sigma_2^* \sqcup \Sigma_3} \to \mathcal{H}_{\Sigma_1 \sqcup \Sigma_3}$ is the natural gluing pairing of state spaces obtained by (A2) and (A3).

When $\Sigma = \emptyset$ (empty), we suppose $\mathcal{H}_\emptyset = \mathbb{C}$, and so we suppose $Z_M = Z(M) \in \mathbb{C}$ when M is closed. We also suppose $Z_{\Sigma \times [0,1]} = \mathrm{id}_{\mathcal{H}_\Sigma}$.

Now we construct the TQFT structure for Dijkgraaf–Witten theory, following [67]. We refer to [47], [55], [239], [247] for other constructions. We fix a finite group and a 3-cocycle $c \in Z^3(G, \mathbb{R}/\mathbb{Z})$ once and for all. Let X be an oriented compact manifold X with a fixed finite triangulation $\mathcal{T}$. Let $\mathcal{T}^{(n)}$ denote the set of n-simplices in $\mathcal{T}$. Each $\sigma_{i_0 \cdots i_n} \in \mathcal{T}^{(n)}$ for $i_0 < \cdots < i_n$ has the orientation determined by the numberings $i_0 \cdots i_n$ assigned to vertices and $\sigma_{\pi(i_0) \cdots \pi(i_n)}$ for a permutation π of $i_0 \cdots i_n$ is defined by $\mathrm{sgn}(\pi)\sigma_{i_0 \cdots i_n}$. We then define the gauge group $\mathcal{G}_X$ on X associated to G by

$$\mathcal{G}_X := \{\gamma : \mathcal{T}^{(0)} \longrightarrow G\}$$

and define the space $\mathcal{F}_X$ of gauge fields on X associated to G by

$$\mathcal{F}_X := \{\varrho : \mathcal{T}^{(1)} \longrightarrow G \mid \varrho(\sigma_{i_0 i_1})\varrho(\sigma_{i_1 i_2}) = \varrho(\sigma_{i_0 i_2}) \text{ for } i_0 < i_1 < i_2\},$$

on which $\mathcal{G}_X$ acts from the right by

$$(\varrho.\gamma)(\sigma_{ij}) := \gamma(\sigma_i)^{-1}\varrho(\sigma_{ij})\gamma(\sigma_j).$$

Note that $\mathcal{F}_X$ and $\mathcal{G}_X$ are finite sets. We remark that the quotient space $\mathcal{F}_X/\mathcal{G}_X$ is identified with the set $\mathcal{M}_X$ of isomorphism classes principal G-bundles (G-torsors) on X by the parallel transport along 1-simplices σ_{ij} and that the holonomy gives the bijection between $\mathcal{M}_X$ and $\mathrm{Hom}(\pi_1(X), G)/G$ if X is connected, where $\mathrm{Hom}(\pi_1(X), G)/G$ is the quotient of $\mathrm{Hom}(\pi_1(X), G)$ by the conjugate action of G from the right: $\mathrm{Hom}(\pi_1(X), G) \times G \to \mathrm{Hom}(\pi_1(X), G)$; $(\varrho, g) \mapsto g^{-1}\varrho g$.

First, we construct the classical theory in the sense of physics. The key ingredient is the *transgression homomorphism* for an oriented compact d-manifold X and $m \geq d$

$$\mathrm{trans}_X^m : C^m(G, \mathbb{R}/\mathbb{Z}) \longrightarrow C^{m-d}(\mathcal{G}_X, \mathrm{Map}(\mathcal{F}_X, \mathbb{R}/\mathbb{Z})),$$

where $\mathrm{Map}(\mathcal{F}_X, \mathbb{R}/\mathbb{Z})$ is the additive group of maps $\mathcal{F}_X \to \mathbb{R}/\mathbb{Z}$, on which $\mathcal{G}_X$ acts from the left by $(\gamma.\psi)(\varrho) := \psi(\varrho.\gamma)$ for $\gamma \in \mathcal{G}_X$, $\psi \in \mathrm{Map}(\mathcal{F}_X, \mathbb{R}/\mathbb{Z})$ and $\varrho \in \mathcal{F}_X$. The explicit expressions of trans_X^m for the cases that $d = m, m + 1$ are given as follows. For $\gamma \in \mathcal{G}_X, \varrho \in \mathcal{F}_X$ and $\sigma_i \in \mathcal{T}^{(0)}, \sigma_{ij} \in \mathcal{T}^{(1)}$, we set $\gamma_i :=$ $\gamma(\sigma_i), \varrho_{ij} := \varrho(\sigma_{ij})$.

Case that $d = m$: For $\alpha \in C^m(G, \mathbb{R}/\mathbb{Z}), \varrho \in \mathcal{F}_X$,

$$\mathrm{trans}_X^m(\alpha)(\varrho) = \sum_{\sigma_{01 \cdots m} \in \mathcal{T}^{(m)}} \varepsilon_{\sigma_{01 \cdots m}} \alpha(\varrho_{01}, \ldots, \varrho_{m-1, m}),$$

where $\varepsilon_{\sigma_{01 \cdots m}} = 1$ if the orientations of $\sigma_{01 \cdots m}$ and X coincides, and $\varepsilon_{\sigma_{01 \cdots m}} := -1$ otherwise.

Case that $d = m + 1$: For $\alpha \in C^m(G, \mathbb{R}/\mathbb{Z}), \gamma \in \mathcal{G}_X$ and $\varrho \in \mathcal{F}_X$,

$$(\mathrm{trans}_X^m(\alpha))(\gamma)(\varrho) = \sum_{\sigma_{01 \cdots m-1} \in \mathcal{T}^{(m-1)}} \varepsilon_{\sigma_{01 \cdots m-1}} \sum_{j=0}^{m-1} (-1)^j$$
$$\alpha(\gamma_0^{-1} \varrho_{01} \gamma_1, \ldots, \gamma_{j-1}^{-1} \varrho_{j-1\, j} \gamma_j, \gamma_j, \varrho_{j\, j+1}, \ldots, \varrho_{m-2\, m-1}),$$

where the signature $\varepsilon_{\sigma_{01 \cdots m-1}}$ is defined as above. It can be shown that the following Stokes-type formula holds:

$$\mathrm{trans}_X^{m+1} \circ \delta = (-1)^d \delta \circ \mathrm{trans}_X^m + \mathrm{res}^*(\mathrm{trans}_{\partial X}^m),$$

where δ denotes the coboundary map of group cochains and res denotes the map on the cochain induced by the restriction $\mathcal{F}_X \to \mathcal{F}_{\partial X}$ (resp. $\mathcal{G}_X \to \mathcal{G}_{\partial X}$).

Now, let $m = 3$ and consider the cases that X is an oriented closed surface Σ ($d = 2$) or X is an oriented compact 3-manifold M ($d = 3$), and we set

$$\begin{aligned} \lambda_\Sigma &:= \mathrm{trans}_\Sigma^3(c) \in C^1(\mathcal{G}_\Sigma, \mathrm{Map}(\mathcal{F}_\Sigma, \mathbb{R}/\mathbb{Z})), \\ CS_M &:= \mathrm{trans}_M^3(c) \in \mathrm{Map}(\mathcal{F}_M, \mathbb{R}/\mathbb{Z}), \end{aligned} \tag{16.3}$$

which are explicitly given as follows:

$$\begin{aligned} \lambda_\Sigma(\gamma)(\varrho) &= \sum_{\sigma_{012} \in \mathcal{T}^{(2)}} \varepsilon_{\sigma_{012}} \{c(\gamma_0, \varrho_{01}, \varrho_{12}) - c(\gamma_0 \varrho_{01} \gamma_1^{-1}, \gamma_1, \varrho_{12}) \\ &\quad + c(\gamma_0 \varrho_{01} \gamma_1^{-1}, \gamma_1 \varrho_{12} \gamma_2^{-1}, \gamma_2)\} \qquad (\gamma \in \mathcal{G}_\Sigma, \varrho \in \mathcal{F}_\Sigma), \\ CS_M(\varrho) &= \sum_{\sigma_{0123} \in \mathcal{T}^{(3)}} \varepsilon_{\sigma_{0123}} c(\varrho_{01}, \varrho_{12}, \varrho_{23}) \qquad (\varrho \in \mathcal{F}_M), \end{aligned} \tag{16.4}$$

where $\varepsilon_\sigma := 1$ if the orientations of σ and X coincides, and $\varepsilon_\sigma := -1$ otherwise. We suppose that the triangulation $\mathcal{T}_{\partial M}$ on ∂M is the restriction of the triangulation $\mathcal{T}_M$ of M. Since c is a 3-cocycle, by (16.3) and (16.4),

$$\delta CS_M = \mathrm{res}^* \lambda_{\partial M}, \quad \delta \lambda_\Sigma = 0, \tag{16.5}$$

where res : $\mathcal{F}_M \to \mathcal{F}_{\partial M}$ (resp. res : $\mathcal{G}_M \to \mathcal{G}_{\partial M}$) is the restriction map. We call λ_Σ the *Chern–Simons 1-cocycle* associated to c for an oriented closed surface Σ. The cohomology class of λ_Σ is independent of the choice of a finite triangulation $\mathcal{T}$. We call CS_M the *Chern–Simons functional* associated to c for an oriented compact 3-manifold M.

Using λ_Σ, we define a $\mathcal{G}_\Sigma$-equivariant principal $\mathbb{R}/\mathbb{Z}$-bundle $\mathcal{L}_\Sigma$ by the product bundle

$$\mathcal{L}_\Sigma := \mathcal{F}_\Sigma \times \mathbb{R}/\mathbb{Z}, \tag{16.6}$$

on which $\mathcal{G}_\Sigma$ acts by $(\varrho, m).\gamma = (\varrho.\gamma, m + \lambda_\Sigma(\gamma)(\varrho))$ for $\varrho \in \mathcal{F}_\Sigma, m \in \mathbb{R}/\mathbb{Z}$ and $\gamma \in \mathcal{G}_\Sigma$. It depends on the cohomology class of λ_Σ up to isomorphism of $\mathcal{G}_\Sigma$-equivariant principal $\mathbb{R}/\mathbb{Z}$-bundles. Taking the quotient by the action of $\mathcal{G}_\Sigma$, let $\overline{\mathcal{L}}_\Sigma$ denote the principal $\mathbb{R}/\mathbb{Z}$-bundle over $\mathcal{M}_\Sigma = \mathcal{F}_\Sigma/\mathcal{G}_\Sigma$, induced by $\mathcal{L}_\Sigma \to \mathcal{F}_\Sigma$. We call $\mathcal{L}_\Sigma \to \mathcal{F}_\Sigma$ or $\overline{\mathcal{L}}_\Sigma \to \mathcal{M}_\Sigma$ the *prequantization principal $\mathbb{R}/\mathbb{Z}$-bundle* for Σ. Let L_Σ be the complex line bundle associated to $\mathcal{L}_\Sigma$ via the homomorphism $\mathbb{R}/\mathbb{Z} \hookrightarrow \mathbb{C}^\times; m \mapsto e^{2\pi\sqrt{-1}m}$, and we have the complex line bundle $\overline{L}_\Sigma$ over $\mathcal{M}_\Sigma$. The line bundle $L_\Sigma \to \mathcal{F}_\Sigma$ or $\overline{L}_\Sigma \to \mathcal{M}_\Sigma$ is called the *prequantization complex line bundle* for a surface Σ. Let $\Gamma_{\mathcal{G}_M}(\mathcal{F}_M, \text{res}^*(\mathcal{L}_{\partial M}))$ (resp. $\Gamma_{\mathcal{G}_M}(\mathcal{F}_M, \text{res}^*(L_{\partial M}))$) denote the space of $\mathcal{G}_M$-equivariant sections of $\text{res}^*\mathcal{L}_{\partial M}$ (resp. $\text{res}^*L_{\partial M}$), which coincides with $\Gamma(\mathcal{M}_M, \text{res}^*(\overline{\mathcal{L}}_{\partial M}))$ (resp. $\Gamma(\mathcal{M}_M, \text{res}^*(\overline{L}_{\partial M}))$). By (16.5), we see that CS_M (resp. $e^{2\pi\sqrt{-1}CS_M}$) is in $\Gamma_{\mathcal{G}_M}(\mathcal{F}_M, \text{res}^*(\mathcal{L}_{\partial M}))$ (resp. $\Gamma_{\mathcal{G}_M}(\mathcal{F}_M, \text{res}^*(L_{\partial M}))$).

$$CS_M \in \Gamma_{\mathcal{G}_M}(\mathcal{F}_M, \text{res}^*(\mathcal{L}_{\partial M})) = \Gamma(\mathcal{M}_M, \text{res}^*(\overline{\mathcal{L}}_{\partial M})),$$
$$e^{2\pi\sqrt{-1}CS_M} \in \Gamma_{\mathcal{G}_M}(\mathcal{F}_M, \text{res}^*(L_{\partial M})) = \Gamma(\mathcal{M}_M, \text{res}^*(\overline{L}_{\partial M})). \tag{16.7}$$

Next, we construct the quantum theory in the sense of physics, namely, the correspondences in Definition 6.2.1 of (2+1)-dimensional TQFT. We define the state space $\mathcal{H}_\Sigma$ for an oriented closed surface Σ by the space of sections of the prequantization bundle $\overline{L}_\Sigma$ over $\mathcal{M}_\Sigma$, equivalently, the space of $\mathcal{G}_\Sigma$-equivariant sections of the prequantization line bundle L_Σ over $\mathcal{F}_\Sigma$:

$$\mathcal{H}_\Sigma = \{\vartheta : \mathcal{F}_\Sigma \to \mathbb{C} \,|\, \vartheta(\varrho.\gamma) = e^{2\pi\sqrt{-1}\lambda_\Sigma(\gamma)(\varrho)}\vartheta(\varrho)$$
$$\forall \gamma \in \mathcal{G}_\Sigma, \ \forall \varrho \in \mathcal{F}_\Sigma\} \tag{16.8}$$
$$= \Gamma(\mathcal{M}_\Sigma, \overline{L}_\Sigma).$$

We call $\mathcal{H}_\Sigma$ the *Dijkgraaf–Witten state space* and the above construction is along the line similar to the *geometric quantization*. We define the *Dijkgraaf–Witten partition function* Z_M by the following finite sum fixing the boundary condition

$$Z_M(\varrho_{\partial M}) = \frac{1}{\#'\mathcal{T}_M^{(0)}} \sum_{\substack{\varrho \in \mathcal{F}_M \\ \text{res}(\varrho) = \varrho_{\partial M}}} e^{2\pi\sqrt{-1}CS_M(\varrho)} \tag{16.9}$$

for $\varrho_{\partial M} \in \mathcal{F}_{\partial M}$, where $\#'\mathcal{T}_M^{(0)}$ is the number of 0-simplices in the interior of M. By (16.5), we see $Z_M \in \mathcal{H}_{\partial M}$. The value $Z_M(\varrho_{\partial M})$ is called the *Dijkgraaf–Witten invariant* of $\varrho_{\partial M} \in \mathcal{F}_{\partial M}$.

Theorem 16.2.2 *The above correspondences*

$$\text{oriented closed surface } \Sigma \rightsquigarrow \text{Dijkgraaf–Witten state space } \mathcal{H}_\Sigma,$$
$$\text{oriented 3-manifold } M \quad \rightsquigarrow \text{Dijkgraaf–Witten partition function}$$
$$Z_M \in \mathcal{H}_{\partial M},$$

satisfy the axioms (A1)–(A4) in the Definition 16.2.1 of the $(2 + 1)$*-dimensional TQFT.*

For the proof of Theorem 16.2.2, we consult [47], [55], [67], [239], [247].

Remark 16.2.3

(1) The above constructions depend only on the cohomology class of the fixed cocycle c. So we may take c to be normalized. Furthermore, the above constructions turn out to be independent of the choice of triangulations of Σ and M. Suppose that another choice of triangulation $\mathcal{T}'_\Sigma$ of Σ yields $\mathcal{H}_\Sigma^{\mathcal{T}'_\Sigma}$ as above. Then it can be shown that there is an isomorphism $\Theta_{\mathcal{T}_\Sigma, \mathcal{T}'_\Sigma} : \mathcal{H}_\Sigma^{\mathcal{T}_\Sigma} \xrightarrow{\sim} \mathcal{H}_\Sigma^{\mathcal{T}'_\Sigma}$. So taking the colimit of $\mathcal{H}_\Sigma^{\mathcal{T}_\Sigma}$'s with respect to triangulations $\mathcal{T}_\Sigma$ of Σ, we obtain the state space $\mathcal{H}_\Sigma$ which is independent of $\mathcal{T}_\Sigma$. Suppose that another choice of triangulation $\mathcal{T}'_M$ of M yields $Z_M^{\mathcal{T}'_M} \in \mathcal{H}_{\partial M}^{\mathcal{T}_{\partial M}}$. Then we can show $\Theta_{\mathcal{T}_{\partial M}, \mathcal{T}'_{\partial M}}(Z_M^{\mathcal{T}_M}) = Z_M^{\mathcal{T}'_M}$ and so we have a topological invariant $Z_M \in \mathcal{H}_{\partial M}$ [67].

(2) For Chern–Simons theory with a compact Lie gauge group, it is known that the state space $\mathcal{H}_\Sigma$ is isomorphic to the space of conformal blocks [115] and its element is called an non-Abelian theta function [14]. The dimension of $\mathcal{H}_\Sigma$ is given by Verlinde's formula [236]. Dijkgraaf–Witten theory is a finite analogue and an element of $\mathcal{H}_\Sigma$ may be regarded as a sort of non-Abelian finite theta function or non-Abelian Gaussian sum.

16.3   Arithmetic Chern–Simons Functionals and Arithmetic Dijkgraaf–Witten Invariants for Number Fields

Let k be a finite algebraic number field. Let $\mathcal{O}_k$ be the ring of integers of k and set $X := \mathrm{Spec}(\mathcal{O}_k)$. Let S_k^∞ be the set of infinite primes of k and set $\overline{X} := X \sqcup S_k^\infty$. Let N be a natural number > 1 and let $\mu_N (\subset \overline{\mathbb{Q}})$ denote the group of N-th roots of unity. We choose a fixed primitive N-th root of unity ζ_N which induces the isomorphism $\mu_N \simeq \mathbb{Z}/N\mathbb{Z}$. We assume that k contains μ_N. We fix a finite group G and a 3-

cocycle c of G with coefficients in $\mathbb{Z}/N\mathbb{Z}$, $c \in Z^3(G, \mathbb{Z}/N\mathbb{Z})$. These are given data.

Let $\pi_1(\overline{X})$ be the modified étale fundamental group of $\overline{X}$ as defined in Example 2.2.20. It is the Galois group of the maximal extension of k which is unramified over all finite and infinite primes. We consider a Grothendieck topology (site) on $\overline{X}$, called the *Artin–Verdier topology* (*site*), and the modified étale cohomology groups of $\overline{X}$ which takes the infinite primes S_k^∞ into account [3], [15], [85]. Let $\tilde{X} := \varprojlim \overline{Y}_i$ and $H^j(\tilde{X}, \mathbb{Z}/N\mathbb{Z}) := \varprojlim H^j(\overline{Y}_i, \mathbb{Z}/N\mathbb{Z})$, where $\overline{Y}_i \to \overline{X}$ runs over finite Galois coverings. Then we have the Hochschild–Serre spectral sequence $H^i(\pi_1(\overline{X}), H^j(\tilde{X}, \mathbb{Z}/N\mathbb{Z})) \Rightarrow H^{i+j}(\overline{X}, \mathbb{Z}/N\mathbb{Z})$ and the edge homomorphism

$$j^3 : H^3(\pi_1(\overline{X}), \mathbb{Z}/N\mathbb{Z}) \longrightarrow H^3(\overline{X}, \mathbb{Z}/N\mathbb{Z}).$$

Here we note that $H^3(\overline{X}, \mathbb{Z}/N\mathbb{Z})$ coincides with the modified étale cohomology group $\hat{H}^3(X, \mathbb{Z}/N\mathbb{Z})$ in Sect. 2.3.3 ([15]). By the canonical isomorphism (called the *invariant isomorphism* in class field theory) $H^3(\overline{X}, \mu_N) \simeq \mathbb{Z}/N\mathbb{Z}$ and the fixed isomorphism $\mu_N \simeq \mathbb{Z}/N\mathbb{Z}$ on $\overline{X}$, we have the isomorphism, denoted by inv_k,

$$\mathrm{inv}_k : H^3(\overline{X}, \mathbb{Z}/N\mathbb{Z}) \xrightarrow{\sim} \mathbb{Z}/N\mathbb{Z}.$$

Let $\mathcal{F}_{\overline{X}} := \mathrm{Hom}_{\mathrm{cont}}(\pi_1(\overline{X}), G)$ be the set of continuous homomorphisms $\pi_1(\overline{X}) \to G$, which may be regarded as an arithmetic analogue of gauge fields on $\overline{X}$ for G. It is a finite set, since $\pi_1(\overline{X})$ is a profinite group. For $\rho \in \mathcal{F}_{\overline{X}}$, we have the homomorphism $f_{\overline{X}}^\rho$ obtained as the composite of maps

$$f_{\overline{X}}^\rho : H^3(G, \mathbb{Z}/N\mathbb{Z}) \xrightarrow{\rho^*} H^3(\pi_1(\overline{X}), \mathbb{Z}/N\mathbb{Z}) \xrightarrow{j^3} H^3(\overline{X}/\mathbb{Z}/N\mathbb{Z}) \xrightarrow{\mathrm{inv}_k} \mathbb{Z}/N\mathbb{Z}.$$

We then define the mod N *arithmetic Chern–Simons functional* over $\overline{X}$ associated to c, $CS_{\overline{X}} : \mathcal{F}_{\overline{X}} \longrightarrow \mathbb{Z}/N\mathbb{Z}$, by

$$CS_{\overline{X}}(\rho) := f_{\overline{X}}^\rho([c]). \tag{16.10}$$

It is easy to see that the value $CS_{\overline{X}}(\rho)$, called the *arithmetic Chern–Simons invariant*, depends on the cohomology class $[c]$ of c and the conjugacy class of ρ under the conjugate action of G on $\mathcal{F}_{\overline{X}}$.

Now the *arithmetic Dijkgraaf–Witten invariant* $Z(\overline{X})$ of $\overline{X}$ is then defined by

$$Z(\overline{X}) := \frac{1}{\#G} \sum_{\rho \in \mathcal{F}_{\overline{X}}} \zeta_N^{CS_{\overline{X}}(\rho)}. \tag{16.11}$$

(16.10) and (16.11) may be regarded as arithmetic analogues of (16.1) and (16.2), respectively.

Example 16.3.1

(1) As in Example 16.1.2 (1), when $[c]$ is the trivial class in $H^3(G, \mathbb{Z}/N\mathbb{Z})$, by (16.10) and (16.11), we have $CS_{\overline{X}}(\rho) = 0$ for any $\rho \in \mathcal{F}_{\overline{X}}$ and

$$Z(\overline{X}) = \frac{1}{\#G} \#\mathrm{Hom}_{\mathrm{cont}}(\pi_1(\overline{X}), G),$$

which is the arithmetic invariant of a number field k.

(2) As in Example 16.1.2 (2), for $G = \mathbb{Z}/N\mathbb{Z}$ with $N \geq 2$, consider the normalized 3-cocycle $c \in Z^3(\mathbb{Z}/N\mathbb{Z}, \mathbb{Z}/N\mathbb{Z})$ defined by

$$c(g_1, g_2, g_3) := \overline{g_1}(\overline{g_2} + \overline{g_3} - \overline{g_2 + g_3}) \bmod N\mathbb{Z},$$

where $\overline{g} \in \{0, 1, \ldots, N-1\}$ and $\overline{g} \bmod N = g \in \mathbb{Z}/N\mathbb{Z}$. In fact, it can be shown that the class $[c]$ generates $H^3(\mathbb{Z}/n\mathbb{Z}, \mathbb{Z}/N\mathbb{Z}) \simeq \mathbb{Z}/N\mathbb{Z}$.

Let us consider the case $N = 2$, $G = \mathbb{Z}/2\mathbb{Z}$ and take $c \in Z^3(G, \mathbb{Z}/2\mathbb{Z})$ to be the above cocycle. Let $S := \{(p_1), \ldots, (p_r)\}$ be a finite set of primes of $\mathbb{Q}$ $(r \geq 2)$ such that $p_i \equiv 1 \bmod 4$ and let $k := \mathbb{Q}(\sqrt{p_1 \cdots p_r})$ be the quadratic extension of $\mathbb{Q}$ ramified over $(p_1), \ldots, (p_r)$. In order to describe the arithmetic Dijkgraaf–Witten invariant $Z^c(\overline{X})$, we introduce the mod 2 arithmetic linking diagram D_S of S, following the linking diagram in Example 16.1.2 (2) and the analogy between the linking number and the Legendre symbol in Chap. 4. The mod 2 *arithmetic linking diagram D_S* of S consists of r vertices and edges. Each vertex represents each prime (p_i) and two vertices (p_i) and (p_j) are adjacent by an edge if and only if $\left(\dfrac{p_i}{p_j}\right) = -1$. Since $p_i \equiv 1 \bmod 4$, D_S is well defined by the quadratic reciprocity law. The following formula is an arithmetic analogue of Theorem 16.1.2.1 and it can also be proved by using genus theory for k (cf. Corollary 6.3.2).

Theorem 16.3.1.1 ([39], [85]) *Notations being as above, we have*

$$Z(\overline{X}) = \begin{cases} 2^{r-2} & \text{if any connected component of } D_S \text{ is a circuit,} \\ 0 & \text{otherwise.} \end{cases}$$

Example 16.3.1.2 Let $S = \{5, 13, 37, 101\}$ so that $\left(\frac{5}{13}\right) = \left(\frac{5}{37}\right) = \left(\frac{13}{37}\right) = -1$, $\left(\frac{5}{101}\right) = \left(\frac{13}{101}\right) = \left(\frac{37}{101}\right) = 1$. Then the mod 2 arithmetic linking diagram D_S is given by the following figure (Fig. 16.2). Let $k := \mathbb{Q}(\sqrt{5 \cdot 13 \cdot 37 \cdot 101}) = \mathbb{Q}(\sqrt{242905})$ and $\overline{X} := \mathrm{Spec}(\mathbb{Z}[\frac{1+\sqrt{242905}}{2}])$.

By Theorem 16.3.1.1, we have $Z(\overline{X}) = 2^2 = 4$.

Fig. 16.2 The mod 2
arithmetic linking diagram of
a finite set of primes

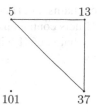

16.4 Arithmetic Dijkgraaf–Witten TQFT for S-Integer Number Rings

In this section, we shall construct an arithmetic analogue of Dijkgraaf-Witten TQFT in a special situation, which corresponds to the topological case that M is a link complement and Σ is the boundary tori of a tubular neighborhood of a link. For the details of the proofs, we refer to [88].

As in Sect. 16.3, let k a finite algebraic number field containing the group μ_N ($N > 1$). Let $\mathcal{O}_k$ be the ring of integers of k and set $X := \operatorname{Spec}(\mathcal{O}_k)$. Let S_k^∞ denote the set of infinite primes of k and set $\overline{X} := X \sqcup S_k^\infty$. We choose a fixed primitive N-th root of unity ζ_N which induces the isomorphism $\mu_N \simeq \mathbb{Z}/N\mathbb{Z}$. By the analogies presented in Chap. 3, we see X, S_k^∞ and $\overline{X}$ as analogues of a non-compact 3-manifold M, the set of ends and the end-compactification $\overline{M}$, respectively. A maximal ideal $\mathfrak{p}$ of $\mathcal{O}_k$ is identified with the residue field $\operatorname{Spec}(\mathcal{O}_k/\mathfrak{p}) = K(\widehat{\mathbb{Z}}, 1)$, which is seen as an analogue of the circle $S^1 = K(\mathbb{Z}, 1)$. We see the mod $\mathfrak{p}$ reduction map $\operatorname{Spec}(\mathcal{O}_k/\mathfrak{p}) \hookrightarrow X$ as an analogue of an embedding $S^1 \hookrightarrow M$, namely, a knot. Let $\mathcal{O}_\mathfrak{p}$ be the ring of $\mathfrak{p}$-adic integers and let $k_\mathfrak{p}$ be the $\mathfrak{p}$-adic field. We denote $\operatorname{Spec}(\mathcal{O}_\mathfrak{p})$ and $\operatorname{Spec}(k_\mathfrak{p})$ by $V_\mathfrak{p}$ and $\partial V_\mathfrak{p}$, respectively, which may be seen as analogues of a tubular neighborhood of a knot and its boundary torus, respectively. Let $\Pi_\mathfrak{p}$ denote the étale fundamental group of $\operatorname{Spec}(k_\mathfrak{p})$, $\Pi_\mathfrak{p} := \pi_1(\operatorname{Spec}(k_\mathfrak{p}))$, which is the absolute Galois group $\operatorname{Gal}(\overline{k}_\mathfrak{p}/k_\mathfrak{p})$ ($\overline{k}_\mathfrak{p}$ being an algebraic closure of $k_\mathfrak{p}$). We see $\Pi_\mathfrak{p}$ as an analogue of the peripheral group of a knot. To be precise, the tame quotient of $\Pi_\mathfrak{p}$ may be seen as a closer analogue of the peripheral group (cf. (3.3)). For a finite set $S = \{\mathfrak{p}_1, \ldots, \mathfrak{p}_r\}$ of maximal ideals of $\mathcal{O}_k$, let $\overline{X}_S := \overline{X} \setminus S$. We see S and $\overline{X}_S$ as an analogue of a link in a 3-manifold and the link complement, respectively. We may also see $\overline{X}_S$ as an analogue of a compact 3-manifold with boundary (union of tori), where $\partial V_S := \operatorname{Spec}(k_{\mathfrak{p}_1}) \sqcup \cdots \sqcup \operatorname{Spec}(k_{\mathfrak{p}_r})$ plays an analogous role of the boundary of $\overline{X}_S$, "$\partial \overline{X}_S = \partial V_S$". Let $\pi_1(\overline{X}_S)$ be the modified étale fundamental group of $\overline{X}_S$, which is the Galois group of the maximal extension k_S of k which is unramified at any (finite and infinite) prime outside S (cf. Example 2.2.20). We see $\pi_1(\overline{X}_S)$ as an analogue of the link group (cf. (3.11)).

Based on the analogies recalled above, we construct an arithmetic analogue of Dijkgraaf–Witten TQFT in the situation, which corresponds to the case that a 3-manifold M is a link complement and a surface Σ is the boundary tori of a tubular neighborhood of a link in Definition 16.2.1. As in the topological case in Sect. 16.2, one of the key ingredients is the transgression homomorphism for profinite group cochains. Let Π be a profinite group and let A be an additive discrete group on which Π acts continuously from the left. Let $C^n(\Pi, A)$ denote the group of continuous n-cochains of Π with coefficients in A. Note that Π acts on $C^n(\Pi, A)$ from the left by

$$(g.\alpha)(\gamma_1, \ldots, \gamma_n) := g\alpha(g^{-1}\gamma_1 g, \ldots, g^{-1}\gamma_n g)$$

for $\alpha \in C^n(\Pi, A)$, $g, \gamma_1, \ldots, \gamma_n \in \Pi$. For $g \in \Pi$, $0 \leq i \leq m - 1$, we define the map $s_g^i : \Pi^{m-1} \to \Pi^m$ by

$$s_g^i(\gamma_1, \ldots, \gamma_{m-1}) := (\gamma_1, \ldots, \gamma_i, g, g^{-1}\gamma_{i+1}g, \ldots, g^{-1}\gamma_{m-1}g).$$

and then define the *transgression homomorphism* $\mathrm{trans}_g^m : C^m(\Pi, A) \to C^{m-1}(\Pi, A)$ by

$$\mathrm{trans}_g^m(\alpha) := \sum_{0 \leq i \leq m-1} (-1)^i (\alpha \circ s_g^i)$$

for $\alpha \in C^m(\Pi, A)$. For example, explicit expressions of $\mathrm{trans}_g^m(\alpha)$ $(m = 2, 3)$ are given as follows:

$$\mathrm{trans}_g^2(\alpha)(\gamma) = \alpha(g, g^{-1}\gamma g) - \alpha(\gamma, g),$$
$$\mathrm{trans}_g^3(\alpha)(\gamma_1, \gamma_2) = \alpha(g, g^{-1}\gamma_1 g, g^{-1}\gamma_2 g) - \alpha(\gamma_1, g, g^{-1}\gamma_2 g) + \alpha(\gamma_1, \gamma_2, g)$$

$$(16.12)$$

for $\gamma, \gamma_1, \gamma_2 \in \Pi$. By straightforward computations, we can show the following formulas:

$$\sigma.\alpha - \alpha = \mathrm{trans}_g^{m+1} \circ \delta(\alpha) + \delta \circ \mathrm{trans}^m(\alpha),$$
$$\mathrm{trans}_{g_1 g_2}^m(\alpha) = g_1.\mathrm{trans}_{g_2}^m(\alpha) + \mathrm{trans}_{g_1}^m(\alpha) = 0 \bmod B^{m-1}(\Pi, A),$$

$$(16.13)$$

where $g, g_1, g_2 \in G$ and δ denotes the coboundary homomorphisms in group cochains.

Another key ingredient for our construction is the canonical isomorphism (called the *invariant isomorphism*) for $k_\mathfrak{p}$

$$\mathrm{inv}_\mathfrak{p} : H^2(\Pi_\mathfrak{p}, \mathbb{Z}/N\mathbb{Z}) \xrightarrow{\sim} \mathbb{Z}/N\mathbb{Z}$$

in local class field theory. (Cf. 2.3.2. Note that k is assumed to contain μ_N.) We note that the isomorphism $\mathrm{inv}_\mathfrak{p}$ tells us that $\partial V_\mathfrak{p}$ is "orientable" and we choose (implicitly) the "orientation" of $\partial V_\mathfrak{p}$ corresponding to $1 \in \mathbb{Z}/N\mathbb{Z}$.

Now, as in the topological case in Sect. 16.2, we firstly construct the classical theory in our arithmetic TQFT. For this, we start to develop the local theory at a finite prime $\mathfrak{p}$. We fix, once and for all, a finite group G and a 3-cocycle $c \in Z^3(G, \mathbb{Z}/N\mathbb{Z})$.

For an arithmetic analogue of the space of gauge fields, we consider the finite set

$$\mathcal{F}_\mathfrak{p} := \mathrm{Hom}_c(\Pi_\mathfrak{p}, G)$$

consisting of continuous homomorphisms of $\Pi_\mathfrak{p}$ to G. We take G simply as an arithmetic analogue of the gauge group, which acts on $\mathcal{F}_\mathfrak{p}$ from the right by the conjugation: $\rho_\mathfrak{p}.g := g^{-1}\rho_\mathfrak{p}g$ for $\rho_\mathfrak{p} \in \mathcal{F}_\mathfrak{p}, g \in G$, and let $\mathcal{M}_\mathfrak{p}$ denote the quotient space by this action: $\mathcal{M}_\mathfrak{p} := \mathcal{F}_\mathfrak{p}/G$. Let $\mathrm{Map}(\mathcal{F}_\mathfrak{p}, \mathbb{Z}/N\mathbb{Z})$ denote the additive group consisting of maps from $\mathcal{F}_\mathfrak{p}$ to $\mathbb{Z}/N\mathbb{Z}$, on which G acts from the left by $(g.\psi_\mathfrak{p})(\rho_\mathfrak{p}) := \psi_\mathfrak{p}(\rho_\mathfrak{p}.g)$ for $g \in G$, $\psi_\mathfrak{p} \in \mathrm{Map}(\mathcal{F}_\mathfrak{p}, \mathbb{Z}/N\mathbb{Z})$ and $\rho_\mathfrak{p} \in \mathcal{F}_\mathfrak{p}$. For $\rho_\mathfrak{p} \in \mathcal{F}_\mathfrak{p}$ and $\alpha \in C^n(G, \mathbb{Z}/N\mathbb{Z})$, $\alpha \circ \rho_\mathfrak{p} \in C^n(\Pi, \mathbb{Z}/N\mathbb{Z})$ is defined by $(\alpha \circ \rho_\mathfrak{p})(\gamma_1, \dots, \gamma_n) := \alpha(\rho_\mathfrak{p}(\gamma_1), \dots, \rho_\mathfrak{p}(\gamma_n))$.

For $\rho_\mathfrak{p} \in \mathcal{F}_\mathfrak{p}$, we define $\mathcal{L}_\mathfrak{p}(\rho_\mathfrak{p})$ by the quotient set

$$\mathcal{L}_\mathfrak{p}(\rho_\mathfrak{p}) := \delta^{-1}(c \circ \rho_\mathfrak{p})/B^2(\Pi_\mathfrak{p}, \mathbb{Z}/N\mathbb{Z}), \qquad (16.14)$$

where δ denotes the coboundary map $C^2(\Pi_\mathfrak{p}, \mathbb{Z}/N\mathbb{Z}) \to C^3(\Pi_\mathfrak{p}, \mathbb{Z}/N\mathbb{Z})$. Note that $\delta^{-1}(c \circ \rho_\mathfrak{p})$ is non-empty, because the cohomological dimension of $\Pi_\mathfrak{p}$ is 2 [176, Theorem 7.1.8], [206, Chapitre II, 5.3, Proposition 15] and so $H^3(\Pi_\mathfrak{p}, \mathbb{Z}/N\mathbb{Z}) = 0$. Thus $\mathcal{L}_\mathfrak{p}(\rho_\mathfrak{p})$ is an $H^2(\Pi_\mathfrak{p}, \mathbb{Z}/N\mathbb{Z})$-torsor in the obvious manner and hence it is a $\mathbb{Z}/N\mathbb{Z}$-torsor via the invariant isomorphism $\mathrm{inv}_\mathfrak{p}$. In the following, we write $n = \alpha_\mathfrak{p} - \beta_\mathfrak{p}$ if $\alpha_\mathfrak{p} = \beta_\mathfrak{p}.n$ for $\alpha_\mathfrak{p}, \beta_\mathfrak{p} \in \mathcal{L}(\rho_\mathfrak{p}), n \in \mathbb{Z}/N\mathbb{Z}$. Choosing $\beta_\mathfrak{p} \in \mathcal{L}(\rho_\mathfrak{p})$ gives the bijection, called the *trivialization* at $\beta_\mathfrak{p}$,

$$\varphi_{\beta_\mathfrak{p}} : \mathcal{L}(\rho_\mathfrak{p}) \overset{\sim}{\longrightarrow} \mathbb{Z}/N\mathbb{Z}; \ \alpha_\mathfrak{p} \longmapsto \alpha_\mathfrak{p} - \beta_\mathfrak{p}.$$

Let $\mathcal{L}_\mathfrak{p}$ be the disjoint union of $\mathcal{L}_\mathfrak{p}(\rho_\mathfrak{p})$ over all $\rho_\mathfrak{p} \in \mathcal{F}_\mathfrak{p}$:

$$\mathcal{L}_\mathfrak{p} := \bigsqcup_{\rho_\mathfrak{p} \in \mathcal{F}_\mathfrak{p}} \mathcal{L}_\mathfrak{p}(\rho_\mathfrak{p})$$

and consider the projection $\pi_\mathfrak{p} : \mathcal{L}_\mathfrak{p} \to \mathcal{F}_\mathfrak{p}; \alpha_\mathfrak{p} \mapsto \rho_\mathfrak{p}$ if $\alpha_\mathfrak{p} \in \mathcal{L}_\mathfrak{p}(\rho_\mathfrak{p})$. Since each fiber $\pi_\mathfrak{p}^{-1}(\rho_\mathfrak{p}) = \mathcal{L}_\mathfrak{p}(\rho_\mathfrak{p})$ is a $\mathbb{Z}/N\mathbb{Z}$-torsor, we may regard $\mathcal{L}_\mathfrak{p}$ as a principal $\mathbb{Z}/N\mathbb{Z}$-bundle over $\mathcal{F}_\mathfrak{p}$.

Let $g \in G$. Using the transgression homomorphism, we define $h_g \in C^2(G, \mathbb{Z}/N\mathbb{Z})/B^2(G, \mathbb{Z}/N\mathbb{Z})$ by

$$h_g := \mathrm{trans}_g^3(c) \bmod B^2(G, \mathbb{Z}/N\mathbb{Z}),$$

which is given explicitly by (16.12) as

$$h_g(\gamma_1, \gamma_2) = c(g, g^{-1}\gamma_1 g, g^{-1}\gamma_2 g) - c(\gamma_1, g, g^{-1}\gamma_2 g) + c(\gamma_1, \gamma_2, g)$$
$$\bmod B^2(G, \mathbb{Z}/N\mathbb{Z}),$$

where $\gamma_1, \gamma_2 \in G$. By (16.13),

$$\delta h_g = g.c - c \mod B^2(G, \mathbb{Z}/N\mathbb{Z}) \tag{16.15}$$

and the 1-cocycle relation

$$h_{gg'} = h_g + g.h_{g'} \tag{16.16}$$

for $g, g' \in G$. By (16.15), we have the isomorphism of $\mathbb{Z}/N\mathbb{Z}$-torsors

$$h_{\mathrm{p}}(g, \rho_{\mathrm{p}}) : \mathcal{L}_{\mathrm{p}}(\rho_{\mathrm{p}}) \xrightarrow{\sim} \mathcal{L}_{\mathrm{p}}(\rho_{\mathrm{p}}.g); \quad \alpha_{\mathrm{p}} \mapsto \alpha_{\mathrm{p}} + h_g \circ \rho_{\mathrm{p}},$$

and, by (16.16), G acts on $\mathcal{L}_{\mathrm{p}}$ from the right by $\alpha_{\mathrm{p}}.g := h(g, \rho_{\mathrm{p}})(\alpha_{\mathrm{p}})$ for $\alpha_{\mathrm{p}} \in \mathcal{L}_{\mathrm{p}}, g \in G$, which satisfies $(\alpha_{\mathrm{p}}.m).g = (\alpha_{\mathrm{p}}.g).m$, $\quad \pi_{\mathrm{p}}(\alpha_{\mathrm{p}}.g) = \pi_{\mathrm{p}}(\alpha_{\mathrm{p}}).g$ for $\alpha_{\mathrm{p}} \in \mathcal{F}_{\mathrm{p}}, m \in \mathbb{Z}/N\mathbb{Z}$ and $g \in G$. Thus $\mathcal{L}_{\mathrm{p}}$ is a G-equivariant principal $\mathbb{Z}/N\mathbb{Z}$-bundle over $\mathcal{F}_{\mathrm{p}}$. Taking the quotient by the action of G, we obtain the principal $\mathbb{Z}/N\mathbb{Z}$-bundle $\overline{\pi}_{\mathrm{p}} : \overline{\mathcal{L}}_{\mathrm{p}} \to \mathcal{M}_{\mathrm{p}}$. We call $\pi_{\mathrm{p}} : \mathcal{L}_{\mathrm{p}} \to \mathcal{F}_{\mathrm{p}}$ or $\overline{\pi}_{\mathrm{p}} : \overline{\mathcal{L}}_{\mathrm{p}} \to \mathcal{M}_{\mathrm{p}}$ the *arithmetic prequantization* $\mathbb{Z}/N\mathbb{Z}$-*bundle* for $\partial V_{\mathrm{p}} = \mathrm{Spec}(k_{\mathrm{p}})$.

Since $\mathcal{F}_{\mathrm{p}}$ and $\mathcal{L}_{\mathrm{p}}$ are finite sets, we may choose a section $x_{\mathrm{p}} \in \Gamma(\mathcal{F}_{\mathrm{p}}, \mathcal{L}_{\mathrm{p}})$, namely, the map $x_{\mathrm{p}} : \mathcal{F}_{\mathrm{p}} \to \mathcal{L}_{\mathrm{p}}$ such that $\pi_{\mathrm{p}} \circ x_{\mathrm{p}} = \mathrm{id}_{\mathcal{F}_{\mathrm{p}}}$. Then we have the trivialization at $x_{\mathrm{p}}(\rho_{\mathrm{p}})$:

$$\varphi_{x_{\mathrm{p}}(\rho_{\mathrm{p}})} : \mathcal{L}_{\mathrm{p}}(\rho_{\mathrm{p}}) \xrightarrow{\sim} \mathbb{Z}/N\mathbb{Z}; \quad \alpha_{\mathrm{p}} \mapsto \alpha_{\mathrm{p}} - x_{\mathrm{p}}(\alpha_{\mathrm{p}}).$$

For $g \in G$ and $\rho_{\mathrm{p}} \in \mathcal{F}_{\mathrm{p}}$, we let

$$\lambda_{\mathrm{p}}^{x_{\mathrm{p}}}(g, \rho_{\mathrm{p}}) := h_{\mathrm{p}}(g, \rho_{\mathrm{p}})(x_{\mathrm{p}}(\rho_{\mathrm{p}})) - x_{\mathrm{p}}(\rho_{\mathrm{p}}.g) = x_{\mathrm{p}}(\rho_{\mathrm{p}}).g - x_{\mathrm{p}}(\rho_{\mathrm{p}}.g)$$

so that we get the following commutative diagram:

$$
\begin{array}{ccc}
\mathcal{L}_{\mathrm{p}}(\rho_{\mathrm{p}}) & \xrightarrow{h_{\mathrm{p}}(g,\rho_{\mathrm{p}})} & \mathcal{L}_{\mathrm{p}}(\rho_{\mathrm{p}}.g) \\
\varphi_{x_{\mathrm{p}}(\rho_{\mathrm{p}})} \downarrow & & \downarrow \varphi_{x_{\mathrm{p}}(\rho_{\mathrm{p}}.g)} \\
\mathbb{Z}/N\mathbb{Z} & \xrightarrow[+\lambda_{\mathrm{p}}^{x_{\mathrm{p}}}(g,\rho_{\mathrm{p}})]{} & \mathbb{Z}/N\mathbb{Z}.
\end{array}
$$

We define the map $\lambda_{\mathrm{p}}^{x_{\mathrm{p}}} : G \to \mathrm{Map}(\mathcal{F}_{\mathrm{p}}, \mathbb{Z}/N\mathbb{Z})$ by $\lambda_{\mathrm{p}}^{x_{\mathrm{p}}}(g)(\rho_{\mathrm{p}}) := \lambda_{\mathrm{p}}^{x_{\mathrm{p}}}(g, \rho_{\mathrm{p}})$ for $g \in G$ and $\rho_{\mathrm{p}} \in \mathcal{F}_{\mathrm{p}}$. By (16.16), the map $\lambda_{\mathrm{p}}^{x_{\mathrm{p}}}$ satisfies the 1-cocycle relation

$$\lambda_{\mathrm{p}}^{x_{\mathrm{p}}}(gg') = \lambda_{\mathrm{p}}^{x_{\mathrm{p}}}(g) + (g.\lambda_{\mathrm{p}}^{x_{\mathrm{p}}})(g')$$

for $g, g' \in G$. We call $\lambda_{\mathrm{p}}^{x_{\mathrm{p}}}$ the *arithmetic Chern–Simons 1-cocycle* for ∂V_{p} with respect to the section x_{p}. We see easily that the cohomology class of $\lambda_{\mathrm{p}}^{x_{\mathrm{p}}}$ is independent of the choice of x_{p}. As a local arithmetic analogue for ∂V_{p} of the topological prequantization bundle $\mathcal{L}_\Sigma$ in (16.6), we define $\mathcal{L}_p^{x_p}$ by the product (trivial) principal $\mathbb{Z}/N\mathbb{Z}$-bundle over $\mathcal{F}_p$:

$$\mathcal{L}_p^{x_p} := \mathcal{F}_p \times \mathbb{Z}/N\mathbb{Z},$$

on which G acts from the right by $(\rho_p, m).g := (\rho_p.g, m + \lambda_p^{x_p}(g, \rho_p))$ for $\rho_p \in \mathcal{F}_p$, $m \in \mathbb{Z}/N\mathbb{Z}$ and $g \in G$. Then the projection $\pi_p^{x_p} : \mathcal{L}_p^{x_p} \to \mathcal{F}_p$ is a G-equivariant principal $\mathbb{Z}/N\mathbb{Z}$-bundle. Taking the quotient by G, we have the principal $\mathbb{Z}/N\mathbb{Z}$-bundle $\overline{\pi}_p^{x_p} : \overline{\mathcal{L}}_p^{x_p} \to \mathcal{M}_p$. We call $\pi_p^{x_p} : \mathcal{L}_p \to \mathcal{F}_p$ or $\overline{\pi}_p^{x_p} : \overline{\mathcal{L}}_p^{x_p} \to \mathcal{M}_p$ the *arithmetic prequantization $\mathbb{Z}/N\mathbb{Z}$-bundle for ∂V_p with respect to x_p*. We then have the following isomorphism of G-equivariant principal $\mathbb{Z}/N\mathbb{Z}$-bundles over $\mathcal{F}_p$:

$$\Phi_p^{x_p} : \mathcal{L}_p \xrightarrow{\sim} \mathcal{L}_p^{x_p}; \quad \alpha_p \longmapsto (\pi_p(\alpha_p), \alpha_p - x_p(\pi_p(\alpha_p))).$$

For another section $x_p' \in \Gamma(\mathcal{F}_p, \mathcal{L}_p)$, we have the isomorphism

$$\Phi_p^{x_p, x_p'} : \mathcal{L}_p^{x_p} \xrightarrow{\sim} \mathcal{L}_p^{x_p'}; \quad (\rho_p, m) \longmapsto (\rho_p, m + (x_p(\rho_p) - x_p'(\rho_p))),$$

where $\Phi_p^{x_p'} = \Phi_p^{x_p, x_p'} \circ \Phi_p^{x_p}$.

Let F be a field containing μ_N. Let L_p be the F-line bundle over $\mathcal{F}_p$ associated to the principal $\mathbb{Z}/N\mathbb{Z}$-bundle $\mathcal{L}_p$ and the homomorphism $\mathbb{Z}/N\mathbb{Z} \hookrightarrow F^\times; m \mapsto \zeta_N^m$, namely,

$$\begin{aligned}
L_p &:= \mathcal{L}_p \times_{\mathbb{Z}/N\mathbb{Z}} F \\
&:= (\mathcal{L}_p \times F)/(\alpha_p, z) \sim (\alpha_p.m, \zeta_N^{-m} z) \quad (\alpha_p \in \mathcal{L}_p, m \in \mathbb{Z}/N\mathbb{Z}, z \in F),
\end{aligned}$$

on which G acts from the right by $[(\alpha_p, z)].g := [(\alpha_p.g, z)]$ for $[(\alpha_p, z)] \in \mathcal{L}_p, g \in G$. The projection $\pi_{p,F} : L_p \longrightarrow \mathcal{F}_p; [(\alpha_p, z)] \mapsto \pi_p(\alpha_p)$ is a G-equivariant F-line bundle. We denote the fiber $\pi_{p,F}^{-1}(\rho_p)$ over ρ_p by $L_p(\rho_p)$:

$$\begin{aligned}
L_p(\rho_p) &:= \{[(\alpha_p, z)] \in L_p \mid \pi_p(\alpha_p) = \rho_p, z \in F\} \\
&\simeq F \quad ([(\alpha_p, z)] \longmapsto z).
\end{aligned}$$

Taking the quotient by the action of G, we obtain the F-line bundles $\overline{\pi}_{p,F} : \overline{L}_p \to \mathcal{M}_p$. We call $\pi_{p,F} : L_p \to \mathcal{F}_p$ or $\overline{\varpi}_{p,F} : \overline{L}_p \to \mathcal{M}_p$ the *arithmetic prequantization F-line bundle* for ∂V_p.

For $x_p \in \Gamma(\mathcal{F}_p, \mathcal{L}_p)$, let $L_p^{x_p}$ be the product F-line bundle over $\mathcal{F}_p$:

$$L_p^{x_p} := \mathcal{F}_p \times F,$$

on which G acts from the right by $(\rho_{\mathfrak{p}}, z).g := (\rho_{\mathfrak{p}}.g, z\zeta_N^{\lambda_{\mathfrak{p}}^{x_{\mathfrak{p}}}(g,\rho_{\mathfrak{p}})})$ for $(\rho_{\mathfrak{p}}, z) \in$ $L_{\mathfrak{p}}^{x_{\mathfrak{p}}}, g \in G$, and the projection $\pi_{\mathfrak{p},F}^{x_{\mathfrak{p}}} : L_{\mathfrak{p}}^{x_{\mathfrak{p}}} \longrightarrow \mathcal{F}_{\mathfrak{p}}$ is a G-equivariant F-line bundle. Taking the quotient of $\varpi_{\mathfrak{p},F}^{x_{\mathfrak{p}}} : L_{\mathfrak{p}}^{x_{\mathfrak{p}}} \to \mathcal{F}_{\mathfrak{p}}$ by the action of G, we have the F-line bundle $\overline{\varpi}_{\mathfrak{p},F}^{x_{\mathfrak{p}}} : \overline{L}_{\mathfrak{p}}^{x_{\mathfrak{p}}} \to \mathcal{M}_{\mathfrak{p}}$. We call $\varpi_{\mathfrak{p},F}^{x_{\mathfrak{p}}} : L_{\mathfrak{p}}^{x_{\mathfrak{p}}} \to \mathcal{F}_{\mathfrak{p}}$ or $\overline{\varpi}_{\mathfrak{p},F}^{x_{\mathfrak{p}}} : \overline{L}_{\mathfrak{p}}^{x_{\mathfrak{p}}} \to \mathcal{M}_{\mathfrak{p}}$ the *arithmetic prequantization F-line bundle* for $\partial V_{\mathfrak{p}}$ with respect to the section $x_{\mathfrak{p}}$. We then have the following isomorphism of G-equivariant F-line bundles over $\mathcal{F}_{\mathfrak{p}}$

$$\Phi_{\mathfrak{p},F}^{x_{\mathfrak{p}}} : L_{\mathfrak{p}} \xrightarrow{\sim} L_{\mathfrak{p}}^{x_{\mathfrak{p}}}; [(\alpha_{\mathfrak{p}}, z)] \longmapsto (\pi_{\mathfrak{p}}(\alpha_{\mathfrak{p}}), z\zeta_N^{\alpha_{\mathfrak{p}} - x_{\mathfrak{p}}(\pi_{\mathfrak{p}}(\alpha_{\mathfrak{p}}))}).$$

For another section $x_{\mathfrak{p}}'$, we have the following isomorphism of G-equivariant F-line bundles over $\mathcal{F}_{\mathfrak{p}}$

$$\Phi_{\mathfrak{p},F}^{x_{\mathfrak{p}},x_{\mathfrak{p}}'} : L_{\mathfrak{p}}^{x_{\mathfrak{p}}} \xrightarrow{\sim} L_{\mathfrak{p}}^{x_{\mathfrak{p}}'} : (\rho_{\mathfrak{p}}, z) \longmapsto (\rho_{\mathfrak{p}}, z\zeta_N^{x_{\mathfrak{p}}(\rho_{\mathfrak{p}}) - x_{\mathfrak{p}}'(\rho_{\mathfrak{p}})}),$$

where $\Phi_{\mathfrak{p},F}^{x_{\mathfrak{p}}'} = \Phi_{\mathfrak{p},F}^{x_{\mathfrak{p}},x_{\mathfrak{p}}'} \circ \Phi_{\mathfrak{p},F}^{x_{\mathfrak{p}}}$. *for* $x_{\mathfrak{p}}, x_{\mathfrak{p}}', x_{\mathfrak{p}}'' \in \Gamma(\mathcal{F}_{\mathfrak{p}}, \mathcal{L}_{\mathfrak{p}})$.

Let $S = \{\mathfrak{p}_1, \ldots, \mathfrak{p}_r\}$ be a finite set of finite primes of k and let $\partial V_S := \partial V_{\mathfrak{p}_1} \sqcup \cdots \sqcup \partial V_{\mathfrak{p}_r}$. Let $\mathcal{F}_S$ be the direct product of $\mathcal{F}_{\mathfrak{p}_i}$'s:

$$\mathcal{F}_S := \mathcal{F}_{\mathfrak{p}_1} \times \cdots \times \mathcal{F}_{\mathfrak{p}_r}.$$

It is a finite set on which G acts diagonally from the right: $\mathcal{F}_S \times G \to \mathcal{F}_S$; $(\rho_S, g) \mapsto \rho_S.g := (\rho_{\mathfrak{p}_1}.g, \ldots, \rho_{\mathfrak{p}_r}.g)$ for $\rho_S = (\rho_{\mathfrak{p}_1}, \ldots, \rho_{\mathfrak{p}_r}) \in \mathcal{F}_S$ and let $\mathcal{M}_S$ denote the quotient space by this action: $\mathcal{M}_S := \mathcal{F}_S/G$. Let $\mathrm{Map}(\mathcal{F}_S, \mathbb{Z}/N\mathbb{Z})$ be the additive group of maps from $\mathcal{F}_S$ to $\mathbb{Z}/N\mathbb{Z}$, on which G acts from the left by $(g.\psi_S)(\rho_S) := \psi_S(\rho_S.g)$ for $\psi_S \in \mathrm{Map}(\mathcal{F}_S, \mathbb{Z}/N\mathbb{Z})$, $g \in G$ and $\rho_S \in \mathcal{F}_S$.

For $\rho_S = (\rho_{\mathfrak{p}_1}, \ldots, \rho_{\mathfrak{p}_r}) \in \mathcal{F}_S$, let $\mathcal{L}_S(\rho_S)$ be the quotient space of the product $\mathcal{L}_{\mathfrak{p}_1}(\rho_{\mathfrak{p}_1}) \times \cdots \times \mathcal{L}_{\mathfrak{p}_r}(\rho_{\mathfrak{p}_r})$:

$$\mathcal{L}_S(\rho_S) := (\mathcal{L}_{\mathfrak{p}_1}(\rho_{\mathfrak{p}_1}) \times \cdots \times \mathcal{L}_{\mathfrak{p}_r}(\rho_{\mathfrak{p}_r}))/\sim, \tag{16.17}$$

where $(\alpha_{\mathfrak{p}_1}, \ldots, \alpha_{\mathfrak{p}_r}) \sim (\alpha_{\mathfrak{p}_1}', \ldots, \alpha_{\mathfrak{p}_r}') \Leftrightarrow \sum_{i=1}^r (\alpha_{\mathfrak{p}_i} - \alpha_{\mathfrak{p}_i}') = 0$. We see easily that $\mathcal{L}_S(\rho_S)$ is equipped with the simply transitive action of $\mathbb{Z}/N\mathbb{Z}$ defined by $[\alpha_S].m := [(\alpha_{\mathfrak{p}_1}.m, \ldots, \alpha_{\mathfrak{p}_r})] = \cdots = [(\alpha_{\mathfrak{p}_1}, \ldots, \alpha_{\mathfrak{p}_r}.m)]$ for $\alpha_S = (\alpha_{\mathfrak{p}_1}, \ldots, \alpha_{\mathfrak{p}_r})$ and $m \in \mathbb{Z}/N\mathbb{Z}$, and hence $\mathcal{L}_S(\rho_S)$ is a $\mathbb{Z}/N\mathbb{Z}$-torsor. We write $n = [\alpha_S] - [\beta_S]$ $(n \in \mathbb{Z}/N\mathbb{Z}, [\alpha_S], [\beta_S] \in \mathcal{L}_S(\rho_S))$ if $[\alpha_S] = [\beta_S].n$.

Let $\mathcal{L}_S$ the disjoint union of $\mathcal{L}_{\mathfrak{p}}(\rho_S)$ for $\rho_S \in \mathcal{F}_S$:

$$\mathcal{L}_S := \bigsqcup_{\rho_S \in \mathcal{F}_S} \mathcal{L}_S(\rho_S), \tag{16.18}$$

on which G acts from the right by $[(\alpha_{\mathfrak{p}_1}, \ldots, \alpha_{\mathfrak{p}_r})].g = [(\alpha_{\mathfrak{p}_1}.g, \ldots, \alpha_{\mathfrak{p}_r}.g)]$ for $\alpha_{\mathfrak{p}_i} \in \mathcal{L}_{\mathfrak{p}_i}, g \in G$. Consider the projection $\pi_S : \mathcal{L}_S \to \mathcal{F}_S$; $\alpha_S = (\alpha_{\mathfrak{p}_i}) \mapsto$

$(\pi_{\mathfrak{p}_i}(\alpha_{\mathfrak{p}_i}))$, which is G-equivariant. Since each fiber $\pi_S^{-1}(\rho_S) = \mathcal{L}_S(\rho_S)$ is a $\mathbb{Z}/N\mathbb{Z}$-torsor, we may regard $\pi_S : \mathcal{L}_S \to \mathcal{F}_S$ as a G-equivariant principal $\mathbb{Z}/N\mathbb{Z}$-bundle. Taking the quotient by the action of G, we have the principal $\mathbb{Z}/N\mathbb{Z}$-bundle $\overline{\pi}_S : \overline{\mathcal{L}}_S \to \mathcal{M}_S$. We call $\pi_S : \mathcal{L}_S \to \mathcal{F}_S$ or $\overline{\pi}_S : \overline{\mathcal{L}}_S \to \mathcal{M}_S$ the *arithmetic prequantization $\mathbb{Z}/N\mathbb{Z}$-bundle* for $\partial V_S := \mathrm{Spec}(k_{\mathfrak{p}_1}) \sqcup \cdots \sqcup \mathrm{Spec}(k_{\mathfrak{p}_r})$.

Let x_S be a section of π_S, $x_S \in \Gamma(\mathcal{F}_S, \mathcal{L}_S)$. By (16.17) and (16.18), x_S is written as $x_S = [(x_{\mathfrak{p}_1}, \ldots, x_{\mathfrak{p}_r})]$, where $x_{\mathfrak{p}_i} \in \Gamma(\mathcal{F}_{\mathfrak{p}_i}, \mathcal{L}_{\mathfrak{p}_i})$ for $1 \le i \le r$. For $g \in G$ and $\rho_S = (\rho_{\mathfrak{p}_i}) \in \mathcal{F}_S$, we set

$$\lambda_S^{x_S}(g, \rho_S) := \lambda_{\mathfrak{p}_1}^{x_{\mathfrak{p}_1}}(g, \rho_{\mathfrak{p}_1}) + \cdots + \lambda_{\mathfrak{p}_r}^{x_{\mathfrak{p}_r}}(g, \rho_{\mathfrak{p}_r})$$

and define the map $\lambda_S^{x_S} : G \to \mathrm{Map}(\mathcal{F}_S, \mathbb{Z}/N\mathbb{Z})$ by $\lambda_S^{x_S}(g)(\rho_S) := \lambda_S^{x_S}(g, \rho_S)$ for $g \in G$ and $\rho_S \in \mathcal{F}_S$. Since each $\lambda_{\mathfrak{p}_i}^{x_{\mathfrak{p}_i}}$ is a 1-cocycle, $\lambda_S^{x_S}$ satisfies the 1-cocycle relation:

$$\lambda_S^{x_S} \in Z^1(G, \mathrm{Map}(\mathcal{F}_S, \mathbb{Z}/N\mathbb{Z})).$$

We see easily that the cohomology class of $\lambda_S^{x_S}$ is independent of the choice of a section x_S. We call λ_S the *arithmetic Chern–Simons 1-cocycle* for ∂V_S with respect to x_S. As an arithmetic analogue for ∂V_S of the topological prequantization bundle $\mathcal{L}_\Sigma$ in (16.6), we define $\mathcal{L}_S^{x_S}$ as the product principal $\mathbb{Z}/N\mathbb{Z}$-bundle over $\mathcal{F}_S$:

$$\mathcal{L}_S^{x_S} := \mathcal{F}_S \times \mathbb{Z}/N\mathbb{Z},$$

on which G acts from the right by $(\rho_S, m).g := (\rho_S.g, m + \lambda_S^{x_S}(g, \rho_S))$ for $\rho_S \in \mathcal{F}_S, m \in \mathbb{Z}/N\mathbb{Z}$ and $g \in G$. The projection $\pi_S^{x_S} : \mathcal{L}_S^{x_S} \to \mathcal{F}_S$ is a G-equivariant principal $\mathbb{Z}/N\mathbb{Z}$-bundle. Taking the quotient by the action of G, we have the principal $\mathbb{Z}/N\mathbb{Z}$-bundle $\overline{\pi}_S^{x_S} : \overline{\mathcal{L}}_S \to \mathcal{M}_S$. We call $\pi_S^{x_S} : \mathcal{L}_S^{x_S} \to \mathcal{F}_S$ or $\overline{\pi}_S^{x_S} : \overline{\mathcal{L}}_S^{x_S} \to \mathcal{M}_S$ the *arithmetic prequantization $\mathbb{Z}/N\mathbb{Z}$-bundle* for ∂V_S with respect to x_S. We then have the following isomorphism of G-equivariant principal $\mathbb{Z}/N\mathbb{Z}$-bundles over $\mathcal{F}_S$:

$$\Phi_S^{x_S} : \mathcal{L}_S \overset{\sim}{\longrightarrow} \mathcal{L}_S^{x_S}; \quad [\alpha_S] \longmapsto \left(\pi_S([\alpha_S]), \sum_{i=1}^{r}(\alpha_{\mathfrak{p}_i} - x_{\mathfrak{p}_i}(\pi_{\mathfrak{p}_i}(\alpha_{\mathfrak{p}_i})))\right),$$

where $\alpha_S = (\alpha_{\mathfrak{p}_1}, \ldots, \alpha_{\mathfrak{p}_r})$. For another section $x_S' \in \Gamma(\mathcal{F}_S, \mathcal{L}_S)$, we have the isomorphism

$$\Phi_S^{x_S, x_S'} : \mathcal{L}_S^{x_S} \overset{\sim}{\longrightarrow} \mathcal{L}_S^{x_S'}; \quad (\rho_S, m) \longmapsto \left(\rho_S, m + \sum_{i=1}^{r}(x_{\mathfrak{p}_i}(\rho_{\mathfrak{p}_i}) - x_{\mathfrak{p}_i}'(\rho_{\mathfrak{p}_i}))\right),$$

where $\Phi_S^{x_S'} = \Phi_S^{x_S, x_S'} \circ \Phi_S^{x_S}$.

Let F be a field containing μ_N. Let L_S be the F-line bundle associated to the principal $\mathbb{Z}/N\mathbb{Z}$-bundle $\mathcal{L}_S$ over $\mathcal{F}_S$ via the homomorphism $\mathbb{Z}/N\mathbb{Z} \to F^\times; m \mapsto \zeta_N^m$. Namely,

$$
\begin{aligned}
L_S &:= \mathcal{L}_S \times_{\mathbb{Z}/N\mathbb{Z}} F \\
&:= (\mathcal{L}_S \times F)/([\alpha_S], z) \sim ([\alpha_S].m, \zeta_N^{-m} z) \quad ([\alpha_S] \in \mathcal{L}_S, m \in \mathbb{Z}/N\mathbb{Z}, z \in F),
\end{aligned}
$$

on which G acts from the right by $([([\alpha_S], z)], g) \mapsto [([\alpha_S].g, z)]$ for $g \in G$. The projection $\pi_{S,F} : L_S \longrightarrow \mathcal{F}_S; [([\alpha_S], z)] \mapsto \pi_S([\alpha_S])$ is a G-equivariant F-line bundle. Taking the quotient by the action of G, we obtain the F-line bundle $\overline{\pi}_{S,F} : \overline{L}_S \to \mathcal{M}_S$. We call $\pi_{S,F} : L_S \to \mathcal{F}_S$ or $\overline{\pi}_{S,F} : \overline{L}_S \to \mathcal{M}_S$ the *arithmetic prequantization F-line bundle* for ∂V_S.

For a section $x_S \in \Gamma(\mathcal{F}_S, \mathcal{L}_S)$, $L_S^{x_S}$ be the trivial F-line bundle over $\mathcal{F}_S$:

$$
L_S^{x_S} := \mathcal{F}_S \times F,
$$

on which G acts from the right by $(\rho_S, z).g := (\rho_S.g, z\zeta_N^{\lambda_S^{x_S}(g, \rho_S)})$ for $\rho_S \in \mathcal{F}_S, z \in F, g \in G$, and the projection $L_S^{x_S} \to \mathcal{F}_S$ is a G-equivariant F-line bundle. Taking the quotient by the action of G, we obtain the principal $\mathbb{Z}/N\mathbb{Z}$-bundle $\overline{\pi}_{S,F} : L_S^{x_S} \to \mathcal{M}_S$. We call $\pi_{S,F}^{x_S} : L_S^{x_S} \to \mathcal{F}_S$ or $\overline{\pi}_{S,F} : \overline{L}_S^{x_S} \to \mathcal{M}_S$ the *arithmetic prequantization F-line bundle* for ∂V_S. We then have the following isomorphism of G-equivariant F-line bundles over $\mathcal{F}_S$:

$$
\Phi_{S,F}^{x_S} : L_S \xrightarrow{\sim} L_S^{x_S}; \quad [(\alpha_S, z)] \longmapsto (\pi_S(\alpha_S), z\zeta_N^{\sum_{i=1}^{r}(\alpha_{\mathfrak{p}_i} - x_{\mathfrak{p}_i}(\pi_{\mathfrak{p}_i}(\alpha_{\mathfrak{p}_i})))}).
$$

For another section $x_S' \in \Gamma(\mathcal{F}_S, \mathcal{L}_S)$, we have the isomorphism

$$
\Phi_{S,F}^{x_S, x_S'} : L_S^{x_S} \xrightarrow{\sim} L_S^{x_S'}; \quad [(\rho_S, z)] \longmapsto [(\rho_S, z\zeta_N^{d_S^{x_S, x_S'}(\rho_S)})],
$$

where $\Phi_{S,F}^{x_S'} = \Phi_{S,F}^{x_S, x_S'} \circ \Phi_{S,F}^{x_S}$.

We may also give the description of L_S in terms of the tensor product of F-line bundles. Let $p_i : \mathcal{F}_S \to \mathcal{F}_{\mathfrak{p}_i}$ be the i-th projection. Let $p_i^*(L_{\mathfrak{p}_i})$ be the F-line bundle over $\mathcal{F}_S$ induced from $L_{\mathfrak{p}_i}$ by p_i:

$$
p_i^*(L_{\mathfrak{p}_i}) := \{(\rho_S, [(\alpha_{\mathfrak{p}_i}, z_i)]) \in \mathcal{F}_S \times L_{\mathfrak{p}_i} \mid p_i(\rho_S) = \varpi_{\mathfrak{p}_i}(\alpha_{\mathfrak{p}_i})\},
$$

and let

$$
p_i^*(\pi_{\mathfrak{p}_i}) : p_i^*(L_{\mathfrak{p}_i}) \longrightarrow \mathcal{F}_S; \quad (\rho_S, [(\alpha_{\mathfrak{p}_i}, z_i)]) \mapsto \rho_S
$$

be the induced projection. The fiber over $\rho_S = (\rho_{\mathfrak{p}_i})$ is given by

$$p_i^*(\pi_{\mathfrak{p}_i})^{-1}(\rho_S) = \{\rho_S\} \times \{[(\alpha_{\mathfrak{p}_i}, z_i)] \in L_{\mathfrak{p}_i} \mid \rho_{\mathfrak{p}_i} = \pi_{\mathfrak{p}_i}(\alpha_{\mathfrak{p}_i}), z_i \in F\}$$
$$\simeq L_{\mathfrak{p}_i}(\rho_{\mathfrak{p}_i}) \simeq F.$$

Let $L_{\mathfrak{p}_1} \boxtimes \cdots \boxtimes L_{\mathfrak{p}_r}$ be the tensor product of $p_i^*(L_{\mathfrak{p}_i})$'s:

$$L_{\mathfrak{p}_1} \boxtimes \cdots \boxtimes L_{\mathfrak{p}_r} := p_1^*(L_{\mathfrak{p}_1}) \otimes \cdots \otimes p_r^*(L_{\mathfrak{p}_r}),$$

which is an F-line bundle over $\mathcal{F}_S$. An element of $L_{\mathfrak{p}_1} \boxtimes \cdots \boxtimes L_{\mathfrak{p}_r}$ is written as $(\rho_S, [(\alpha_{\mathfrak{p}_1}, z_1)] \otimes \cdots \otimes [(\alpha_{\mathfrak{p}_r}, z_r)])$, where $\rho_S = (\rho_{\mathfrak{p}_i}) \in \mathcal{F}_S$, $[(\alpha_{\mathfrak{p}_i}, z_i)] \in L_{\mathfrak{p}_i}(\rho_{\mathfrak{p}_i})$. Let $\varpi_S^{\boxtimes} : L_{\mathfrak{p}_1} \boxtimes \cdots \boxtimes L_{\mathfrak{p}_r} \to \mathcal{F}_S$ be the projection. For fibers over ρ_S,

$$(\varpi_S^{\boxtimes})^{-1}(\rho_S) \xrightarrow{\sim} F; (\rho_S, [(\alpha_{\mathfrak{p}_1}, z_1)] \otimes \cdots \otimes [(\alpha_{\mathfrak{p}_r}, z_r)]) \longmapsto \prod_{i=1}^{r} z_i.$$

The right action of G on $L_{\mathfrak{p}_1} \boxtimes \cdots \boxtimes L_{\mathfrak{p}_r}$ is given by

$$L_{\mathfrak{p}_1} \boxtimes \cdots \boxtimes L_{\mathfrak{p}_r} \times G \to L_{\mathfrak{p}_1} \boxtimes \cdots \boxtimes L_{\mathfrak{p}_r};$$

$$((\rho_S, [(\alpha_{\mathfrak{p}_1}, z_1)] \otimes \cdots \otimes [(\alpha_{\mathfrak{p}_r}, z_r)]), g)$$

$$\mapsto (\rho_S.g, [(\alpha_{\mathfrak{p}_1}.g, z_1)] \otimes \cdots \otimes [(\alpha_{\mathfrak{p}_r}.g, z_r)]).$$

The projection $\pi_S^{\boxtimes}$ is G-equivariant. We then have the following isomorphism of G-equivariant F-line bundles over $\mathcal{F}_S$

$$\Phi_{S,F}^{\boxtimes} : L_{\mathfrak{p}_1} \boxtimes \cdots \boxtimes L_{\mathfrak{p}_r} \xrightarrow{\sim} L_S;$$

$$(\rho_S, [(\alpha_{\mathfrak{p}_1}, z_1)] \otimes \cdots \otimes [(\alpha_{\mathfrak{p}_r}, z_r)]) \longmapsto [([\alpha_S], \textstyle\prod_{i=1}^{r} z_i)],$$

where $\rho_S = (\rho_{\mathfrak{p}_i}) \in \mathcal{F}_S$, $[(\alpha_{\mathfrak{p}_i}, z_i)] \in L_{\mathfrak{p}_i}(\rho_{\mathfrak{p}_i})$, and $\alpha_S = (\alpha_{\mathfrak{p}_1}, \ldots, \alpha_{\mathfrak{p}_r})$.

The arithmetic prequantization $\mathbb{Z}/N\mathbb{Z}$-bundles $\mathcal{L}_S$, $\mathcal{L}_S^{\chi_S}$ and arithmetic prequantization bundles L_S, $L_S^{\chi_S}$ defined above may be regarded as arithmetic analogues for ∂V_S of the prequantization $\mathbb{R}/\mathbb{Z}$-bundle $\mathcal{L}_\Sigma$ and prequantization complex line bundle L_Σ, respectively, in the topological case in Sect. 16.2.

Now we shall introduce an arithmetic analogue of the Chern–Simons functional for $\overline{X}_S := \overline{X} \setminus S$. Let $\pi_1(\overline{X}_S)$ be the modified étale fundamental group of $\overline{X}_S$ (Example 2.2.20), namely, the Galois group of the maximal extension of k such that any prime outside S is unramified. We view $\overline{X}_S$ as an arithmetic analogue of the link exterior so that the boundary of $\overline{X}_S$ is $\partial V_S := \text{Spec}(k_{\mathfrak{p}_1}) \sqcup \cdots \sqcup \text{Spec}(k_{\mathfrak{p}_r})$. In the following, we assume that all maximal ideals of $\mathcal{O}_k$ dividing N are contained in S (in particular, S is non-empty). As an arithmetic analogue of the gauge fields over $\mathcal{F}_{\overline{X}_S}$, we consider the set $\mathcal{F}_{\overline{X}_S}$ of continuous representations of $\pi_1(\overline{X}_S)$ to G:

$$\mathcal{F}_{\overline{X}_S} := \mathrm{Hom}_{\mathrm{cont}}(\pi_1(\overline{X}_S), G),$$

on which G acts from the right by $\mathcal{F}_{\overline{X}_S} \times G \to \mathcal{F}_{\overline{X}_S}$; $(\rho, g) \mapsto \rho.g := g^{-1}\rho g$, and let $\mathcal{M}_{\overline{X}_S}$ denote the quotient set by this action: $\mathcal{M}_{\overline{X}_S} := \mathcal{F}_{\overline{X}_S}/G$. Let $\mathrm{Map}(\mathcal{F}_{\overline{X}_S}, \mathbb{Z}/N\mathbb{Z})$ be the additive group of maps from $\mathcal{F}_{\overline{X}_S}$ to $\mathbb{Z}/N\mathbb{Z}$, on which G acts from the left by $(g.\psi)(\rho) := \psi(\rho.g)$ for $g \in G$, $\psi \in \mathrm{Map}(\mathcal{F}_{\overline{X}_S}, \mathbb{Z}/N\mathbb{Z})$ and $\rho \in \mathcal{F}_{\overline{X}_S}$.

We fix an embedding $\overline{k} \hookrightarrow \overline{k}_{\mathfrak{p}_i}$, which induces the continuous homomorphism $\iota_{\mathfrak{p}_i} : \Pi_{\mathfrak{p}_i} \longrightarrow \pi_1(\overline{X}_S)$ for each $1 \le i \le r$. Let $\mathrm{res}_{\mathfrak{p}_i}$ and res_S denote the restriction maps (the pull-backs by $\iota_{\mathfrak{p}_i}$) defined by

$$\mathrm{res}_{\mathfrak{p}_i} : \mathcal{F}_{\overline{X}_S} \longrightarrow \mathcal{F}_{\mathfrak{p}_i}; \quad \rho \longmapsto \rho \circ \iota_{\mathfrak{p}_i},$$
$$\mathrm{res}_S := (\mathrm{res}_{\mathfrak{p}_i}) : \mathcal{F}_{\overline{X}_S} \longrightarrow \mathcal{F}_S,$$

which are G-equivariant. We denote by $\mathrm{Res}_{\mathfrak{p}_i}$ and Res_S the homomorphisms on cochains defined by

$$\mathrm{Res}_{\mathfrak{p}_i} : C^n(\pi_1(\overline{X}_S), \mathbb{Z}/N\mathbb{Z}) \longrightarrow C^n(\Pi_{\mathfrak{p}_i}, \mathbb{Z}/N\mathbb{Z}); \quad \alpha \mapsto \alpha \circ \iota_{\mathfrak{p}_i},$$
$$\mathrm{Res}_S := (\mathrm{Res}_{\mathfrak{p}_i}) : C^n(\Pi_S, \mathbb{Z}/N\mathbb{Z}) \longrightarrow \prod_{i=1}^r C^n(\Pi_{\mathfrak{p}_i}, \mathbb{Z}/N\mathbb{Z}).$$

The key fact to define the Chern–Simons functional over $\mathcal{F}_{\overline{X}_S}$ is

$$H^3(\pi_1(\overline{X}_S), \mathbb{Z}/N\mathbb{Z}) = 0, \tag{16.19}$$

which follows from the fact that the cohomological p-dimension of $\Pi_S \le 2$ for any prime number p dividing N (cf. [176, Proposition 8.3.18] for $p > 2$ and [176, Theorem 10.6.7] for $N = p = 2$). For $\rho \in \mathcal{F}_{\overline{X}_S}$, we have $c \circ \rho \in Z^3(\pi_1(\overline{X}_S), \mathbb{Z}/N\mathbb{Z})$. By (16.19), there is $\beta_\rho \in C^2(\pi_1(\overline{X}_S), \mathbb{Z}/N\mathbb{Z})/B^2(\pi_1(\overline{X}_S), \mathbb{Z}/N\mathbb{Z})$ such that

$$c \circ \rho = \delta\beta_\rho, \tag{16.20}$$

where $\delta : C^2(\pi_1(\overline{X}_S), \mathbb{Z}/N\mathbb{Z}) \to C^3(\pi_1(\overline{X}_S), \mathbb{Z}/N\mathbb{Z})$ is the coboundary homomorphism, and so

$$c \circ \mathrm{res}_{\mathfrak{p}_i}(\rho) = \delta \, \mathrm{Res}_{\mathfrak{p}_i}(\beta_\rho)$$

for $1 \le i \le r$. Therefore, by (16.14) and (16.17),

$$[\mathrm{Res}_S(\beta_\rho)] \in \mathcal{L}_S(\mathrm{res}_S(\rho)).$$

Let $\mathrm{res}_S^*(\mathcal{L}_S)$ be the G-equivariant principal $\mathbb{Z}/N\mathbb{Z}$-bundle over $\mathcal{F}_{\overline{X}_S}$ induced from $\mathcal{L}_S$ by res_S:

$$\mathrm{res}_S^*(\mathcal{L}_S) := \{(\rho, \alpha_S) \in \mathcal{F}_{\overline{X}_S} \times \mathcal{L}_S \mid \mathrm{res}_S(\rho) = \pi_S(\alpha_S)\}.$$

and let $\mathrm{res}_S^*(\pi_S)$ be the projection $\mathrm{res}_S^*(\mathcal{L}_S) \to \mathcal{F}_{\overline{X}_S}$. The quotient by the action of G is the principal $\mathbb{Z}/N\mathbb{Z}$-bundle $\mathrm{res}^*(\overline{\mathcal{L}}_S)$ over $\mathcal{M}_{\overline{X}_S}$ induced from $\overline{\mathcal{L}}_S$ by res_S. A section of $\mathrm{res}_S^*(\pi_S)$ is naturally identified with a map $y_S : \mathcal{F}_{\overline{X}_S} \to \mathcal{L}_S$ satisfying $\pi_S \circ y_S = \mathrm{res}_S$:

$$\Gamma(\mathcal{F}_{\overline{X}_S}, \mathrm{res}_S^*(\mathcal{L}_S)) = \{y_S : \mathcal{F}_{\overline{X}_S} \longrightarrow \mathcal{L}_S \mid \pi_S \circ y_S = \mathrm{res}_S\},$$

on which G acts by $(y_S.g)(\rho) := y_S(\rho.g)$ for $\rho \in \mathcal{F}_{\overline{X}_S}, g \in G$. Let $\Gamma_G(\mathcal{F}_{\overline{X}_S}, \mathrm{res}_S^*(\mathcal{L}_S))$ be the set of G-equivariant sections of $\mathrm{res}_S^*(\mathcal{L}_S)$, which is identified with $\Gamma(\mathcal{M}_{\overline{X}_S}, \mathrm{res}_S^*(\overline{\mathcal{L}}_S))$. We define the *arithmetic Chern–Simons functional* $CS_{\overline{X}_S} : \mathcal{F}_{\overline{X}_S} \to \mathcal{L}_S$ by

$$CS_{\overline{X}_S}(\rho) := [\mathrm{Res}_S(\beta_\rho)]$$

for $\rho \in \mathcal{F}_{\overline{X}_S}$. The value $CS_{\overline{X}_S}(\rho) \in \mathcal{L}_S$ is called the *arithmetic Chern–Simons invariant* of ρ. The following may be regarded as an arithmetic analogue of (16.7).

Proposition 16.4.1

(1) $CS_{\overline{X}_S}(\rho)$ is independent of the choice of β_ρ.

(2) $CS_{\overline{X}_S}$ (resp. $\zeta_N^{CS_{\overline{X}_S}}$) is a G-equivariant section of $\mathrm{res}_S^*(\mathcal{L}_S)$ (resp. $\mathrm{res}_S^*(L_S)$):

$$CS_{\overline{X}_S} \in \Gamma_G(\mathcal{F}_{\overline{X}_S}, \mathrm{res}^*(\mathcal{L}_S)) = \Gamma(\mathcal{M}_{\overline{X}_S}, \mathrm{res}^*(\overline{\mathcal{L}}_S)),$$
$$\zeta_N^{CS_{\overline{X}_S}} \in \Gamma_G(\mathcal{F}_{\overline{X}_S}, \mathrm{res}^*(L_S)) = \Gamma(\mathcal{M}_{\overline{X}_S}, \mathrm{res}^*(\overline{L}_S)).$$

Proof

(1) Let $\beta_\rho' \in C^2(\pi_1(\overline{X}_S), \mathbb{Z}/N\mathbb{Z})/B^2(\pi_1(\overline{X}_S), \mathbb{Z}/N\mathbb{Z})$ be another choice satisfying $c \circ \rho = d\beta_\rho'$. Then $\beta_\rho' = \beta_\rho + z$ for some $z \in H^2(\pi_1(\overline{X}_S), \mathbb{Z}/N\mathbb{Z})$ and so

$$\mathrm{Res}_{\mathfrak{p}_i}(\beta_\rho') - \mathrm{Res}_{\mathfrak{p}_i}(\beta_\rho) = \mathrm{inv}_{\mathfrak{p}_i}(\mathrm{Res}_{\mathfrak{p}_i}(z)) \quad (1 \le i \le r).$$

Noting that any primes dividing N is contained in S, the Tate–Poitou exact sequence [176, 8.6.10] implies that the composite of the following maps:

$$H^2(\Pi_S, \mathbb{Z}/N\mathbb{Z}) \xrightarrow{\prod_{\mathfrak{p} \in \overline{S}} \mathrm{Res}_{\mathfrak{p}}} \prod_{\mathfrak{p} \in S \cup S_k^\infty} H^2(\Pi_{\mathfrak{p}}, \mathbb{Z}/N\mathbb{Z}) \xrightarrow{\sum_{\mathfrak{p} \in \overline{S}} \mathrm{inv}_{\mathfrak{p}}} \mathbb{Z}/N\mathbb{Z}$$

is the zero map, where $\overline{S} = S \cup S_k^\infty$. Here we note that for any infinite prime $v \in S_k^\infty$, the restriction map $\Pi_v := \mathrm{Gal}(\overline{k}_v/k_v) \to \Pi_S$ is the trivial homomorphism and so $H^2(\Pi_S, \mathbb{Z}/2\mathbb{Z}) \to H^2(\Pi_v, \mathbb{Z}/2\mathbb{Z})$ is the zero map. Hence $\sum_{i=1}^r \mathrm{inv}_{\mathfrak{p}_i}(\mathrm{Res}_{\mathfrak{p}_i}(z)) = 0$. By (16.18), we obtain

$$[\mathrm{Res}_S(\beta'_\rho)] = [\mathrm{Res}_S(\beta_\rho)].$$

(2) By the definition of $CS_{\overline{X}_S}$, we see $CS_{\overline{X}_S} \in \Gamma(\mathcal{F}_{\overline{X}_S}, \mathrm{res}_S^*(\mathcal{L}_S))$. So it suffices to show that $CS_{\overline{X}_S}$ is G-equivariant. By (16.15) and (16.4.1),

$$\delta\beta_{\rho.g} = c \circ (\rho.g) = (g.c) \circ \rho = (c + \delta h_g) \circ \rho = \delta(\beta_\rho + h_g \circ \rho).$$

for $g \in G$ and $\rho \in \mathcal{F}_{\overline{X}_S}$. Therefore there is $z \in H^2(\pi_1(\overline{X}_S), \mathbb{Z}/\mathbb{Z})$ such that $\beta_{\rho.g} = \beta_\rho + h_g \circ \rho + z$ and so

$$\begin{aligned}
\mathrm{Res}_S(\beta_{\rho.g}) &= \mathrm{Res}_S(\beta_\rho) + h_g \circ \mathrm{res}_S(\rho) + \mathrm{Res}_S(z) \\
&= \mathrm{Res}_S(\beta_\rho).g + \mathrm{Res}_S(z).
\end{aligned}$$

By the same argument as in (1), we obtain

$$CS_{\overline{X}_S}(\rho.g) = [\mathrm{Res}_S(\beta_{\rho.g})] = [\mathrm{Res}_S(\beta_\rho)].g = CS_{\overline{X}_S}(\rho).g.$$

The assertion for $\zeta_N^{CS_{\overline{X}_S}}$ follows from the definition of L_S.

Let $x_S = [(x_{\mathfrak{p}_1}, \ldots, x_{\mathfrak{p}_r})] \in \Gamma(\mathcal{F}_S, \mathcal{L}_S)$ be a section and let $\mathcal{L}_S^{x_S}$ be the arithmetic prequantization principal $\mathbb{Z}/N\mathbb{Z}$-bundle over $\mathcal{F}_S$ with respect to x_S. Let $\mathrm{res}_S^*(\mathcal{L}_S^{x_S})$ be the G-equivariant principal $\mathbb{Z}/N\mathbb{Z}$-bundle over $\mathcal{F}_{\overline{X}_S}$ induced from $\mathcal{L}_S^{x_S}$ by res_S:

$$\begin{aligned}
\mathrm{res}_S^*(\mathcal{L}_S^{x_S}) &= \{(\rho, (\rho_S, m)) \in \mathcal{F}_{\overline{X}_S} \times \mathcal{L}_S^{x_S} \mid \mathrm{res}_S(\rho) = \rho_S\} \\
&= \mathcal{F}_{\overline{X}_S} \times \mathbb{Z}/N\mathbb{Z}
\end{aligned}$$

by identifying $(\rho, (\rho_S, m))$ with (ρ, m). So a section of $\mathrm{res}_S^*(\mathcal{L}_S^{x_S})$ over $\mathcal{F}_{\overline{X}_S}$ is identified with a map $\mathcal{F}_{\overline{X}_S} \to \mathbb{Z}/N\mathbb{Z}$:

$$\Gamma(\mathcal{F}_{\overline{X}_S}, \mathrm{res}_S^*(\mathcal{L}_S^{x_S})) = \mathrm{Map}(\mathcal{F}_{\overline{X}_S}, \mathbb{Z}/N\mathbb{Z}).$$

Let $\Gamma_G(\mathcal{F}_{\overline{X}_S}, \mathrm{res}_S^*(\mathcal{L}_S^{x_S}))$ be the set of G-equivariant section of $\mathrm{res}_S^*(\mathcal{L}_S^{x_S})$ over $\mathcal{F}_{\overline{X}_S}$. It coincides with the set of G-equivariant maps $\mathcal{F}_{\overline{X}_S} \to \mathbb{Z}/N\mathbb{Z}$, denoted by $\mathrm{Map}_G(\mathcal{F}_{\overline{X}_S}, \mathbb{Z}/N\mathbb{Z})$:

$$\begin{aligned}
\Gamma_G(\mathcal{F}_{\overline{X}_S}, \mathrm{res}_S^*(\mathcal{L}_S^{x_S})) &= \mathrm{Map}_G(\mathcal{F}_{\overline{X}_S}, \mathbb{Z}/N\mathbb{Z}) \\
&= \{\psi : \mathcal{F}_{\overline{X}_S} \longrightarrow \mathbb{Z}/N\mathbb{Z} \mid \psi(\rho.g) = \psi(\rho) + \lambda_S^{x_S}(g, \mathrm{res}_S(\rho)) \\
&\qquad\qquad\qquad\qquad\qquad \text{for } \rho \in \mathcal{F}_{\overline{X}_S}, g \in G\}.
\end{aligned}$$

The isomorphism $\Phi_S^{x_S} : \mathcal{L}_S \xrightarrow{\sim} \mathcal{L}_S^{x_S}$ induces the isomorphism

$$\Psi^{x_S} : \Gamma_G(\mathcal{F}_{\overline{X}_S}, \mathrm{res}_S^*(\mathcal{L}_S)) \xrightarrow{\sim} \Gamma_G(\mathcal{F}_{\overline{X}_S}, \mathrm{res}_S^*(\mathcal{L}_S^{x_S})) = \mathrm{Map}_G(\mathcal{F}_{\overline{X}_S}, \mathbb{Z}/N\mathbb{Z})$$
$$y_S \longmapsto \Phi_S^{x_S} \circ y_S.$$

We then define the *arithmetic Chern–Simons functional* $CS_{\overline{X}_S}^{x_S} : \mathcal{F}_{\overline{X}_S} \to \mathbb{Z}/N\mathbb{Z}$ with respect to x_S as the image of $CS_{\overline{X}_S}$ under Ψ^{x_S}:

$$CS_{\overline{X}_S}^{x_S} := \Psi^{x_S}(CS_{\overline{X}_S}).$$

Since $CS_{\overline{X}_S}^{x_S} \in \mathrm{Map}_G(\mathcal{F}_{\overline{X}_S}, \mathbb{Z}/N\mathbb{Z})$,

$$CS_{\overline{X}_S}^{x_S}(\rho.g) = CS_{\overline{X}_S}^{x_S}(\rho) + \lambda_S^{x_S}(g, \mathrm{res}_S(\rho))$$

for $g \in G$ and $\rho \in \mathcal{F}_{\overline{X}_S}$. This means

$$\delta CS_{\overline{X}_S}^{x_S} = \mathrm{res}^*(\lambda_S^{x_S}), \tag{16.21}$$

which may be regarded as an arithmetic analogue of (16.5).

For $x_S, x_S' \in \Gamma(\mathcal{F}_S, \mathcal{L}_S)$, the G-equivariant isomorphism $\Phi_S^{x_S, x_S'} : \mathcal{L}_S^{x_S} \xrightarrow{\sim} \mathcal{L}_S^{x_S'}$ induces the isomorphism

$$\Psi^{x_S, x_S'} : \Gamma_G(\mathcal{F}_{\overline{X}_S}, \mathrm{res}_S^*(\mathcal{L}_S^{x_S})) \xrightarrow{\sim} \Gamma_G(\mathcal{F}_{\overline{X}_S}, \mathrm{res}_S^*(\mathcal{L}_S^{x_S'})); \ \psi^{x_S} \mapsto \Phi_S^{x_S, x_S'} \circ \psi^{x_S}.$$

Then we can define the equivalence relation $\sim$ on the disjoint union of $\Gamma_G(\mathcal{F}_{\overline{X}_S}, \mathrm{res}_S^*(\mathcal{L}_S^{x_S}))$ over $x_S \in \Gamma(\mathcal{F}_S, \mathcal{L}_S)$ by

$$\psi^{x_S} \sim \psi^{x_S'} \iff \Psi^{x_S, x_S'}(\psi^{x_S}) = \psi^{x_S'}$$

for $\psi^{x_S} \in \Gamma_G(\mathcal{F}_{\overline{X}_S}, \mathrm{res}_S^*(\mathcal{L}_S^{x_S}))$ and $\psi^{x_S'} \in \Gamma_G(\mathcal{F}_{\overline{X}_S}, \mathrm{res}_S^*(\mathcal{L}_S^{x_S'}))$. Since $\Phi_S^{x_S'} = \Phi_S^{x_S, x_S'} \circ \Phi_S^{x_S}$, $\Psi^{x_S, x_S'}(CS_{\overline{X}_S}^{x_S}) = CS_{\overline{X}_S}^{x_S'}$. Thus we have the following identification:

$$\Gamma_G(\mathcal{F}_{\overline{X}_S}, \mathrm{res}_S^*(\mathcal{L}_S)) = \bigsqcup_{x_S \in \Gamma(\mathcal{F}_S, \mathcal{L}_S)} \Gamma_G(\mathcal{F}_{\overline{X}_S}, \mathrm{res}_S^*(\mathcal{L}_S^{x_S}))/\sim,$$
$$\psi \longmapsto \Psi^{x_S}(\psi)$$

where $CS_{\overline{X}_S}$ and $[CS_{\overline{X}_S}^{x_S}]$ are identified.

Next, we shall construct the arithmetic Dijkgraaf–Witten state space and the arithmetic Dijkgraaf–Witten partition function over the moduli space of Galois representations, following the topological case (16.8), (16.9). We keep the same notations and assumptions as above, and we assume that F is a subfield of $\mathbb{C}$ such that $\mu_N \subset F$ and $\overline{F} = F$ ($\overline{F}$ being the complex conjugate of F).

We define the *arithmetic Dijkgraaf–Witten state space* $\mathcal{H}_S$ for ∂V_S as the space of G-equivariant sections of the arithmetic prequantization F-line bundle $\pi_{S,F}$: $L_S \to \mathcal{F}_S$:

$$\mathcal{H}_S := \Gamma_G(\mathcal{F}_S, L_S) = \Gamma(\mathcal{M}_S, \overline{L}_S).$$

It is a finite-dimensional F-vector space.

Let $x_S = [(x_{\mathfrak{p}_1}, \ldots, x_{\mathfrak{p}_r})] \in \Gamma(\mathcal{F}_S, \mathcal{L}_S)$ be a section. Let $L_S^{x_S}$ be the arithmetic prequantization F-line bundle over $\mathcal{F}_S$ with respect to x_S and let $\mathcal{H}_S^{x_S}$ denote the set of G-equivariant sections of $L_S^{x_S}$:

$$
\begin{aligned}
\mathcal{H}_S^{x_S} &:= \Gamma_G(\mathcal{F}_S, L_S^{x_S}) \\
&= \{\theta : \mathcal{F}_S \longrightarrow F \mid \theta(\rho_S.g) = \zeta_N^{\lambda_S^{x_S}(g,\rho_S)}\theta(\rho_S) \text{ for } \rho_S \in \mathcal{F}_S, g \in G\},
\end{aligned}
$$

which we call the *arithmetic Dijkgraaf–Witten state space for ∂V_S with respect to* x_S. The isomorphism $\Phi_{S,F}^{x_S} : L_S \xrightarrow{\sim} L_S^{x_S}$ induces the isomorphism

$$\Theta^{x_S} : \mathcal{H}_S \xrightarrow{\sim} \mathcal{H}_S^{x_S}; \quad \theta \longmapsto \Phi_{S,F}^{x_S} \circ \theta.$$

For $x_S, x_S' \in \Gamma(\mathcal{F}_S, \mathcal{L}_S)$, the isomorphism $\Phi_{S,F}^{x_S,x_S'} : L_S^{x_S} \xrightarrow{\sim} L_S^{x_S'}$ induces the isomorphism of F-vector spaces:

$$\Theta^{x_S,x_S'} : \mathcal{H}_S^{x_S} \xrightarrow{\sim} \mathcal{H}_S^{x_S'}; \quad \theta^{x_S} \longmapsto \Phi_S^{x_S,x_S'} \circ \theta^{x_S}.$$

So the equivalence relation $\sim$ is defined on the disjoint union of all $\mathcal{H}_S^{x_S}$ running over $x_S \in \Gamma(\mathcal{F}_S, \mathcal{L}_S)$ by $\theta^{x_S} \sim \theta^{x_S'} \iff \Theta^{x_S,x_S'}(\theta^{x_S}) = \theta^{x_S'}$ for $\theta^{x_S} \in \mathcal{H}_S^{x_S}$ and $\theta^{x_S'} \in \mathcal{H}_S^{x_S'}$. Then we have the following identification:

$$\mathcal{H}_S = \bigsqcup_{x_S \in \Gamma(\mathcal{F}_S, \mathcal{L}_S)} \mathcal{H}_S^{x_S} / \sim . \tag{16.22}$$

Remark 16.4.2 The arithmetic state space $\mathcal{H}_S$ is an arithmetic analogue of the Dijkgraaf–Witten state space $\mathcal{H}_\Sigma$ for a closed surface Σ in (2+1)-dimensional Dijkgraaf–Witten TQFT (cf. Remark 16.2.3 (1)). It would also be an interesting question in number theory to describe the dimension and a canonical basis of $\mathcal{H}_S$ in comparison of Verlinde's formulas [236].

For $\rho_S \in \mathcal{F}_S$, we define the subset $\mathcal{F}_{\overline{X}_S}(\rho_S)$ of $\mathcal{F}_{\overline{X}_S}$ by

$$\mathcal{F}_{\overline{X}_S}(\rho_S) := \{\rho \in \mathcal{F}_{\overline{X}_S} \mid \mathrm{res}_S(\rho) = \rho_S\}.$$

We then define the *arithmetic Dijkgraaf–Witten invariant* $Z_{\overline{X}_S}^{x_S}(\rho_S)$ of ρ_S with respect to x_S by

$$Z_{\overline{X}_S}^{x_S}(\rho_S) := \frac{1}{\#G} \sum_{\rho \in \mathcal{F}_{\overline{X}_S}(\rho_S)} \zeta_N^{CS_{\overline{X}_S}^{x_S}(\rho)}.$$

Proposition 16.4.1 (1) and (16.21) yield the following:

Proposition 16.4.3

(1) $Z_{\overline{X}_S}^{x_S}(\rho_S)$ *is independent of the choice of* β_ρ.
(2) We have

$$Z_{\overline{X}_S}^{x_S} \in \mathcal{H}_S^{x_S}.$$

We call $Z_{\overline{X}_S}^{x_S} \in \mathcal{H}_S^{x_S}$ the *arithmetic Dijkgraaf–Witten partition function for* $\overline{X}_S$ with respect to x_S. For sections $x_S, x_S' \in \Gamma(\mathcal{F}_S, \mathcal{L}_S)$, using $\Psi^{x_S, x_S'}(CS_{\overline{X}_S}^{x_S}) = CS_{\overline{X}_S}^{x_S'}$,

$$\Theta^{x_S, x_S'}(Z_{\overline{X}_S}^{x_S}) = Z_{\overline{X}_S}^{x_S'}.$$

By the identification (16.22), $Z_{\overline{X}_S}^{x_S}$ defines the element $Z_{\overline{X}_S}$ of $\mathcal{H}_S$ which is independent of the choice of x_S. We call it the *arithmetic Dijkgraaf–Witten partition function for* $\overline{X}_S$.

An arithmetic analogue of Theorem 16.2.2 is stated as the following:

Theorem 16.4.4 *The correspondences constructed above:*

$$\partial V_S := \mathrm{Spec}(k_{\mathfrak{p}_1}) \sqcup \cdots \sqcup \mathrm{Spec}(k_{\mathfrak{p}_r}) \rightsquigarrow \textit{arithmetic Dijkgraaf} - \textit{Witten}$$
$$\textit{state space } \mathcal{H}_S,$$
$$\overline{X}_S := \overline{\mathrm{Spec}(\mathcal{O}_k)} \setminus S \qquad \rightsquigarrow \textit{arithmetic Dijkgraaf} - \textit{Witten}$$
$$\textit{partition function } Z_{\overline{X}_S} \in \mathcal{H}_S,$$

satisfy analogues of the axioms (A1)–(A4) in $(2 + 1)$-*dimensional TQFT, which are described below.*

(A1) *Functoriality.* Let k' be another number field that contains μ_N and let $S' = \{\mathfrak{p}_1', \ldots, \mathfrak{p}_{r'}'\}$ be a finite set of finite primes of k' such that any finite prime dividing N is contained in S'. The objects constructed by using k' and S' will be denoted by, for example, $\mathcal{L}_{\mathfrak{p}'}, L_{\mathfrak{p}'}, \mathcal{L}_{S'}, L_{S'}, \ldots$ etc., for simplicity of notation. Assume that

$r = r'$ and there are isomorphisms $\xi_i : k_{\mathfrak{p}_i} \xrightarrow{\sim} k'_{\mathfrak{p}'_i}$ for $1 \le i \le r$. Then ξ_i's induces the isomorphisms of arithmetic prequantization bundles

$$\xi_S : \mathcal{L}_S \xrightarrow{\sim} \mathcal{L}_{S'}, \quad \xi_{S,F} : L_S \xrightarrow{\sim} L_{S'},$$

and for $x_S \in \Gamma(\mathcal{F}_S, \mathcal{L}_S)$ and $x_{S'} \in \Gamma(\mathcal{F}_{S'}, \mathcal{L}_{S'})$, we have the isomorphisms of arithmetic prequantization bundles with respect to sections

$$\mathcal{L}_S^{x_S} \xrightarrow{\sim} \mathcal{L}_{S'}^{x_{S'}}, \quad L_S^{x_S} \xrightarrow{\sim} L_{S'}^{x_{S'}}.$$

Suppose further that there is an isomorphism $\tau : k \xrightarrow{\sim} k'$ of number fields which sends $\mathfrak{p}_i$ to $\mathfrak{p}'_i$ for $1 \le i \le r$ so that we have the isomorphism

$$\xi : \overline{X}_S := \overline{X}_k \setminus S \xrightarrow{\sim} \overline{X}_{k'} \setminus S' =: \overline{X}_{S'}.$$

For example, let $k := \mathbb{Q}(\sqrt[3]{2})$, $k' := \mathbb{Q}(\sqrt[3]{2}\omega)$, $\omega := \exp(\frac{2\pi\sqrt{-1}}{3})$ and so $N = 2$. Let ξ be the isomorphism $k \xrightarrow{\sim} k'$ defined by $\xi(\sqrt[3]{2}) := \sqrt[3]{2}\omega$. Noting $2\mathcal{O}_k = (\sqrt[3]{2})^2$, $X^3 - 2 \equiv (X - 4)(X - 7)(X - 20) \bmod 31$, let $S := \{\mathfrak{p}_1 := (\sqrt[3]{2}), \mathfrak{p}_2 := (31, \sqrt[3]{2} - 4), \mathfrak{p}_2 := (31, \sqrt[3]{2} - 7), \mathfrak{p}_4 := (31, \sqrt[3]{2} - 20)\}$, $S' := \xi(S) = \{\mathfrak{p}'_1 := (\sqrt[3]{2}\omega), \mathfrak{p}'_2 := (31, \sqrt[3]{2}\omega - 4), \mathfrak{p}'_3 := (31, \sqrt[3]{2}\omega - 7), \mathfrak{p}'_4 := (31, \sqrt[3]{2}\omega - 20)\}$, so that we have $k_{\mathfrak{p}_1} = k'_{\mathfrak{p}'_1} = \mathbb{Q}_2$ and $k_{\mathfrak{p}_i} = k'_{\mathfrak{p}'_i} = \mathbb{Q}_{31}$ ($2 \le i \le 4$). So this example satisfies the above conditions.

The isomorphism ξ induces the bijection $\xi^* : \mathcal{F}_{\overline{X}_{S'}} \xrightarrow{\sim} \mathcal{F}_{\overline{X}_S}$.

Proposition 16.4.5 *The isomorphism* $\xi_S : \mathcal{L}_S \xrightarrow{\sim} \mathcal{L}_{S'}$ *induces the bijection*

$$\Gamma_G(\mathcal{F}_{\overline{X}_S}, \mathrm{res}_S^*(\mathcal{L}_S)) \xrightarrow{\sim} \Gamma_G(\mathcal{F}_{\overline{X}_{S'}}, \mathrm{res}_{S'}^*(\mathcal{L}_{S'}))$$

which sends $CS_{\overline{X}_S}$ *to* $CS_{\overline{X}_{S'}}$. *The isomorphism* $\xi_{S,F} : L_S \xrightarrow{\sim} L_{S'}$ *induces the isomorphism*

$$\mathcal{H}_S \xrightarrow{\sim} \mathcal{H}_{S'},$$

which sends $Z_{\overline{X}_S}$ *to* $Z_{\overline{X}_{S'}}$.

(A2) *Multiplicativity.* Let S_1 and S_2 be disjoint sets of finite primes of k and let $S = S_1 \sqcup S_2$. Then

$$\mathcal{F}_S = \mathcal{F}_{S_1} \times \mathcal{F}_{S_2}.$$

For the arithmetic quantization F-line bundles, we let $p_i^*(L_{S_i})$ be the G-equivariant F-line bundle over $\mathcal{F}_S$ induced from L_{S_i} by the projection $p_i : \mathcal{F}_S \to \mathcal{F}_{S_i}$ for $i = 1, 2$:

$$p_i^*(L_{S_i}) := \{(\rho_S, [\alpha_{S_i}]) \mid \rho_{S_i} = \pi_{S_i}([\alpha_{S_i}])\ (i = 1, 2)\}$$

for $\rho_S = (\rho_{S_1}, \rho_{S_2})$. Let

$$p_i^*(\pi_{S_i}) : p_i^*(L_{S_i}) \longrightarrow \mathcal{F}_S$$

be the projection. The fiber over $\rho_S = (\rho_{S_1}, \rho_{S_2})$ is given by

$$\begin{aligned} p_i^*(\pi_{S_i})^{-1}(\rho_S) &= \{\rho_S\} \times \{[([\alpha_{S_i}], z_i) \in L_{S_i} \mid \rho_{S_i} = \pi_{S_i}([\alpha_{S_i}]), z_i \in F\} \\ &= L_{S_i}(\rho_{S_i}) \simeq F. \end{aligned}$$

We set

$$L_{S_1} \boxtimes L_{S_2} := p_1^*(L_{S_1}) \otimes p_2^*(L_{S_2}),$$

which is the F-line bundle over $\mathcal{F}_S$ and whose elements are written as $(\rho_S, [([\alpha_{S_1}], z_1)] \otimes [([\alpha_{S_2}], z_2)])$, where $\rho_S = (\rho_{S_1}, \rho_{S_2}) \in \mathcal{F}_S$, $([\alpha_{S_i}], z_i) \in L_{S_i}(\rho_{S_i})$. The right action on $L_{S_1} \boxtimes L_{S_2}$ is defined by

$$(\rho_S, [([\alpha_{S_1}], z_1)] \otimes [([\alpha_{S_2}], z_2)]).g := (\rho_S.g, [([\alpha_{S_1}].g, z_1)] \otimes [([\alpha_{S_2}].g, z_2)])$$

so that the projection $L_{S_1} \boxtimes L_{S_2} \to \mathcal{F}_S$ is G-equivariant. Then we have the isomorphism of G-equivariant F-line bundles over $\mathcal{F}_S$:

$$L_{S_1} \boxtimes L_{S_2} \xrightarrow{\sim} L_S;\ (\rho_S, [([\alpha_{S_1}], z_1)] \otimes [([\alpha_{S_2}], z_2)]) \longmapsto [([\alpha_S], z_1 z_2)],$$

where $\alpha_S = (\alpha_{S_1}, \alpha_{S_2})$. Choose $x_{S_i} \in \Gamma(\mathcal{F}_{S_i}, \mathcal{L}_{S_i})$ and let $x_S := [(x_{S_1}, x_{S_2})] \in \Gamma(\mathcal{F}_S, \mathcal{L}_S)$. Then we see that

$$\lambda_{S_1}^{x_{S_1}}(g, \rho_{S_1}) + \lambda_{S_2}^{x_{S_2}}(g, \rho_{S_2}) = \lambda_S^{x_S}(g, \rho_S)$$

for $g \in G$ and $\rho_S = (\rho_{S_1}, \rho_{S_2})$.

Proposition 16.4.6 *For $\theta_i \in \mathcal{H}_{S_i}^{x_{S_i}}$ ($i = 1, 2$), we define $\theta_1 \cdot \theta_2 \in \mathcal{H}_S^{x_S}$ by*

$$(\theta_1 \cdot \theta_2)(\rho_S) := \theta_1(\rho_{S_1})\theta_2(\rho_{S_2})$$

for $\rho_S = (\rho_{S_1}, \rho_{S_2})$. Then we have the following isomorphism of F-vector spaces:

$$\mathcal{H}_{S_1}^{x_{S_1}} \otimes \mathcal{H}_{S_2}^{x_{S_2}} \xrightarrow{\sim} \mathcal{H}_S^{x_S};\ (\theta_1, \theta_2) \longmapsto \theta_1 \cdot \theta_2.$$

For $\theta_i \in \mathcal{H}_{S_i}$ $(i = 1, 2)$, we define $\theta_1 \boxtimes \theta_2 \in \mathcal{H}_S$ by

$$(\theta_1 \boxtimes \theta_2)(\rho_S) := p_1^*(\theta_1(\rho_{S_1})) \otimes p_2^*(\theta_2(\rho_{S_2}))$$

for $\rho_S = (\rho_{S_1}, \rho_{S_2})$. Here $p_1^*(\theta_1(\rho_{S_1})) \otimes p_2^*(\theta_2(\rho_{S_2}))$ denotes $[([\alpha_S], z_1 z_2)]$ when $\theta_i(\rho_{S_i}) = [([\alpha_{S_i}], z_i)], \alpha_S = (\alpha_{S_1}, \alpha_{S_2})$. Then we have the following isomorphism of F-vector spaces:

$$\mathcal{H}_{S_1} \otimes \mathcal{H}_{S_2} \xrightarrow{\sim} \mathcal{H}_S; \quad (\theta_1, \theta_2) \longmapsto \theta_1 \boxtimes \theta_2.$$

The above isomorphisms are compatible via the isomorphisms $\Theta^{x_{S_i}} : \mathcal{H}_{S_i} \simeq \mathcal{H}_{S_i}^{x_{S_i}}$ $(i = 1, 2)$ and $\Theta^{x_S} : \mathcal{H}_S \simeq \mathcal{H}_S^{x_S}$.

(A3) *Involutority*. For a finite prime $\mathfrak{p}$ of k, the canonical isomorphism

$$\mathrm{inv}_{\mathfrak{p}} : H_{\text{ét}}^2(\partial V_{\mathfrak{p}}, \mathbb{Z}/N\mathbb{Z}) \xrightarrow{\sim} \mathbb{Z}/N\mathbb{Z}$$

indicates that $\partial V_{\mathfrak{p}}$ is "orientable" and we choose (implicitly) the "orientation" of $\partial V_{\mathfrak{p}}$ corresponding $1 \in \mathbb{Z}/N\mathbb{Z}$. We let $\partial V_{\mathfrak{p}}^* = \partial V_{\mathfrak{p}}$ with the "opposite orientation", namely, $\mathrm{inv}_{\mathfrak{p}}([\partial V_{\mathfrak{p}}^*]) = -1$.

The arithmetic prequantization principal $\mathbb{Z}/N\mathbb{Z}$-bundle for $\partial V_{\mathfrak{p}}^*$, denoted by $\mathcal{L}_{\mathfrak{p}^*}$, is defined (formally) by $\mathcal{L}_{\mathfrak{p}}$ with the opposite action of the structure group $\mathbb{Z}/N\mathbb{Z}$, $(\alpha_{\mathfrak{p}}, m) \mapsto \alpha_{\mathfrak{p}}.(-m)$ for $\alpha_{\mathfrak{p}} \in \mathcal{L}_{\mathfrak{p}^*}$ and $m \in \mathbb{Z}/N\mathbb{Z}$. So the arithmetic prequantization F-line bundle $L_{\mathfrak{p}^*}$ for $\partial V_{\mathfrak{p}}^*$ is the dual bundle of $L_{\mathfrak{p}}$, $L_{\mathfrak{p}^*} = L_{\mathfrak{p}}^*$. Noting $\Gamma(\mathcal{F}_{\mathfrak{p}}, \mathcal{L}_{\mathfrak{p}^*}) = \Gamma(\mathcal{F}_{\mathfrak{p}}, \mathcal{L}_{\mathfrak{p}})$, the arithmetic Chern–Simons 1-cocycle $\lambda_{\mathfrak{p}^*}^{x_{\mathfrak{p}}}$ for $\partial V_{\mathfrak{p}}^*$ is given by $-\lambda_{\mathfrak{p}}^{x_{\mathfrak{p}}}$ for $x_{\mathfrak{p}} \in \Gamma(\mathcal{F}_{\mathfrak{p}}, \mathcal{L}_{\mathfrak{p}^*})$. The actions of G on $\mathcal{L}_{\mathfrak{p}^*}^{x_{\mathfrak{p}}} = \mathcal{F}_{\mathfrak{p}} \times \mathbb{Z}/N\mathbb{Z}$ and $L_{\mathfrak{p}^*}^{x_{\mathfrak{p}}} = \mathcal{F}_{\mathfrak{p}} \times F$ are changed to those via $\lambda_{\mathfrak{p}^*}^{x_{\mathfrak{p}}}$.

For a finite set of finite primes $S = \{\mathfrak{p}_1, \ldots, \mathfrak{p}_r\}$, we set $\partial V_S^* := \partial V_{\mathfrak{p}_1}^* \sqcup \cdots \sqcup \partial V_{\mathfrak{p}_r}^*$. Then the arithmetic prequantization bundles $\mathcal{L}_{S^*}, L_{S^*}, \mathcal{L}_{S^*}^{x_S}$ and $L_{S^*}^{x_S}$ $(x_S \in \Gamma(\mathcal{F}_S, \mathcal{L}_{S^*}) = \Gamma(\mathcal{F}_{\mathfrak{p}}, \mathcal{L}_S))$ are defined in the similar manner. For the arithmetic Chern–Simons 1-cocycle, we have

$$\lambda_{S^*}^{x_S} = -\lambda_S^{x_S}.$$

Let $\mathcal{H}_{S^*}^{x_S}$ be the arithmetic quantum space for ∂V_S^* with respect to x_S. Then we see that

$$\begin{aligned}
\mathcal{H}_{S^*}^{x_S} &= \{\theta^* : \mathcal{F}_S \to F \mid \theta^*(\rho_S.g) = \zeta_N^{\lambda_{S^*}^{x_S}(g, \rho_S)} \theta^*(\rho_S) \text{ for } \rho_S \in \mathcal{F}_S, g \in G\} \\
&= \{\theta^* : \mathcal{F}_S \to F \mid \theta^*(\rho_S.g) = \zeta_N^{-\lambda_S(g, \rho_S)} \theta^*(\rho_S) \text{ for } \rho_S \in \mathcal{F}_S, g \in G\} \\
&= \overline{\mathcal{H}}_S^{x_S},
\end{aligned}$$

where $\overline{\mathcal{H}}_S^{xs}$ is the complex conjugate of $\mathcal{H}_S^{xs}$. Since the pairing

$$\mathcal{H}_{S*}^{xs} \times \mathcal{H}_S^{xs} \longrightarrow F; \ (\theta^*, \theta) \longmapsto \sum_{\rho_S \in \mathcal{F}_S} \theta^*(\rho_S)\theta(\rho_S)$$

is a (Hermitian) perfect pairing ($\overline{F} = F$), together with the isomorphism Θ^{xs}, we have the following:

Proposition 16.4.7 $\mathcal{H}_{S*}^{xs}$ and $\mathcal{H}_{S*}$ are the dual spaces of $\mathcal{H}_S^{xs}$ and $\mathcal{H}_S$, respectively:

$$\mathcal{H}_{S*}^{xs} = (\mathcal{H}_S^{xs})^*, \quad \mathcal{H}_{S*} = (\mathcal{H}_S)^*.$$

So far we have chosen implicitly the orientation of $\overline{X}_S$ so that the boundary $\partial \overline{X}_S$ with induced orientation may be identified with ∂V_S. Let $\overline{X}_S^*$ denote $\overline{X}_S$ with the opposite orientation. Then, the arithmetic Chern–Simons functional and the Dijkgraaf-Witten partition function for $\overline{X}_S^*$ are given as follows:

$$CS_{\overline{X}_S^*}^{xs} = -CS_{\overline{X}_S}^{xs}, \quad Z_{\overline{X}_S^*}^{xs}(\rho_S) = \frac{1}{\#G} \sum_{\rho \in \mathcal{F}_{\overline{X}_S}} \zeta_N^{-CS_{\overline{X}_S}^{xs}(\rho)} = \overline{Z_{\overline{X}_S}^{xs}(\rho_S)}.$$

(A4) *Gluing formula.* We show a decomposition formula for arithmetic Chern–Simons invariants and a gluing formula for arithmetic Dijkgraaf–Witten partition functions. For this, we firstly introduce the arithmetic Chern–Simons functional and the arithmetic Dijkgraaf–Witten partition function for $V_S := V_{\mathfrak{p}_1} \sqcup \cdots \sqcup V_{\mathfrak{p}_r}$, which plays a role analogous to a tubular neighborhood of a link, where $V_{\mathfrak{p}_i} := \mathrm{Spec}(\mathcal{O}_{\mathfrak{p}_i})$, $\mathcal{O}_{\mathfrak{p}_i}$ being the ring of $\mathfrak{p}_i$-adic integers.

Let $\tilde{\Pi}_{\mathfrak{p}}$ be the étale fundamental group of $V_{\mathfrak{p}}$, namely, the Galois group of the maximal unramified extension of $k_{\mathfrak{p}}$ and we set

$$\mathcal{F}_{V_{\mathfrak{p}}} := \mathrm{Hom}_{\mathrm{cont}}(\tilde{\Pi}_{\mathfrak{p}}, G), \quad \mathcal{F}_{V_S} := \mathcal{F}_{V_{\mathfrak{p}_1}} \times \cdots \times \mathcal{F}_{V_{\mathfrak{p}_r}}.$$

Since $\tilde{\Pi}_{\mathfrak{p}} \simeq \hat{\mathbb{Z}}$, $\mathcal{F}_{V_{\mathfrak{p}}} \simeq G$. G acts on $\mathcal{F}_{V_S}$ from the right by $\rho.g := (g^{-1}\tilde{\rho}_{\mathfrak{p}_i}g)_i$ for $\rho = (\tilde{\rho}_{\mathfrak{p}_i})_i$ and $g \in G$, and let $\mathcal{M}_{V_S}$ denote the quotient set by this action: $\mathcal{M}_{V_S} := \mathcal{F}_{V_S}/G$. Let $\tilde{\mathrm{res}}_{\mathfrak{p}_i} : \mathcal{F}_{V_{\mathfrak{p}_i}} \to \mathcal{F}_{\mathfrak{p}}$ and $\tilde{\mathrm{res}}_S := (\tilde{\mathrm{res}}_{\mathfrak{p}_i}) : \mathcal{F}_{V_S} \to \mathcal{F}_S$ denote the restriction maps induced by the natural continuous homomorphisms $v_{\mathfrak{p}_i} : \Pi_{\mathfrak{p}_i} \to \tilde{\Pi}_{\mathfrak{p}_i}$ $(1 \le i \le r)$, which are G-equivariant. We denote by $\tilde{\mathrm{Res}}_{\mathfrak{p}_i}$ and $\tilde{\mathrm{Res}}_S$ the homomorphisms on cochains given as the pull-back by $v_{\mathfrak{p}_i}$:

$$\tilde{\mathrm{Res}}_{\mathfrak{p}_i} : C^n(\tilde{\Pi}_{\mathfrak{p}_i}, \mathbb{Z}/N\mathbb{Z}) \longrightarrow C^n(\Pi_{\mathfrak{p}_i}, \mathbb{Z}/N\mathbb{Z}); \ \alpha_i \longmapsto \alpha_i \circ v_{\mathfrak{p}_i},$$
$$\tilde{\mathrm{Res}}_S := (\tilde{\mathrm{Res}}_{\mathfrak{p}_i}) : \prod_{i=1}^r C^n(\tilde{\Pi}_{\mathfrak{p}_i}, \mathbb{Z}/N\mathbb{Z}) \longrightarrow \prod_{i=1}^r C^n(\Pi_{\mathfrak{p}_i}, \mathbb{Z}/N\mathbb{Z}).$$

For $\tilde{\rho} = (\tilde{\rho}_{\mathfrak{p}_i})_i \in \mathcal{F}_{V_S}$, $c \circ \tilde{\rho}_{\mathfrak{p}_i} \in Z^3(\tilde{\Pi}_{\mathfrak{p}_i}, \mathbb{Z}/N\mathbb{Z})$. Since $H^3(\tilde{\Pi}_{\mathfrak{p}_i}, \mathbb{Z}/N\mathbb{Z}) = 0$, there is $\tilde{\beta}_{\mathfrak{p}_i} \in C^2(\tilde{\Pi}_{\mathfrak{p}_i}, \mathbb{Z}/N\mathbb{Z})/B^2(\tilde{\Pi}_{\mathfrak{p}_i}, \mathbb{Z}/N\mathbb{Z})$ such that $c \circ \tilde{\rho}_{\mathfrak{p}_i} = d\tilde{\beta}_{\mathfrak{p}_i}$. We see that $c \circ \tilde{\mathrm{res}}_{\mathfrak{p}_i}(\tilde{\rho}_{\mathfrak{p}_i}) = d\tilde{\mathrm{Res}}_{\mathfrak{p}_i}(\tilde{\beta}_{\mathfrak{p}_i})$ for $1 \leq i \leq r$ and so

$$[\tilde{\mathrm{Res}}_S((\tilde{\beta}_{\mathfrak{p}_i})_i)] \in \mathcal{L}_S(\tilde{\mathrm{res}}_S(\tilde{\rho})).$$

Let $\tilde{\mathrm{res}}_S^*(\mathcal{L}_S)$ be the G-equivariant principal $\mathbb{Z}/N\mathbb{Z}$-bundle over $\mathcal{F}_{V_S}$ induced from $\mathcal{L}_S$ by $\tilde{\mathrm{res}}_S$:

$$\tilde{\mathrm{res}}_S^*(\mathcal{L}_S) := \{(\tilde{\rho}, \alpha_S) \in \mathcal{F}_{V_S} \times \mathcal{L}_S \mid \tilde{\mathrm{res}}_S(\tilde{\rho}) = \pi_S(\alpha_S)\}$$

and let $\tilde{\mathrm{res}}_S^*(\pi_S)$ be the projection $\tilde{\mathrm{res}}_S^*(\mathcal{L}_S) \to \mathcal{F}_{V_S}$. We define the *arithmetic Chern–Simons functional* $CS_{V_S} : \mathcal{F}_{V_S} \to \mathcal{L}_S$ by

$$CS_{V_S}(\tilde{\rho}) := [\tilde{\mathrm{Res}}_S((\tilde{\beta}_{\mathfrak{p}_i})_i)]$$

for $\tilde{\rho} \in \mathcal{F}_{V_S}$. The value $CS_{V_S}(\tilde{\rho})$ is called the *arithmetic Chern–Simons invariant* of $\tilde{\rho}$. It can be shown that $CS_{V_S}(\tilde{\rho})$ is independent of the choice of $\tilde{\beta}_{\mathfrak{p}_i}$ and CS_{V_S} is a G-equivariant section of $\tilde{\mathrm{res}}_S^*(\varpi_S)$:

$$CS_{V_S} \in \Gamma_G(\mathcal{F}_{V_S}, \tilde{\mathrm{res}}_S^*(\mathcal{L}_S)) = \Gamma(\mathcal{M}_{V_S}, \tilde{\mathrm{res}}_S^*(\overline{\mathcal{L}}_S)).$$

For a section $x_S = [(x_{\mathfrak{p}_1}, \ldots, x_{\mathfrak{p}_r})] \in \Gamma(\mathcal{F}_S, \mathcal{L}_S)$, the isomorphism $\Phi_S^{x_S} : \mathcal{L}_S \xrightarrow{\sim} \mathcal{L}_S^{x_S}$ induces the isomorphism

$$\tilde{\Psi}^{x_S} : \Gamma_G(\mathcal{F}_{V_S}, \tilde{\mathrm{res}}_S^*(\mathcal{L}_S)) \xrightarrow{\sim} \Gamma_G(\mathcal{F}_{V_S}, \tilde{\mathrm{res}}_S^*(\mathcal{L}_S^{x_S})) = \mathrm{Map}_G(\mathcal{F}_{V_S}, \mathbb{Z}/N\mathbb{Z});$$
$$y_S \longmapsto \Phi_S^{x_S} \circ y_S.$$

We define the *arithmetic Chern–Simons functional* $CS_{V_S}^{x_S} : \mathcal{F}_{V_S} \to \mathbb{Z}/N\mathbb{Z}$ with respect to x_S by the image of CS_{V_S} under Ψ^{x_S}. We have the following equality in $C^1(G, \mathrm{Map}(\mathcal{F}_{V_S}, \mathbb{Z}/N\mathbb{Z}))$:

$$\delta CS_{V_S}^{x_S} = \tilde{\mathrm{res}}^*(\lambda_S^{x_S}).$$

For $\rho_S \in \mathcal{F}_S$, we define the subset $\mathcal{F}_{V_S}(\rho_S)$ of $\mathcal{F}_{V_S}$ by

$$\mathcal{F}_{V_S}(\rho_S) := \{\tilde{\rho} \in \mathcal{F}_{V_S} \mid \tilde{\mathrm{res}}_S(\tilde{\rho}) = \rho_S\}.$$

We then define the *arithmetic Dijkgraaf–Witten invariant* $Z_{V_S}(\rho_S)$ of ρ_S with respect to x_S by

$$Z_{V_S}^{x_S}(\rho_S) := \frac{1}{\#G} \sum_{\tilde{\rho} \in F_{V_S}(\rho_S)} \zeta_N^{CS_{V_S}^{x_S}(\tilde{\rho})}.$$

It can be shown that $Z_{V_S}^{x_S}(\rho_S)$ is independent of the choice of $\tilde{\beta}_{\rho_{\mathfrak{p}_i}}$ and

$$Z_{V_S}^{x_S} \in \mathcal{H}_S^{x_S}.$$

We call $Z_{V_S}^{x_S}$ the *arithmetic Dijkgraaf–Witten partition function*) for V_S with respect to x_S.

For sections $x_S, x_S' \in \Gamma(\mathcal{F}_S, \mathcal{L}_S)$ we see that

$$\Theta^{x_S, x_S'}(Z_{V_S}^{x_S}) = Z_{V_S}^{x_S'}.$$

By the identification (16.22), $Z_{V_S}^{x_S}$ defines the element Z_{V_S} of $\mathcal{H}_S$ which is independent of the choice of x_S. We call it the *arithmetic Dijkgraaf–Witten partition function* for V_S.

In the above, the orientation of V_S is chosen so that it is compatible with that of ∂V_S. Let V_S^* denote V_S with opposite orientation. Then the arithmetic Chern–Simons functional and the arithmetic Dijkgraaf–Witten partition function are given by

$$CS_{V_S^*}^{x_S} = -CS_{V_S}^{x_S}, \quad Z_{V_S^*}^{x_S}(\rho_S) = \frac{1}{\#G} \sum_{\rho \in \mathcal{F}_{V_S}} \zeta_N^{-CS_{V_S}^{x_S}(\rho)}.$$

Now let S be a finite set of finite primes of k which contains any prime dividing N. We may think of $\overline{X}$ as the space obtained by gluing $\overline{X}_S$ and V_S^* along ∂V_S. Let $\eta_S : \pi_1(\overline{X}_S) \to \pi_1(\overline{X})$, $\iota_{\mathfrak{p}} : \Pi_{\mathfrak{p}} \to \pi_1(\overline{X}_S)$, $v_{\mathfrak{p}} : \Pi_{\mathfrak{p}} \to \tilde{\Pi}_{\mathfrak{p}}$, and $u_{\mathfrak{p}} : \tilde{\Pi}_{\mathfrak{p}} \to \pi_1(\overline{X})$ be the natural homomorphisms, where $\mathfrak{p} \in S$, so that $\eta_S \circ \iota_{\mathfrak{p}} = u_{\mathfrak{p}} \circ v_{\mathfrak{p}}$ for $\mathfrak{p} \in S$:

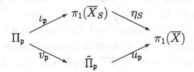

Then we have the following decomposition formula for arithmetic Chern–Simons invariants [31].

Theorem 16.4.8 (Decomposition Formula) *For* $\rho \in \mathrm{Hom}_{\mathrm{cont}}(\pi_1(\overline{X}), G)$,

$$CS_{\overline{X}_S}(\rho \circ \eta_S) - CS_{V_S}((\rho \circ u_{\mathfrak{p}})_{\mathfrak{p} \in S}) = CS_{\overline{X}}(\rho).$$

For a section $x_S \in \Gamma(\mathcal{F}_S, \mathcal{L}_S)$,

$$CS_{\overline{X}}(\rho) + CS_{V_S}^{x_S}((\rho \circ u_{\mathfrak{p}})_{\mathfrak{p} \in S}) = CS_{\overline{X}_S}^{x_S}(\rho \circ \eta_S).$$

We define the pairing $\langle \ , \ \rangle : \mathcal{H}_S^{xs} \times \mathcal{H}_{S*}^{xs} \to F$ by

$$\langle \theta_S, \theta_{S*} \rangle := \#G \sum_{\rho_S \in \mathcal{F}_S} \theta_S(\rho_S)\theta_{S*}(\rho_S)$$

for $\theta_S \in \mathcal{H}_S^{xs}, \theta_{S*} \in \mathcal{H}_{S*}^{xs}$. This induces the pairing $\langle , \rangle : \mathcal{H}_S \times \mathcal{H}_{S*} \to F$ by (16.22). Then we can show the following gluing formula for arithmetic Dijkgraaf–Witten partition functions [88].

Theorem 16.4.9 (Gluing Formula) *Notations being as above, we have the following equality:*

$$\langle Z_{\overline{X}_S}, Z_{V_S^*} \rangle = Z(\overline{X}).$$

For the proofs of Theorems 16.4.8 and 16.4.9, we refer to [88, Section 5].

Remark 16.4.10

(1) In [31] and [129], the authors used the decomposition formula (Theorem 16.4.8) in order to compute arithmetic Chern–Simons invariants $CS_{\overline{X}}(\rho)$ for various examples. In [17] and [3], computations of $CS_{\overline{X}}(\rho)$ have also been carried out by number-theoretic methods.

(2) In [30], the authors computed arithmetic Dijkgraaf–Witten correlation functions for finite cyclic gauge groups in terms of arithmetic linking numbers of primes. Their formula may be regarded as an arithmetic finite analogue of the path integral for linking numbers in abelian Chern–Simons gauge theory (cf. Remark 4.2.3).

(3) In [103], Kim introduced arithmetic Chern–Simons functionals for the case of p-adic Lie gauge group. In [26], an arithmetic analogue of topological BF theory was studied and, in [27], the authors showed an arithmetic path integral formula for the Kubota–Leopoldt p-adic L-function.

(4) A deep aspect of the 3-dimensional Chern–Simons TQFT with compact connected gauge group is a relation with 2-dimensional conformal field theory [115]. For Dijkgraaf–Witten TQFT, Brylinski and McLaughlin [22], [23] studied the analogue for a finite gauge group of the Segal–Witten reciprocity law in conformal field theory [21], [22], [202].

We may find an analogous feature between central extensions of loop groups in conformal field theory and metaplectic coverings in number theory such as the Segal–Witten reciprocity law and the Hilbert reciprocity law [20], [119], [242]. It would be interesting to pursue this analogy in connection with (arithmetic) Chern–Simons TQFT.

Summary

oriented connected 3-manifold M with boundary ∂M	a complement $\overline{X}_S$ of a finite set S of finite primes in $\overline{X} = \overline{\mathrm{Spec}(\mathcal{O}_k)}$ with $\partial \overline{X}_S = \partial V_S = \bigsqcup_{\mathfrak{p} \in S} \mathrm{Spec}(k_{\mathfrak{p}})$
Chern–Simons 1-cocycle for an oriented closed surface Σ $\Sigma \rightsquigarrow \lambda_\Sigma$	arithmetic Chern–Simons 1-cocycle for ∂V_S $\partial V_S \rightsquigarrow \lambda_S^{x_S}$
Chern–Simons functional for M $M \rightsquigarrow CS_M$ $\delta CS_M = \mathrm{res}^* \lambda_{\partial M}$	arithmetic Chern–Simons functional for $\overline{X}_S$ $\overline{X}_S \rightsquigarrow CS_{\overline{X}_S}$ $\delta CS_{\overline{X}_S}^{x_S} = \mathrm{res}^* \lambda_S^{x_S}$
gauge fields $\mathcal{F}_M, \mathcal{F}_\Sigma$ gauge groups $\mathcal{G}_M, \mathcal{G}_\Sigma$	arithmetic gauge fields $\mathcal{F}_{\overline{X}_S}, \mathcal{F}_S$ arithmetic gauge fields $\mathcal{G}_{\overline{X}_S}, \mathcal{G}_S$
prequantization line bundles $\mathcal{L}_\Sigma, L_\Sigma$ $e^{2\pi\sqrt{-1}CS_M} \in \Gamma(\mathcal{F}_M/\mathcal{G}_M, \mathrm{res}^*(L_{\partial M}))$	arithmetic prequantization line bundles $\mathcal{L}_S, L_S$ $\zeta_N^{CS_{\overline{X}_S}} \in \Gamma(\mathcal{F}_{\overline{X}_S}/\mathcal{G}_{\overline{X}_S}, \mathrm{res}^*(L_S))$
Dijkgraaf–Witten state space $\Sigma \rightsquigarrow \mathcal{H}_\Sigma$ $\mathcal{H}_\Sigma := \Gamma(\mathcal{F}_\Sigma/\mathcal{G}_\Sigma, l_\Sigma)$	arithmetic Dijkgraaf–Witten state space $\partial V_S \rightsquigarrow \mathcal{H}_S$ $\mathcal{H}_S := \Gamma(\mathcal{F}_S/\mathcal{G}_S, L_S)$
Dijkgraaf–Witten partition function $M \rightsquigarrow Z_M \in \mathcal{H}_{\partial M}$ $Z_M(\rho) := \dfrac{1}{\#G} \displaystyle\sum_{\substack{\tilde{\rho} \in \mathcal{F}_M \\ \mathrm{res}(\tilde{\rho})=\rho}} e^{2\pi\sqrt{-1}CS_M(\tilde{\rho})}$	arithmetic Dijkgraaf–Witten partition function $\overline{X}_S \rightsquigarrow Z_{\overline{X}_S} \in \mathcal{H}_S$ $Z_{\overline{X}_S}(\rho) := \dfrac{1}{\#G} \displaystyle\sum_{\substack{\tilde{\rho} \in \mathcal{F}_{\overline{X}_S} \\ \mathrm{res}(\tilde{\rho})=\rho}} \zeta_N^{CS_{\overline{X}_S}(\tilde{\rho})}$
When $\partial M = \emptyset$, Chern–Simons invariant $CS_M(\varrho)$ Dijkgraaf–Witten invariant $Z(M)$	When $S = \emptyset$, arithmetic Chern–Simons invariant $CS_{\overline{X}}(\rho)$ arithmetic Dijkgraaf–Witten invariant $Z(\overline{X})$

Bibliography

1. Ahlfors, L.: Complex Analysis: An Introduction of the Theory of Analytic Functions of One Complex Variable, 3rd edn. McGraw-Hill Book Co., New York-Toronto-London (1979)
2. Ahlqvist, E., Carlson, M.: The étale cohomology ring of the ring of a punctured arithmetic curve (2022). arXiv:2110.01597v3
3. Ahlqvist, E., Carlson, M.: The étale cohomology ring of the ring of integers of a number field. Res. Number Theory **9**, Article number: 58 (2023)
4. Amano, F.: On Rédei's dihedral extension and triple reciprocity law. Proc. Jpn. Acad. **90**, Ser. A, 1–5 (2014)
5. Amano, F.: On a certain nilpotent extension over $\mathbb{Q}$ of degree 64 and the 4-th multiple residue symbol. Tohoku Math. J. (2) **66**(4), 501–522 (2014)
6. Amano, F.: Arithmetic of certain nilpotent extensions and multiple residue symbols. Doctoral thesis, Kyushu University (2014)
7. Amano, F., Kodani, H., Morishita, M., Sakamoto, T., Yoshida, T.: Rédei's triple symbols and modular forms, wish Appendix by T. Ogasawara. Tokyo J. Math. **36**(2), 405–427 (2013)
8. Amano, F., Mizusawa, Y., Morishita, M.: On mod 3 triple Milnor invariants and triple cubic residue symbols in the Eisenstein number field. Res. Number Theory **4**, Article number: 7 (2018)
9. Artin, M., Mazur, B.: Etale Homotopy. Lecture Notes in Mathematics, vol. 100. Springer-Verlag, Berlin-New York (1969)
10. Artin, E., Tate, J.: Class Field Theory, Advanced Book Classics. Addison-Wesley, Redwood City (1990)
11. Atiyah, M.: Topological quantum field theories. Inst. Hautes Etudes Sci. Publ. Math. No. **68**, 175–186 (1988)
12. Atiyah, M.: The Geometry and Physics of Knots. Cambridge University Press, Cambridge (1990)
13. Bayer, P., Neukirch, J.: On values of zeta functions and l-adic Euler characteristics. Invent. Math. **50**(1), 35–64 (1978)
14. Beauville, A., Laszlo, Y.: Conformal blocks and generalized theta functions. Commun. Math. Phys. **164**, 385–419 (1994)
15. Bienenfeld, M.: An étale cohomology duality theorem for number fields with a real embedding. Trans. Am. Math. Soc. **303**(1), 71–96 (1987)
16. Birman, J.: Braids, Links, and Mapping Class Groups. Annals of Mathematics Studies, vol. 82. Princeton University Press, Princeton/University of Tokyo Press, Tokyo (1974)
17. Bleher, F.M., Chinburg, T., Greenberg, R., Kakde, M., Pappas, G., Taylor, M.J.: Cup products in the étale cohomology of number fields. New York J. Math. **24**, 514–542 (2018)

© The Author(s), under exclusive license to Springer Nature Singapore Pte Ltd. 2024   245
M. Morishita, *Knots and Primes*, Universitext,
https://doi.org/10.1007/978-981-99-9255-3

18. Boileau, M., Porti, J.: Geometrization of 3-orbifolds of cyclic type. Astérisque **272** (2001)
19. Boston, N.: Galois p-groups unramified at p—a survey. In: Primes and Knots. Contemporary Mathematics, vol. 416, pp. 31–40. American Mathematical Society, Providence (2006)
20. Brylinski, J.-L., Deligne, P.: Central extensions of reductive groups by K_2. Publ. Math. l'IHES, Tome **94**, 5–85 (2001)
21. Brylinski, J.-L., McLaughlin, D.A.: The geometry of degree-4 characteristic classes and of line bundles on loop groups I. Duke Math. J. **75**(3), 603–683 (1994)
22. Brylinski, J.-L., McLaughlin, D.A.: The geometry of degree-4 characteristic classes and of line bundles on loop groups II. Duke Math. J. **83**(1), 105–139 (1996)
23. Brylinski, J.-L,, McLaughlin, D.A.: Non-commutative reciprocity laws associated to finite groups. In: Contemporary Mathematics, vol. 202, pp. 421–438. American Mathematical Society, Providence (1997)
24. Burde, G., Zieschang, H.: Knots, de Gruyter Studies in Mathematics, vol. 5, 2nd edn. de Gruyter, Berlin (2003)
25. Bush, M.R., Gärtner, J., Labute, J., Vogel, D.: Mild 2-relator pro-p-groups. New York J. Math. **17**, 281–294 (2011)
26. Carlson, M., Kim, M.: A note on abelian arithmetic BF-theory. Bull. Lond. Math. Soc. **54**(4), 1299–1307 (2022)
27. Carlson, M., Chung, H.-J., Kim, D., Kim, M., Park, J., Yoo, H.: Path integrals and p-adic L-functions (2022). arXiv:2207.03732
28. Cassels, J.W.S., Fröhlich, A.: Algebraic Number Theory, 2nd revised version. London Mathematical Society, London (2010)
29. Chen, K.-T., Fox, R., Lyndon, R.: Free differential calculus. IV. The quotient groups of the lower central series. Ann. Math. (2) **68**, 81–95 (1958)
30. Chung, H.-J., Kim, D., Kim, M., Pappas, G., Park, J., Yoo, H.: Abelian arithmetic Chern–Simons theory and arithmetic linking numbers. Int. Math. Res. Not. **18**, 5674–5702 (2019)
31. Chung, H.-J., Kim, D., Kim, M., Park, J., Yoo, H.: Arithmetic Chern–Simons theory II. In: p-Adic Hodge Theory (2017 Simons Symposium on p-adic Hodge theory), pp. 81–128. Springer International Publishing, New York (2020)
32. Coates, J., Perrin-Riou, B.: On p-adic L-functions attached to motives over $\mathbb{Q}$. In: Algebraic Number Theory. Advanced Studies in Pure Mathematics, vol. 17, pp. 23–54. Academic Press, Boston (1989)
33. Coates, J., Sujatha, R.: Cyclotomic Fields and Zeta Values. Springer Monographs in Mathematics. Springer-Verlag, Berlin (2006)
34. Coleman, R.: Dilogarithms, regulators and p-adic L-functions. Invent. Math. **69**(2), 171–208 (1982)
35. Crowell, R.: The derived module of a homomorphism. Adv. Math. **6**, 210–238 (1971)
36. Crowell, R., Fox, R.: Introduction to Knot Theory, Reprint of the 1963 Original. Graduate Texts in Mathematics, vol. 57. Springer-Verlag, New York-Heidelberg (1977)
37. Culler, M., Shalen, P.: Varieties of group representations and splittings of 3-manifolds. Ann. Math. **117**, 109–146 (1983)
38. Deligne, P.: Formes modulaires et représentations l-adiques. Lecture Notes in Mathematics, vol. 179, pp. 136–172. Springer, Berlin (1971)
39. Deng, Y., Kurimaru, R., Matsusaka, T.: Arithmetic Dijkgraaf-Witten invariants for real quadratic fields, quadratic residue graphs, and density formulas. Res. Number Theory **9**, Article number: 70 (2023)
40. Deninger, C.: Some analogies between number theory and dynamical systems on foliated spaces. Doc. Math. J. DMV. Extra Volume ICM I, 23–46 (1998)
41. Deninger, C.: On dynamical systems and their possible significance for arithmetic geometry. In: Regulators in Analysis, Geometry and Number Theory. Progress in Mathematics, vol. 171, pp. 29–87. Birkhauser Boston, Boston (2000)
42. Deninger, C.: Number theory and dynamical systems on foliated spaces. Jahresber. Deutsch. Math.-Verein. **103**(3), 79–100 (2001)

43. Deninger, C.: A note on arithmetic topology and dynamical systems. In: Algebraic Number Theory and Algebraic Geometry. Contemporary Mathematics, vol. 300, pp. 99–114. AMS, Providence (2002)
44. Deninger, C.: A dynamical systems analogue of Lichtenbaum's conjectures on special values of Hasse-Weil zeta functions (2007). arXiv:math/0605724
45. Deninger, C.: Analogies between analysis on foliated spaces and arithmetic geometry. In: Groups and Analysis, London Mathematical Society Lecture Note Series, vol. 354, pp. 174–190. Cambridge University Press, Cambridge (2008)
46. Deninger, C.: Dynamical systems for arithmetic schemes (2022). arXiv:1807.06400v2
47. Dijkgraaf, R., Witten, E.: Topological gauge theories and group cohomology. Commun. Math. Phys. **129**, 393–429 (1990)
48. Dimofte, T., Gukov, S., Lenells, J., Zagier, D.: Exact results for purturbative Chern–Simons theory with complex gauge group. Commun. Number Theory Phys. **3**(2), 363–443 (2009)
49. Dion, C., Ray, A.: Topological Iwasawa invariants and arithmetic statistics (2022). arXiv:2203.11422
50. Dunnington, G.W.: Carl Friedrich Gauss, Titan of Science. The Mathematical Association of America, Washington (2004)
51. Eilenberg, S., MacLane, S.: Cohomology theory in abstract groups. I. Ann. Math. **48**, 51–78 (1947)
52. Fox, R.H.: Free differential calculus. I: Derivation in the free group ring. Ann. Math. **57**, 547–560 (1953)
53. Fox, R.H.: Free differential calculus III. Subgroups. Ann. Math. **64**(3), 407–419 (1956)
54. Fox, R.H.: Covering spaces with singularities. In: A Symposium in Honor of S. Lefschetz, pp. 243–257. Princeton University Press, Princeton (1957)
55. Freed, D., Quinn, F.: Chern–Simons theory with finite gauge group. Commun. Math. Phys. **156**(3), 435–472 (1993)
56. Fried, D.: Analytic torsion and closed geodesics on hyperbolic manifolds. Invent. Math. **84**(3), 523–540 (1986)
57. Friedlander, E.: Étale Homotopy of Simplicial Schemes. Annals of Mathematics Studies, vol. 104. Princeton University Press, Princeton/University of Tokyo Press, Tokyo (1982)
58. Fujiwara, K.: Algebraic number theory and low dimensional topology. Report of the 47-th Algebra Symposium at Muroran Inst. Univ., pp. 172–185 (2002)
59. Fuluta, K.: On capitulation theorems for infinite groups. In: Primes and Knots. Contemporary Mathematics, vol. 416, pp. 41–47. American Mathematical Society, Providence (2006). Preprint (2003)
60. Furuta, Y.: The genus field and genus number in algebraic number fields. Nagoya Math. J. **29**, 281–285 (1967)
61. Gärtner, J.: Rédei symbols and arithmetical mild pro-2-groups. Annales mathématiques du Québec **38**, 13–36 (2014)
62. Gärtner, J.: Higher Massey products in the cohomology of mild pro-p-groups. J. Algebra **422**, 788–820 (2015)
63. Gauss, C.F.: Zur mathematischen Theorie der electrodynamischen Wirkungen. Werke **V** (1833)
64. Gauss, C.F.: Disquisitiones arithmeticae. Yale University, New Haven (1966)
65. Ghrist, R.: Flows on S^3 supporting all links as orbits. Electron. Res. Announc. Am. Math. Soc. **1**(2), 91–97 (1995)
66. Ghrist, R., Kin, E.: Flowlines transverse to knot and link fibrations. Pac. J. Math. **217**, 61–86 (2004)
67. Gomi, K.: Extended topological quantum field theory: a toy model (Japanese). In: Proceedings of the 4th Kinosaki Seminar, 2007, 18 pp. Available at Homepage of Kiyonori Gomi
68. González-Acuña, F., Short, H.: Cyclic branched coverings of knots and homology spheres. Rev. Mat. Univ. Complut. Madrid **4**(1), 97–120 (1991)
69. Gordon, C., Luecke, J.: Knots are determined by their complements. J. Am. Math. Soc. **2**, 371–415 (1989)

70. Greenberg, R.: Iwasawa theory for p-adic representations. In: Algebraic Number Theory. Advanced Studies in Pure Mathematics, vol. 17, pp. 97–137. Academic Press, Boston (1989)
71. Greenberg, R., Stevens, G.: p-Adic L-functions and p-adic periods of modular forms. Invent. Math. 111(2), 407–447 (1993)
72. Grothendieck, A.: Revêtements étales et groupe fondamental. Séminaire de Géometrie Algébrique du Bois Marie 1960–1961 (SGA 1). Lecture Notes in Mathematics, vol. 224. Springer-Verlag, Berlin-New York (1971)
73. Grothendieck, A. (with M. Artin and J. L. Verdier): Théorie des topos et cohomologie étale des schémas, Tome 1,2, 3. (French) Séminaire de Géométrie Algébrique du Bois-Marie 1963–1964 (SGA 4). Lecture Notes in Mathematics, vol. 269, 270, 305. Springer-Verlag, Berlin-New York (1972–1973)
74. Haberland, K.: Galois Cohomology of Algebraic Number Fields, 145 pp. VEB Deutscher Verlag der Wissenschaften, Berlin (1978)
75. Habibi, A.E., Mizusawa, Y.: On pro-p-extensions of number fields with restricted ramification over intermediate $\mathbb{Z}_p$-extensions. J. Number Theory 231, 214–238 (2022)
76. Harada, S.: Hasse-Weil zeta function of absolutely irreducible SL_2 representations of the figure 8 knot group. Proc. Am. Math. Soc. 139(9), 3115–3125 (2011)
77. Hempel, J.: 3-Manifolds. Annals of Mathematics Studies, vol. 86. Princeton University Press, Princeton/University of Tokyo Press, Tokyo (1976)
78. Hida, H.: Iwasawa modules attached to congruences of cusp forms. Ann. Sci. École Norm. Sup. (4) 19, 231–273 (1986)
79. Hida, H.: Galois representations into $GL_2(\mathbb{Z}_p[[X]])$ attached to ordinary cusp forms. Invent. Math. 85, 545–613 (1986)
80. Hida, H.: Modular Forms and Galois Cohomology. Cambridge Studies in Advanced Mathematics, vol. 69. Cambridge University Press, Cambridge (2000)
81. Hida, H.: Hilbert Modular Forms and Iwasawa Theory. Oxford Mathematical Monographs. Oxford University Press, Oxford (2006)
82. Hilbert, D.: Die Theorie der algebraischen Zahlkörper, Jahresbericht der Deutschen Mathematikervereinigung Bd. 4 (1897), pp. 175–546. In: Gesammelte Abhandlungen. Band I: Zahlentheorie. (German) Zweite Auflage Springer-Verlag, Berlin-New York (1970)
83. Hillman, J.: Algebraic Invariants of Links. Series on Knots and Everything, vol. 32. World Scientific Publishing Co., Singapore (2002)
84. Hillman, J., Matei, D., Morishita, M.: Pro-p link groups and p-homology groups. In: Primes and Knots. Contemporary Mathematics, vol. 416, pp. 121–136. American Mathematical Society, Providence (2006)
85. Hirano, H.: On mod 2 arithmetic Dijkgraaf-Witten invariants for certain real quadratic fields. Osaka J. Math. 60(4), 933–954 (2023)
86. Hirano, H.: Arithmetic Dijkgraaf-Witten theory for number rings. Doctoral Thesis, Kyushu University (2023)
87. Hirano, H., Morishita, M.: Arithmetic topology in Ihara theory II: Milnor invariants, dilogarithmic Heisenberg coverings and triple power residue symbols. J. Number Theory 198, 211–238 (2019)
88. Hirano, H., Kim, J., Morishita, M.: On arithmetic Dijkgraaf-Witten theory. Commun. Number Theory Phys. 17(1), 1–61 (2023)
89. Hironaka, E.: Alexander stratifications of character varieties. Ann. Inst. Fourier (Grenoble) 47, 555–583 (1997)
90. Ihara, Y.: On Galois representations arising from towers of coverings of $\mathbb{P}^1 \setminus \{0, 1, \infty\}$. Invent. Math. 86, 427–459 (1986)
91. Ihara, Y.: On beta and gamma functions associated with the Grothendieck-Teichmüller groups. In: Aspects of Galois Theory (Gainesville, FL, 1996). London Mathematical Society Lecture Note Series, vol. 256, pp. 144–179. Cambridge University Press, Cambridge (1999)
92. Iwasawa, K.: On a certain analogy between algebraic number fields and function fields (Japanese). Sugaku 15, 65–67 (1963)
93. Iwasawa, K.: On $\mathbb{Z}_l$-extensions of algebraic number fields. Ann. Math. (2) 98, 246–326 (1973)

94. Iyanaga, S., Tamagawa, T.: Sur la theorie du corps de classes sur le corps des nombres rationnels. J. Math. Soc. Jpn. **3**, 220–227 (1951)
95. Kadokami, T., Mizusawa, Y.: Iwasawa type formulas for covers of a link in a rational homology sphere. J. Knot Theory Ramif. **17**(10), 1199–1221 (2008)
96. Kadokami, T., Mizusawa, Y.: On the Iwasawa invariants of a link in the 3-sphere. Kyushu J. Math. **67**(1), 215–226 (2013)
97. Kapranov, M.: Analogies between the Langlands correspondence and topological quantum field theory. In: Progress in Mathematics, vol. 131, pp. 119–151. Birkhäuser, Basel (1995)
98. Kapranov, M.: Analogies between number fields and 3-manifolds, unpublished note (1996)
99. Kato, K.: A Hasse principle for two-dimensional global fields. J. Reine Angew. Math. **366**, 142–183 (1986)
100. Kato, K.: Lectures on the approach to Iwasawa theory for Hasse-Weil L-functions via B_{dR}. In: Arithmetic Algebraic Geometry. Lecture Notes in Mathematics, vol. 1553, pp. 50–163. Springer, Berlin (1993)
101. Kawauchi, A.: A Survey of Knot Theory. Translated and revised from the 1990 Japanese original by the author. Birkhäuser Verlag, Basel (1996)
102. Kim, M.: Arithmetic gauge theory: a brief introduction. Mod. Phys. Lett. A **33**(29), 1830012 (2018)
103. Kim, M.: Arithmetic Chern-Simons theory I. In: Galois Cover, Grothendieck-Teichmüller Theory and Dessins d'Enfants – Interactions between Geometry, Topology, Number Theory and Algebra. Springer Proceedings in Mathematics & Statistics, vol. 330, pp. 155–180 (2020)
104. Kim, J.: On dynamical zeta functions for 3-dimensional Riemannian foliated dynamical systems. Doctoral Thesis, Kyushu University (2021)
105. Kim, J., Morishita, M., Noda, T., Terashima, Y.: On 3-dimensional foliated dynamical systems and Hilbert type reciprocity law. Münster J. Math. **14**(2), 323–348 (2021)
106. Kirk, P., Livingston, C.: Twisted Alexander invariants, Reidemeister torsion, and Casson-Gordon invariants. Topology **38**(3), 635–661 (1999)
107. Kitanao, T., Goda, H., Morifuji, T.: Twisted Alexander Invariants (Japanese). MSJ Memoir, vol. 5. Mathematical Society of Japan (2006)
108. Kitayama, T., Morishita, M., Tange, R., Terashima, Y.: On certain L-functions for deformations of knot group representations. Trans. AMS **370**, 3171–3195 (2018)
109. Kitayama, T., Morishita, M., Tange, R., Terashima, Y.: On adjoint homological Selmer modules for SL_2-representations of knot groups. Int. Math. Res. Not. **2023**(23), 19801–19826 (2023)
110. Knudsen, F., Mumford, D.: The projectivity of the moduli space of stable curves. I. Preliminaries on "det" and "Div". Math. Scand. **39**(1), 19–55 (1976)
111. Koch, H.: Galoissche Theorie der p-Erweiterungen. Springer, Berlin–New York/VEB Deutscher Verlag der Wissenschaften, Berlin (1970)
112. Koch, H.: On p-extensions with given ramification, Appendix 1 in [74], pp. 89–126
113. Kodani, H.: Arithmetic topology on braid and absolute Galois groups. Doctoral Thesis, Kyushu University (2017)
114. Kodani, H., Morishita, M., Terashima, Y.: Arithmetic topology in Ihara theory. Publ. RIMS Kyoto Univ. **53**, 629–688 (2017)
115. Kohno, T.: Conformal field theory and topology. In: Translations of Mathematical Monographs, vol. 210. Iwanami Series in Modern Mathematics. AMS, Providence (2002)
116. Kontsevich, M.: Vassiliev's Knot Invariants, I. M. Gelfand Seminar. Advances in Soviet Mathematics, vol. 16, Part 2, pp. 137–150. American Mathematical Society, Providence (1993)
117. Kopei, F.: A remark on a relation between foliations and number theory. In: Walczak, P., et al. (eds.) Foliation 2005, pp. 245–249. World Scientific, Singapore (2006)
118. Kopei, F.: A foliated analogue of one- and two-dimensional Arakelov theory. Abh. Math. Semin. Univ. Hambg. **81**(2), 141–189 (2011)
119. Kubota, T.: Topological coverings of SL(2) over a local field. J. Math. Soc. Jpn. **19**, 114–121 (1967)

120. Kubota, T., Leopoldt, H.: Eine p-adische Theorie der Zetawerte. I. Einführung der p-adischen Dirichletschen L-Funktionen. J. Reine Angew. Math. **214/215**, 328–339 (1964)
121. Kurihara, M.: Iwasawa theory and Fitting ideals. J. Reine Angew. Math. **561**, 39–86 (2003)
122. Kurihara, M.: Refined Iwasawa theory and Kolyvagin systems of Gauss sum type. Proc. Lond. Math. Soc. **104**, 728–769 (2012)
123. Labute, J.: The Lie algebra associated to the lower central series of a link group and Murasugi's conjecture. Proc. Am. Math. Soc. **109**(4), 951–956 (1990)
124. Labute, J.: Mild pro-p-groups and Galois groups of p-extensions of **Q**. J. Reine Angew. Math. **596**, 155–182 (2006)
125. Labute, J., Mináč, J.: Mild pro-2-groups and 2-extensions of $\mathbb{Q}$ with restricted ramification. J. Algebra **332**, 136–158 (2011)
126. Lang, S.: Cyclotomic Fields I and II, combined second edition. With an appendix by Karl Rubin. Graduate Texts in Mathematics, vol. 121. Springer-Verlag, New York (1990)
127. Lang, S.: Algebraic Number Theory, 2nd edn. Graduate Texts in Mathematics, vol. 110. Springer-Verlag, New York (1994)
128. Le, T.: Varieties of representations and their subvarieties of homology jumps for certain knot groups. Russ. Math. Surv. **46**, 250–251 (1991)
129. Lee, J., Park, J.: Arithmetic Chern-Simons theory with real places. J. Knot Theory Ramif. **32**(04), 2350027 (2023)
130. Lück, W.: Analytic and topological torsion for manifolds with boundary and symmetry. J. Diff. Geom. **37**, 263–322 (1993)
131. Manin, Y., Marcolli, M.: Holography principle and arithmetic of algebraic curves. Adv. Theor. Math. Phys. **5**(3), 617–650 (2001)
132. Marcolli, M., Xu, Y.: Quantum statistical mechanics in arithmetic topology. J. Geom. Phys. **114**, 153–183 (2017)
133. Massey, W.: Algebraic Topology: An Introduction, Reprint of the 1967 edition. Graduate Texts in Mathematics, vol. 56. Springer-Verlag, New York-Heidelberg (1977)
134. Matsusaka, T., Ueki, J.: Modular knots, automorphic forms, and the Rademacher symbols for triangle groups. Res. Math. Sci. **10**(1) (2023), Paper no. 4
135. Mazur, B.: Remarks on the Alexander polynomials, 1963 or 1964. Available at Homepage of Barry Mazur
136. Mazur, B.: Notes on étale cohomology of number fields. Ann. Sci. École Norm. Sup. (4) **6**, 521–552 (1973)
137. Mazur, B.: Deforming Galois representations, Galois groups over $\mathbb{Q}$. Math. Sci. Res. Inst. Publ. **16**, 385–437 (1989). Springer
138. Mazur, B.: Two-dimensional p-adic Galois representations unramified away from p. Compos. Math. **74**, 115–133 (1990)
139. Mazur, B.: The theme of p-adic variation. In: Mathematics: Frontiers and Perspectives, pp. 433–459. AMS, Providence (2000)
140. Mazur, B.: Primes, knots and Po, 2012. Available at Homepage of Barry Mazur
141. Mazur, B., Wiles, A.: Class fields of Abelian extensions of $\mathbb{Q}$. Invent. Math. **76**(2), 179–330 (1984)
142. Mazur, B., Wiles, A.: On p-adic analytic families of Galois representations. Compos. Math. **59**(2), 231–264 (1986)
143. McMullen, C.: From dynamics on surfaces to rational points on curves. Bull. Am. Math. Soc. (N.S.) **37**(2), 119–140 (2000)
144. McMullen, C.: Knots which behaves like the prime numbers. Compos. Math. **149**, 1235–1244 (2013)
145. Mihara, T.: Cohomological approach to class field theory in arithmetic topology. Can. J. Math. **71**(4), 891–935 (2019)
146. Milne, J.: Étale Cohomology. Princeton Mathematical Series, vol. 33. Princeton University Press, Princeton (1980)
147. Milne, J.: Arithmetic Duality Theorems. Perspectives in Mathematics, vol. 1. Academic Press, Inc., Boston (1986)

148. Milnor, J.: Isotopy of links. In: Fox, R.H., Spencer, D.S., Tucker, W. (eds.) Algebraic Geometry and Topology, A Symposium in Honour of S. Lefschetz, pp. 280–306. Princeton University Press, Princeton (1957)

149. Milnor, J.: Infinite cyclic coverings. In: 1968 Conference on the Topology of Manifolds (Michigan State Univ., E. Lansing, Mich., 1967), pp. 115–133. Prindle, Weber & Schmidt, Boston

150. Mizusawa, Y.: On pro-p link groups of number fields. Trans. Am. Math. Soc. **372**(10), 7225–7254 (2019)

151. Mizusawa, Y., Yamamoto, G.: Iwasawa invariants and linking numbers of primes, "Development of Iwasawa theory – the Centennial of K. Iwasawa's Birth". Adv. Stud. Pure Math. **86**, 639–654 (2020)

152. Morin, B.: Utilisation d'une cohomologie étale equivariante en topologie arithmétique. Compos. Math. **144**(1), 32–60 (2008)

153. Morin, B.: Sur le topos Weil-étale d'un corps de nombres. Thèse, Université Bordeaux I (2008)

154. Morishita, M.: Milnor's link invariants attached to certain Galois groups over $\mathbb{Q}$. Proc. Jpn. Acad. **76**(2), 18–21 (2000)

155. Morishita, M.: Knots and prime numbers, 3-dimensional manifolds and algebraic number fields (Japanese). In: Algebraic Number Theory and Related Topics (Japanese) (Kyoto, 2000). Surikaisekikenkyusho Kokyuroku **1200**, 103–115 (2001)

156. Morishita, M.: A theory of genera for cyclic coverings of links. Proc. Jpn. Acad. **77**(7), 115–118 (2001)

157. Morishita, M.: On certain analogies between knots and primes. J. Reine Angew. Math. **550**, 141–167 (2002)

158. Morishita, M.: Primes and Knots –On analogies between algebraic number theory and 3-dimensional topology– (Japanese), Report of the 47-th Algebra Symposium at Muroran Inst. Univ., pp. 157–171 (2002)

159. Morishita, M.: On capitulation problem for 3-manifolds. In: Galois Theory and Modular Forms, Developments in Mathematgics, vol. 11, pp. 305–313. Kluwer, Boston (2003)

160. Morishita, M.: Milnor invariants and Massey products for prime numbers. Compos. Math. **140**, 69–83 (2004)

161. Morishita, M.: Analogies between prime numbers and knots (Japanese). Sugaku **58**(1), 40–63 (2006)

162. Morishita, M.: On the Alexander stratification in the deformation space of Galois characters. Kyushu J. Math. **60**, 405–414 (2006)

163. Morishita, M.: Milnor invariants and l-class groups. In: Geometry and Dynamics of Groups and Spaces, In Memory of Alexander Reznikov. Progress in Mathematics, vol. 265, pp. 589–603. Birkäuser, Basel (2007)

164. Morishita, M.: Knots and Primes (Japanese), xiii + 201 pp. Springer-Japan (2009)

165. Morishita, M.: Analogies between knots and primes, 3-manifolds and number rings. Sugaku Expo. **23**(1), 1–30 (2010). AMS

166. Morishita, M.: Galois Categories and Arithmetic Dualities for Number Rings. Springer (in preparation)

167. Morishita, M., Terashima, Y.: Arithmetic topology after Hida theory. In: Intelligence of Low Dimensional Topology 2006. World Scientific Publishing Co. in the Knots and Everything Book Series, vol. 40, pp. 213–222

168. Morishita, M., Terashima, Y.: Chern-Simons variation and Deligne cohomology. In: Spectral Analysis in Geometry and Number Theory on Professor Toshikazu Sunada's 60th birthday. Contemporary Mathematics, vol. 484, pp. 127–134. AMS, Providence (2007)

169. Morishita, M., Takakura, Y., Terashima, Y., Ueki, J.: On the universal deformations of SL_2-representations of knot groups. Tohoku Math. J. (2) **69**(1), 67–84 (2017)

170. Murasugi, K.: On Milnor's invariant for links. Trans. Am. Math. Soc. **124**, 94–110 (1966)

171. Murasugi, K.: Nilpotent coverings of links and Milnor's invariant. In: Low-Dimensional Topology, London Mathematical Society. Lecture Note Series, vol. 95, pp. 106–142. Cambridge University Press, Cambridge-New York (1985)

172. Murasugi, K.: Classical knot invariants and elementary number theory. In: Primes and Knots. Contemporary Mathematics, vol. 416, pp. 167–196. American Mathematical Society, Providence (2006)

173. Murre, J.P.: Lectures on an introduction to Grothendieck's theory of the fundamental group, Notes by S. Anantharaman. Tata Institute of Fundamental Research Lectures on Mathematics, No 40. Tata Institute of Fundamental Research, Bombay (1967)

174. Neukirch, J.: Class Field Theory. Grundlehren der Mathematischen Wissenschaften, vol. 280. Springer-Verlag, Berlin (1986)

175. Neukirch, J.: Algebraic number theory, Translated from the 1992 German original and with a note by Norbert Schappacher. With a foreword by G. Harder. Grundlehren der Mathematischen Wissenschaften [Fundamental Principles of Mathematical Sciences], vol. 322. Springer-Verlag, Berlin (1999)

176. Neukirch, J., Schmidt, A., Wingberg, K.: Cohomology of Number Fields, 2nd edn. Grundlehren der Mathematischen Wissenschaften, vol. 323. Springer-Verlag, Berlin (2008)

177. Neumann, W., Zagier, D.: Volumes of hyperbolic three-manifolds. Topology **24**(3), 307–332 (1985)

178. Nguyen Quang Do, T.: Formations de classes et modules d'Iwasawa, Number theory, Noordwijkerhout 1983 (Noordwijkerhout, 1983). Lecture Notes in Mathematics, vol. 1068, pp. 167–185. Springer, Berlin (1984)

179. Niibo, H.: Idèlic class field theory for 3-manifolds. Kyushu J. Math. **68**, 421–436 (2014)

180. Niibo, H.: Idèlic class field theory for 3-manifolds. Doctoral Thesis, Kyushu University (2017)

181. Niibo, H., Ueki, J.: Idèlic class field theory for 3-manifolds and very admissible links. Trans. AMS **371**(12), 8467–8488 (2019)

182. Noguchi, A.: A functional equation for the Lefschetz zeta functions of infinite cyclic coverings with an application to knot theory, Spring Topology and Dynamical Systems Conference. Topology Proc. **29**(1), 277–291 (2005)

183. Noguchi, A.: Zeros of the Alexander polynomial of knot. Osaka J. Math. **44**(3), 567–577 (2007)

184. Oda, T.: Note on meta-Abelian quotients of pro-l free groups (1985). Preprint

185. Ohtani, S.: An analogy between representations of knot groups and Galois groups. Osaka J. Math. **48**(4), 857–872 (2011)

186. Ono, T.: An Introduction to Algebraic Number Theory. University Series in Mathematics. Plenum Publishers, New York (1990)

187. Pappas, G.: Volume and symplectic structure for ℓ-adic local systems. Adv. Math. **387** (2021), Paper No. 107836

188. Parry, W., Pollicott, M.: Zeta functions and the periodic orbit structure of hyperbolic dynamics. Astérisque, tome **187–188** (1990)

189. Porter, R.: Milnor's $\bar{\mu}$-invariants and Massey products. Trans. Am. Math. Soc. **275**, 39–71 (1980)

190. Ramachandran, N.: A note on arithmetic topology. C. R. Math. Acad. Sci. Soc. R. Can. **23**(4), 130–135 (2001)

191. Rédei, L.: Arithmetischer Beweis des Satzes über die Anzahl der durch vier teilbaren Invarianten der absoluten Klassengruppe im quadratischen Zahlkörper. J. Reine Angew. Math. **171**, 55–60 (1934)

192. Rédei, L.: Ein neues zahlentheoretisches Symbol mit Anwendungen auf die Theorie der quadratischen Zahlkörper. I J. Reine Angew. Math. **180**, 1–43 (1939)

193. Reznikov, A.: Three-manifolds class field theory (Homology of coverings for a nonvirtually b_1-positive manifold). Sel. Math. New Ser. **3**, 361–399 (1997)

194. Reznikov, A.: Embedded incompressible surfaces and homology of ramified coverings of three-manifolds. Sel. Math. New Ser. **6**, 1–39 (2000)

195. Reznikov, A., Moree, P.: Three-manifold subgroup growth, homology of coverings and simplicial volume. Asian J. Math. **1**(4), 764–768 (1997)

196. Rolfsen, D.: Knots and Links. Mathematics Lecture Series, no. 7. Publish or Perish, Inc., Berkeley (1976)

197. Sakuma, M.: The homology groups of abelian coverings of links. Math. Sem. Notes Kobe Univ. **7**(3), 515–530 (1979)

198. Schmidt, K.: Dynamical Systems of Algebraic Origin. Progress in Mathematics, vol. 128, xviii+310 pp. Birkhäuser Verlag, Basel (1995)

199. Schmidt, A.: Circular sets of prime numbers and p-extensions of the rationals. J. Reine Angew. Math. **596**, 115–130 (2006)

200. Schmidt, A.: Singular homology of arithmetic schemes. Algebra Number Theory **1**(2), 183–222 (2007)

201. Schwarz, A.S.: The partition function of a degenerate quadratic functional. Commun. Math. Phys. **67**, 1–16 (1979)

202. Segal, G.: The definition of conformal field theory. In: Differential Geometrical Methods in Theoretical Physics, Part of the NATO ASI Series book series (ASIC, vol. 250), pp. 165–171

203. Selberg, A.: Harmonic analysis and discontinuous subgroups in weakly symmetric Riemannian spaces with applications to Dirichlet series. J. Indian Math. Soc. **20**, 47–87 (1956)

204. Serre, J.-P.: A Course in Arithmetic, Translated from the French. Graduate Texts in Mathematics, vol. 7. Springer-Verlag, New York-Heidelberg (1973)

205. Serre, J.-P.: Local Fields, Translated from the French by Marvin Jay Greenberg. Graduate Texts in Mathematics, vol. 67. Springer-Verlag, New York-Berlin (1979)

206. Serre, J.-P.: Cohomologie Galoisienne. Lecture Notes in Mathematics, vol. 5. Springer (2nd Corrected Printing), Berlin (1986)

207. Sharifi, R.: Massey products and ideal class groups. J. Reine Angew. Math. **603**, 1–33 (2007)

208. Shiraishi, D.: On ℓ-adic polylogarithms and triple ℓ-th power residue symbols. Kyushu J. Math. **75**, 95–113 (2021)

209. Sikora, A.: Analogies between group actions on 3-manifolds and number fields. Comment. Math. Helv. **78**, 832–844 (2003)

210. Silver, D., Williams, S.: Mahler measure, links and homology growth. Topology **41**, 979–991 (2002)

211. Soma, T.: Hyperbolic, fibred links and fibre-concordances. Math. Proc. Camb. Philos. Soc. **96**, 283–294 (1984)

212. Stallings, J.: Homology and central series of groups. J. Algebra **2**, 170–181 (1965)

213. Sugiyama, K.: An analog of the Iwasawa conjecture for a compact hyperbolic threefold. J. Reine Angew. Math. **613**, 35–50 (2007)

214. Sugiyama, K.: The geometric Iwasawa conjecture from a viewpoint of the arithmetic topology. In: Proceedings of the Symposium on Algebraic Number Theory and Related Topics, pp. 235–247. RIMS Kôkyûroku Bessatsu, B4, Res. Inst. Math. Sci. (RIMS), Kyoto (2007)

215. Sugiyama, K.: On geometric analogues of Iwasawa main conjecture for a hyperbolic threefold. In: Noncommutativity and Singularities. Advanced Studies in Pure Mathematics, vol. 55, pp. 117–135. Mathematical Society of Japan, Tokyo (2009)

216. Sunada, T.: Fundamental Groups and Laplacians (Japanese). Kinokuniya, Tokyo (1988)

217. Takakura, Y.: Gauss' genus theory from the viewpoint of arithmetic topology (Japanese). Master thesis, Kyushu University (2007)

218. Tamme, G.: Introduction to étale cohomology. Translated from the German by Manfred Kolster. Universitext. Springer-Verlag, Berlin (1994)

219. Tange, R.: Fox formulas for twisted Alexander invariants associated to representations of knot groups over rings of S-integers. J. Knot Theory Ramif. **27**(5), 1850033 (2018)

220. Tange, R.: Iwasawa theory for representations of knot groups. Doctoral Thesis, Kyushu University (2018)

221. Tange, R., Ueki, J.: Twisted Iwasawa invariants of knots (2022). arXiv:2203.03239

222. Tange, R., Tran, A.T., Ueki, J.: Non-acyclic SL_2-representations of twist knots, -3-Dehn Surgeries, and L-functions. Int. Math. Res. Not., 42 pp. (2021). rnab034

223. Taylor, R., Wiles, A.: Ring-theoretic properties of certain Hecke algebras. Ann. Math. (2) **141**(3), 553–572 (1995)

224. Thurston, W.: The Geometry and Topology of 3-Manifolds. Lecture Notes, Princeton (1977)

225. Traldi, L.: Milnor's invariants and the completions of link modules. Trans. Am. Math. Soc. **284**, 401–424 (1984)

226. Turaev, V.G.: Milnor's invariants and Massey products. English transl. J. Soviet Math. **12**, 128–137 (1979)

227. Ueki, J.: On the homology of branched coverings of 3-manifolds. Nagoya Math. J. **213**, 21–39 (2014)

228. Ueki, J.: Arithmetic topology on branched covers of 3-manifolds. Doctoral Thesis, Kyushu University (2015)

229. Ueki, J.: On the Iwasawa μ-invariants of branched $\mathbb{Z}_p$-covers. Proc. Jpn. Acad. Ser. A **92**(6), 67–72 (2016)

230. Ueki, J.: On the Iwasawa invariants for links and Kida's formula. Int. J. Math. **28**(6), 1750035 (30 pp.) (2017)

231. Ueki, J.: On the Iwasawa invariants of branched $\mathbb{Z}_p$-covers, a survey. In: Algebraic Number Theory and Related Topics 2015. RIMS Kokyu-roku Bessatsu, vol. B72, pp. 331–342. Res. Inst. Math. Sci. Kyoto University (2018)

232. Ueki, J.: p-Adic Mahler measure and $\mathbb{Z}$-covers of links. Ergodic Theory Dyn. Syst. **40**, 272–288 (2020)

233. Ueki, J.: Olympic links is a Chebotarev link. Proc. Int. Geom. Cent. **13**(4), 40–49 (2020)

234. Ueki, J.: Chebotarev link is stably generic. Bull. Lond. Math. Soc. **53**(1), 82–91 (2021)

235. Ueki, J.: Modular knots obey the Chebotarev law (2022). arXiv:2105.10745

236. Verlinde, E.: Fusion rules and modular transformations in 2-D conformal field theory. Nucl. Phys. **B300**, 360–376 (1988)

237. Vogel, D.: Massey products in the Galois cohomology of number fields. Dissertation, Universität Heidelberg (2004)

238. Vogel, D.: On the Galois group of 2-extensions with restricted ramification. J. Reine Angew. Math. **581**, 117–150 (2005)

239. Wakui, M.: On Dijkgraaf-Witten invariant for 3-manifolds. Osaka J. Math. **29**, 675–696 (1992)

240. Waldspurger, J.-L.: Entrelacements sur Spec($\mathbb{Z}$). Bull. Sci. Math. **100**, 113–139 (1976)

241. Washington, L.: Introduction to Cyclotomic Fields. Graduate Texts in Mathematics, vol. 83, 2nd edn. Springer-Verlag, Berlin (1997)

242. Weil, A.: Sur certains groupes d'opérateurs unitaires. Acta Math. **111**, 143–211 (1976)

243. Whitten, W.: Knots complements and groups. Topology **26**, 41–44 (1987)

244. Witten, E.: Quantum field theory and the Jones polynomial. Commun. Math. Phys. **121**, 351–399 (1989)

245. Wiles, A.: Modular elliptic curves and Fermat's last theorem. Ann. Math. (2) **141**(3), 443–551 (1995)

246. Yamamoto, Y.: Class number problems for quadratic fields (concentrating on the 2-part) (Japanese). Sugaku **40**(2), 167–174 (1988)

247. Yetter, D.: Topological quantum field theories associated to finite groups and crossed G-sets. J. Knot Theory Ramif. **1**, 1–20 (1992)

248. Zink, T.: Etale cohomology and duality in number fields, Appendix 2 in [74], pp. 127–145

Index

Printed in the United States
by Baker & Taylor Publisher Services